ETHICS, COMPUTING, and GENOMICS

Edited by
Herman T. Tavani, PhD,
Rivier College

JONES AND BARTLETT PUBLISHERS
Sudbury, Massachusetts
BOSTON   TORONTO   LONDON   SINGAPORE

World Headquarters

Jones and Bartlett
  Publishers
40 Tall Pine Drive
Sudbury, MA 01776
978-443-5000
info@jbpub.com
www.jbpub.com

Jones and Bartlett
  Publishers Canada
6339 Ormindale Way
Mississauga, ON L5V1J2
CANADA

Jones and Bartlett
  Publishers International
Barb House, Barb Mews
London W6 7PA
UK

Jones and Bartlett's books and products are available through most bookstores and online booksellers. To contact Jones and Bartlett Publishers directly, call 800-832-0034, fax 978-443-8000, or visit our website www.jbpub.com.

Substantial discounts on bulk quantities of Jones and Bartlett's publications are available to corporations, professional associations, and other qualified organizations. For details and specific discount information, contact the special sales department at Jones and Bartlett via the above contact information or send an email to specialsales@jbpub.com.

Copyright © 2006 by Jones and Bartlett Publishers, Inc.

Cover Image: © Image100 Ltd, © AbleStock, © LiquidLibrary

Library of Congress Cataloging-in-Publication Data
Tavani, Herman T.
  Ethics, computing, and genomics / by Herman Tavani.
    p. cm.
  Includes bibliographical references.
  ISBN 0-7637-3620-1
  1. Genomics—Moral and ethical aspects. 2. Genetics—Data processing—Moral and ethical aspects. I. Title.
  QH441.2.T38 2006
  174'.957—dc22
                                        2005007396

All rights reserved. No part of the material protected by this copyright notice may be reproduced or utilized in any form, electronic or mechanical, including photocopying, recording, or any information storage or retrieval system, without written permission from the copyright owner.

Production Credits
Acquisitions Editor: Tim Anderson
Production Director: Amy Rose
Production Assistant: Alison Meier
Editorial Assistant: Kate Koch
V.P. of Manufacturing: Therese Connell
Cover Design: Timothy Dziewit
Composition: Northeast Compositors
Printing and Binding: Malloy Inc.
Cover Printing: Malloy Inc.

Printed in the United States of America
09 08 07 06 05   10 9 8 7 6 5 4 3 2 1

In memory of Nicholas Tavani, Jr.

Preface

The purpose of *Ethics, Computing, and Genomics* is to bring together in a single volume a collection of papers that examine ethical issues at the intersection of computing technology and genomic research—an area of inquiry that we refer to as *computational genomics*. Several textbooks that examine ethical issues involving computer/information technology are currently available; and a growing number of bioethics/biotechnology ethics textbooks that examine ethical aspects of the "new genetics," especially from the point of view of genetic testing, have also been published. So, one might ask why there is a need for a book dedicated to ethical concerns at the juncture of computing *and* genomics. For one thing, none of the existing books specifically address the cluster of ethical issues that arise *because of* the role that computers and computational techniques play in genetic/genomic research. But why, exactly, are these issues sufficiently important to warrant special consideration and thus a separate text? The answer that I propose, and defend in Chapter 1, is that ethical issues affecting computational genomics deserve special consideration for at least two reasons: (1) the computational techniques and methods used in genomic research introduce ethical concerns that are not always easy to anticipate, and that also may be difficult to articulate and analyze *as* ethical issues; and (2) these ethical concerns are significant because they can result in serious harm, including discrimination and stigmatization, both to individuals and groups. Several of the papers in this book either directly or indirectly support this thesis.

The papers included in this volume identify and examine a wide range of social and ethical issues affecting computational genomics, thereby contributing to the book's overall objective of providing a comprehensive and in-depth analysis of the impact these issues have both at the micro and macro levels—i.e., at the level of the individual as well as at the societal level. Some macro-level ethical issues that we consider have broad social and public-policy implications. For example, they affect the public health, epidemiologic research, and property laws (such as legislation concerning the patenting of genes and DNA sequences). Other ethical concerns that we examine have a greater impact at the micro-level because they more directly affect individuals than society in general. These concerns can be described in terms of three distinct but related kinds of issues: (1) privacy and confidentiality concerns affecting participants in genomics studies who, as a result of computational tools and methodologies used, are potentially at increased risk for denial of employment and health insurance; (2) concerns affecting the autonomy of research subjects that arise from current informed-consent practices used in collecting personal genetic data; and (3) concerns involving ownership rights to, and thus control over, personal genetic information that resides in computer databases.

At least four distinct constituencies or stakeholder groups stand to benefit from a clear understanding of the impact that these ethical issues have for them:

- human subjects that participate in genetics/genomics studies;
- scientists and administrators in the genomics research community (in the non-profit sector);
- entrepreneurs in the commercial sector;
- lawmakers.

First, human research subjects would be able to comprehend the breadth and depth of the risks (as well as the benefits) that they can expect from their participation in genetics/genomics studies that employ computational techniques. Second, scientists and administrators working in academic environments or in nonprofit organizations, where government funding is used to support their research, could benefit from a clear and comprehensive understanding of the ELSI (Ethical, Legal, and Social Implications) requirements that must be satisfied when conducting genomics research that involves human subjects. Third, entrepreneurs and business executives who direct biotechnology and pharmaceutical companies involved in genomics research and development in the private sector could benefit from an in-depth analysis of genomics-related ethical controversies that particularly affect the commercial sphere. Fourth, lawmakers who debate, and in some cases draft, legislation in the area of genetics stand to benefit from a clear and comprehensive analysis of the issues examined in this book. In addition to these four constituencies, others also could benefit from understanding the ethical issues at stake—especially since the health and welfare of virtually everyone living in the 21st century will be affected in one way or another by developments in the field of genomics that are made possible by computing technology.

The Scope of Ethical Issues Examined in this Volume

While the aim of this book is to identify and analyze *ethical* issues at the intersection of computing and genomics, no final "solutions" to the moral problems we examine are proffered. An important distinction can be drawn between the terms *ethics* and *morality*, which often tend to be used interchangeably. We define ethics as the "study of morality"—more precisely, as the *philosophical* study of morality. Morality, in turn, is a system comprised of two key components: (1) rules that guide our conduct, and (2) principles and values that provide standards for evaluating the action-guiding rules.[1] A *moral system* can be analyzed from multiple vantage points, three of which include legal, religious, and philosophical/ethical perspectives. In this book, we employ the perspective of philosophical ethics, based on logical reasoning and ethical theory, to guide our inquiry into and analysis of the moral issues we examine.[2]

The terms "morals," "values," and "moral values" have recently begun to take on a distinct, and arguably narrow, meaning in their popular or colloquial usage. This trend may have been influenced by the way in which members of some conservative religious organizations and political groups use these expressions. Often, representatives of these groups use the term "moral" in a very limited sense to refer almost exclusively to concerns involving reproductive and sexual issues. These concerns typically include abortion, cloning, and stem cell research, as well as issues involving gay marriage and extramarital sex. While these issues deserve ethical analysis, they do not exhaust the list of moral concerns that warrant serious and sustained consideration. As many ethicists and philosophers would quickly point out, the domain of moral issues is much more robust than the one portrayed by some conservative thinkers and organizations. Thus we need to expand upon the narrow scheme of morality used by these political and religious conservatives—a scheme that also currently tends to be reinforced, perhaps unwittingly, in the popular media.

A wide range of social concerns having to do with equity of, and access to, resources also need to be identified and analyzed as *moral* issues. For example, grossly unequal distributions of wealth—and especially unequal distributions of vital human resources—between and within nations, also generate moral concerns that warrant ethical analysis. When people in developing countries, as well as poor people in developed countries, are

---

[1] These distinctions are developed more fully in Chapter 2 of my book *Ethics and Technology: Ethical Issues in an Age of Information and Communication Technology* (Hoboken, NJ: John Wiley and Sons, 2004).

[2] In Chapters 3, 14, and 15 of this volume, some legal perspectives for analyzing moral issues are also provided.

denied access to life-sustaining medical drugs because of pricing practices tied to genetic patents owned by pharmaceutical companies, a moral issue arises. And this issue, like other moral issues, deserves careful ethical analysis.

The moral controversies examined in this book, however, do not focus on concerns involving equity of vital human resources; nor do they examine directly the set of moral concerns involving reproductive and sexual issues that have been so widely discussed by representatives of conservative organizations. Instead, they concentrate on moral concerns that arise because of policies and practices involving computational genomics research that have the potential to deprive human beings of autonomy, respect, and dignity.

Autonomy issues arise because of concerns having to do with whether research subjects can make fully informed, autonomous decisions when consenting to participate in genetics studies that involve the use of certain kinds of controversial computational tools and techniques. Issues concerning respect for persons emerge because of concerns about protecting the privacy and confidentiality of human research subjects, which are exacerbated by the use of computers in genomics research. Worries about human dignity arise because of the significant level of discrimination and stigmatization that genomics research subjects can experience in being identified with certain groups and subgroups generated by the use of computational techniques such as data mining. Finally, basic issues having to do with fairness arise because of concerns about who should own, and thus control who can and cannot access, personal genetic information that resides in computer databases.

At this point, it is worth drawing an important distinction between the notions of *ethicist* and *moralist*. Unlike many moralists, who may presume to have *the* correct answers to important moral issues, ethicists make no such presumption. Instead, we see the role of ethicist as one who identifies moral issues and then carefully analyzes them via the perspectives of logical reasoning and ethical theory. As ethicists, we do not presume that there is one uniquely correct answer to every moral issue. However, we follow Bernard Gert[3] in recognizing that, even though there may be disagreement among reasonable people as to what the uniquely correct answer should be in the case of a particular moral issue, there can nonetheless be considerable agreement on what the "wrong" answers are to many moral issues. So it is possible to disagree on what the "best" answer is in the case of a specific moral issue, while also acknowledging that not simply any answer will do. Thus, as ethicists, we can avoid the charge of *moral*

---

[3]Bernard Gert, *Common Morality: Deciding What to Do* (New York: Oxford University Press, 2004).

relativism—the naïve, but popular, view that all answers to moral questions are equally acceptable.[4]

Organization and Structure of the Book

This volume is comprised of eighteen chapters that are organized into five sections. Section I, consisting of a single chapter, defines computational genomics and introduces some of the ethical issues that arise in this field. Section II includes four chapters, each providing a distinct framework or methodological perspective—viz., moral, legal, policy, and (scientific) research—for analyzing specific controversies that are examined in detail in Sections III-V. The five chapters included in Section III examine ethical concerns involving privacy and confidentiality, as well as informed-consent policies used for research subjects participating in population-based genomics studies. Section IV includes five chapters that examine two different kinds of property right issues involving genetic/genomic data: (1) questions about ownership rights to *personal* genetic information[5] that resides in computer databases, and (2) questions about the right to patent *non-personal* genetic data, including DNA sequences, genes, and entire genomes. Section V, the concluding section of this volume, includes three chapters that describe, and in some instances speculate about, future directions and challenges pertaining to genomic research. The closing chapter in this section examines some controversies surrounding research in nanotechnology, which has significant implications for the future of computational genomics research.

Comprehensive introductions to the five sections of the book, each aimed at establishing a context for the chapters included in a particular section, are also included. Review questions, designed to quiz students on some key themes and controversies examined in each chapter, are included in each section's introduction.

Ethics, Computing, and Genomics can be used as a stand-alone text in courses that examine ethical aspects of bioinformatics, computational biology, and genome informatics. It can also be used as a supplementary text in computer science, genetics, genomics, and public health courses that include a component on ethical, legal, and/or social issues. This volume should also be of interest to a wider audience, including those outside academia who are eager to examine some ethical and social implications of the "new genetics."

---

[4]This point is developed more fully by Gert in Chapter 2 of this volume.

[5]In this book, the phrase "personal genetic information" is used to refer to genetic data that pertains to specific persons, as opposed to "non-personal genetic data" that refers to DNA in general.

Editing this volume has provided me with the opportunity to work on a project whose topics and themes span my research interests in two fields of applied ethics: computer/information technology ethics and public health ethics. My work in the latter area has benefited from my ongoing research as an ethicist in the Visiting Scholars Program at the Harvard School of Public Health, where my projects have centered on ethical issues at the intersection of environmental health and genomics research. Increasingly, ethical issues in this area are converging with ethical concerns in computer/information technology. A primary objective of this book is to examine some of the ways in which issues in these fields of applied ethics now converge at some points and intersect at others.

The idea for this book was conceived while I was organizing the Conference on Privacy, Informed Consent, and Genomic Research, held at the Harvard School of Public Health in March 2003. Five papers presented at that conference are included in this volume. Other papers in this anthology were originally presented at the Fifth International Conference on Computer Ethics—Philosophical Enquiry (CEPE 2003) and at the Sixth Annual Conference on Ethics and Technology, both hosted by Boston College in June 2003. Some papers in this book originally appeared as articles in journals such as the *Australian Journal of Professional and Applied Ethics*; *Boston University Journal of Science & Technology Law*; *Bioethics*; *Computers and Society*; *Ethics and Information Technology*; the *Journal of Information, Communication, and Ethics in Society*; the *Journal of Occupational and Environmental Medicine*; and *Nature*. Four papers were written especially for this volume. I am grateful to the authors who agreed to contribute their work to this anthology.

Acknowledgments

I am especially grateful to Ann Backus, Director of the Visiting Scholars Program in the Department of Environmental Health at the Harvard School of Public Health, for pointing me in the direction of a research project that eventually resulted in this volume. Ann's support was also critical in securing a Harvard-NIEHS grant (COEP Administrative Grant Supplement ES000239) for a project on ELSI-related aspects of Genetics, Genomics, and Proteomics, which provided funding for the Harvard Conference on Privacy, Informed Consent, and Genomics Research. As previously noted, a number of papers originally presented at that conference are included in this book.

I wish to thank Rivier College for granting me a sabbatical leave during the fall term of 2002, which enabled me to work on the project at Harvard that eventually led to this book. Several of my Rivier colleagues have contributed to this book in one way or another—some by providing comments on drafts of my papers that are included as chapters in this volume, and others by offering feedback on presentations on this topic that I delivered at faculty lecture sessions sponsored by Rivier's Committee on Scholarly Research and Professional Development. In particular, I would like to thank Mark Bolt, Jerry Dolan, and Martin Menke for some very helpful comments on Chapter 1. I am especially grateful to Lloyd Carr for the extraordinary amount of support he provided, both in reviewing multiple drafts of material included in Chapters 1 and 10 of this volume and in suggesting some refinements to formulations of certain views that I defend in this book. I am also grateful to many Rivier students in my Computer Ethics, Contemporary Moral Issues, and Technology, Values, and Society courses, who contributed indirectly to this book.

I would like to thank the editing and production staff at Jones and Bartlett, including Tim Anderson, Lesley Chiller, Alison Meier, and Amy Rose, for their support and attention to details during the production of this book. I am especially grateful to Steve Solomon, formerly of Jones and Bartlett, who believed in this book and who provided considerable support throughout the initial stages of this project.

Finally, I would like to thank my wife, Joanne, and our daughter, Regina, for the extraordinary patience and support they demonstrated during the long, and sometimes tedious, period that was required to compose this volume. Without their support, this book could not have been completed.

This book is dedicated to the memory of Nicholas Tavani, Jr. (1943-2005)—cousin and friend, mathematician and mentor, but most importantly, a gentle and humble person whose kind and generous spirit continues to inspire many who had the good fortune to know him.

Herman T. Tavani
May 2005

Contents

Preface v
Contributors xv
About the Editor xviii

Section I	**Introduction: Mapping the Conceptual Terrain** 1
Chapter 1	Herman T. Tavani, "Ethics at the Intersection of Computing and Genomics" 5
Section II	**Moral, Legal, Policy, and (Scientific) Research Perspectives** 27
Chapter 2	Bernard Gert, "Moral Theory and the Human Genome Project" 33
Chapter 3	Dan L. Burk, "Lex Genetica: The Law and Ethics of Programming Biological Code" 55
Chapter 4	Ruth Chadwick and Antonio Marturano, "Computing, Genetics, and Policy: Theoretical and Practical Considerations" 75
Chapter 5	David C. Christiani, Richard R. Sharp, Gwen W. Collman, and William A. Suk, "Applying Genomic Technologies in Environmental Health Research: Challenges and Opportunities" 85
Section III	**Personal Privacy and Informed Consent** 99
Chapter 6	James H. Moor, "Using Genetic Information While Protecting the Privacy of the Soul" 109
Chapter 7	Judith Wagner DeCew, "Privacy and Policy for Genetic Research" 121
Chapter 8	David Baumer, Julia Brande Earp, and Fay Cobb Payton, "Privacy of Medical Records: IT Implications of HIPAA" 137
Chapter 9	Bart Custers, "The Risks of Epidemiological Data Mining" 153
Chapter 10	Herman T. Tavani, "Environmental Genomics, Data Mining, and Informed Consent" 167

Section IV	**Intellectual Property Rights and Genetic/Genomic Information** 187	
Chapter 11	Adam D. Moore, "Intellectual Property, Genetic Information, and Gene Enhancement Techniques" 197	
Chapter 12	Richard A. Spinello, "Property Rights in Genetic Information" 213	
Chapter 13	Antonio Marturano, "Molecular Biologists as Hackers of Human Data: Rethinking Intellectual Property Rights" 235	
Chapter 14	Dan L. Burk, "Bioinformatics Lessons from the Open Source Movement" 247	
Chapter 15	James Boyle, "Enclosing the Genome: What the Squabbles over Genetic Patents Could Teach Us" 255	
Section V	**Challenges for the Future of Computational Genomics** 279	
Chapter 16	Francis S. Collins, Eric D. Green, Alan E. Guttmacher, and Mark S. Guyer, "A Vision for the Future of Genomics Research: A Blueprint for the Genomic Era" 287	
Chapter 17	Kenneth W. Goodman, "Bioinformatics: Challenges at the Frontier" 317	
Chapter 18	John Weckert, "The Control of Scientific Research: The Case of Nanotechnology" 323	
Index	341	

Contributors

DAVID L. BAUMER, PHD, JD, is Professor of Law and Technology and Director of the Cyberlaw Initiative at North Carolina State University, a licensed attorney in the State of North Carolina, and editor of *The Technology Report for the American Legal Scholars in Business*.

JAMES BOYLE is the William Neal Reynolds Professor of Law and co-founder of the Center for the Study of the Public Domain at Duke Law School, and the author of *Shamans, Software, and Spleens: Law and the Construction of the Information Society* (Harvard University Press 1996).

DAN L. BURK, JD, JSM, is the Oppenheimer, Wolff & Donnelly Professor of Law at the University of Minnesota, where he is a Faculty Associate at the Bioethics Center, teaching in the areas of Patent, Copyright, and Biotechnology Law.

RUTH CHADWICK, DPHIL, LLB, is Professor of Bioethics and Director of the ESRC Centre for Economic and Social Aspects of Genomics (CESAGen), Lancaster University (UK), and the editor of the *Encyclopedia of the Ethics of New Technologies* (Elsevier/Academic Press, 2001).

DAVID C. CHRISTIANI, MD, MPH, is Professor of Occupational Medicine and Epidemiology at the Harvard School of Public Health, and Professor of Medicine at the Harvard Medical School and Massachusetts General Hospital.

FRANCIS S. COLLINS, MD, PHD, is Director of the National Human Genome Research Institute (NHGRI) at the National Institutes of Health, as well as a Senior Investigator in the Genome Technology Branch at NHGRI.

GWEN W. COLLMAN, PHD, is the Scientific Program Administrator, Chemical Exposures and Molecular Biology Branch, Division of Extramural Research, at the National Institute of Environmental Health Sciences, National Institutes of Health.

BART CUSTERS, PHD, is a researcher at Tilburg University (The Netherlands) and the author of *The Power of Knowledge: Ethical, Legal, and Technological Aspects of Data Mining and Group Profiling in Epidemiology* (Wolf Legal Publishers, 2004).

JUDITH WAGNER DECEW, PHD, is Professor of Philosophy and Department Chair at Clark University, and the author of *In Pursuit of Privacy: Law, Ethics and the Rise of Technology* (Cornell University Press, 1997) and *Unionization in the Academy: Visions and Realities* (Rowman & Littlefield, 2003).

JULIA B. EARP, PHD, an Associate Professor of Information Systems in the College of Management at North Carolina State University, is a senior research collaborator with *theprivacyplace.org* and the author of numerous articles on Internet security and privacy.

BERNARD GERT, PHD, the Stone Professor of Intellectual and Moral Philosophy, Dartmouth College and Adjunct Professor of Psychiatry, Dartmouth Medical School, is the author of *Common Morality: Deciding What to Do* (Oxford University Press, 2004) and *Morality: Its Nature and Justification* (Oxford University Press, revised edition, 2005).

KENNETH W. GOODMAN, PHD, is Director of the Bioethics Program at the University of Miami, Associate Professor of medicine, philosophy, epidemiology and public health, nursing and anesthesiology, and the author of *Ethics and Evidence-Based Medicine* (Cambridge University Press, 2004).

ERIC D. GREEN, MD, PHD, is Scientific Director at the National Human Genome Research Institute (NHGRI) at the National Institutes of Health (NIH), Senior Investigator and Chief of the NHGRI Genome Technology Branch, Director of the NIH Intramural Sequencing Center, Editor of the journal *Genome Research*, and Co-Editor of the *Annual Reviews of Genomics and Human Genetics* series.

ALAN E. GUTTMACHER, MD, is Deputy Director of the National Human Genome Research Institute (NHGRI) at the National Institutes of Health, and Director of the NHGRI Office of Policy, Communications and Education.

MARK S. GUYER, PHD, is Director of the Division of Extramural Research at the National Human Genome Research Institute at the National Institutes of Health.

JAMES H. MOOR, PHD, is Professor of Philosophy at Dartmouth College, President of the International Society for Ethics and Information Technology, editor of the journal *Minds and Machines,* and co-author of *The Logic Book* (McGraw Hill, fourth edition, 2004).

ADAM D. MOORE, PHD, is an Assistant Professor of Philosophy at the University of Washington, the author of *Intellectual Property and Information Control* (Transaction Publishers, 2001, 2004), and the editor of *Information Ethics: Privacy, Property, and Power* (The University of Washington Press, 2005).

ANTONIO MARTURANO, PHD, is a Research Fellow and Ethics Officer at the School of Business and Management at the University of Exeter (UK), a member of the Ethics Committee of the Italian Ministry of Defence, and a member of the Editorial Advisory Board of the *Journal of Information, Communication and Ethics in Society*.

FAY COBB PAYTON, PHD, an Associate Professor of Information Systems in the College of Management at North Carolina State University, has worked in industry and/or health care related projects with several organizations including Blue Cross/Blue Shield of Ohio and North Carolina.

RICHARD R. SHARP, PHD, former Director of the Program in Environmental Health Policy at the National Institute of Environmental Health Sciences, is Assistant Professor of Medicine in the Center for Medical Ethics and Health Policy at Baylor College.

RICHARD A. SPINELLO, PHD, is an Associate Research Professor in the Carroll School of Management at Boston College, the author of *Case Studies in Information Technology Ethics* (Prentice Hall, second edition, 2003), and the co-editor of *Readings in CyberEthics* (Jones and Bartlett, second edition, 2004).

WILLIAM A. SUK, PHD, is Director of the Office of Program Development in the Division of Extramural Research and Training at the National Institute of Environmental Health Sciences, National Institutes of Health.

JOHN WECKERT, PHD, is Professor of Information Technology in the School of Information Studies and Professorial Fellow in the Centre for Applied Philosophy and Public Ethics at Charles Sturt University (Australia), and the editor of *Electronic Monitoring in the Workplace: Controversies and Solutions* (Idea Group Publishing, 2005).

About the Editor

HERMAN T. TAVANI, PHD, is Professor of Philosophy and Department Chair at Rivier College. He also holds appointments at the Harvard School of Public Health, where he is a visiting scholar/ethicist in the Department of Environmental Health, and at Boston College where he is an adjunct lecturer (in ethics) in the Carroll School of Management. Currently, he is President of the Northern New England Philosophical Association and Co-Director of the International Society for Ethics and Information Technology. Tavani is the author of *Ethics and Technology* (John Wiley and Sons, 2004) and the co-editor (with Richard Spinello) of *Readings in CyberEthics* (Jones and Bartlett, second edition, 2004) and *Intellectual Property Rights in a Networked World* (Idea Group, 2005).

I: INTRODUCTION: MAPPING THE CONCEPTUAL TERRAIN

Section I, comprised of a single chapter, provides an overview of ethical issues at the intersection of computing and genomics—a relatively new field of inquiry that we refer to as *computational genomics*. This chapter intends to perform two functions: (a) introduce the reader to ethical issues that are examined in much greater detail in subsequent chapters of the book, and (b) serve as a standalone chapter that presents arguments for the view that ethical issues affecting computational genomics deserve special consideration.

Chapter 1 begins with a brief examination of some conceptual and practical connections that link the fields of computer science and genetics. It then defines computational genomics as a field of inquiry, differentiating it from three distinct but related fields: bioinformatics, computational biology, and genome informatics. Next, it identifies and briefly examines a cluster of ethical issues affecting computational genomics. This is followed by a discussion of the ELSI (Ethical, Legal, and Social Implications) Research Program, which was established to guide researchers working on the Human Genome Project. Specific ELSI guidelines are then examined in the context of computational genomics research involving population studies. In particular, the case of deCODE Genetics, a privately owned genomics company in Iceland, is examined from the perspective of ethical concerns identified in the ELSI guidelines. For example, does deCODE's practice of cross-referencing and "mining" personal genetic data violate the ethical guidelines articulated in the ELSI Program? This question, of course, can also be extended to practices involving computational techniques used by other organizations that conduct population genomics research.

The final section of Chapter 1 considers whether ethical issues in computational genomics deserve special normative consideration. Arguments for framing special genetics laws and policies are often based on the thesis of genetic exceptionalism—viz., the view that controversies involving personal genetic data are special or "exceptional" because one's genetic data is different from other kinds of personal data, including medical data. It is then argued that the debate between genetic exceptionalists and "anti-exceptionalists" proceeds on the basis of a questionable assumption that is used as a premise in arguments advanced by both camps. Following an analysis of that assumption, the chapter concludes with an alternative proposal for why personal genetic data deserves special consideration from a normative perspective—i.e., because of the computational tools used to manipulate that data.

Review Questions for Section I
1. Identify and describe some theoretical and practical connections that link the fields of computing and genetics/genomics.
2. What is *computational genomics*, and how is this field both similar to and different from bioinformatics, computational biology, and genome informatics?
3. Which kinds of ethical issues arise at the intersection of computing and genomics?
4. What is the ELSI Research Program? What kinds of functions has it played thus far in guiding researchers working on the Human Genome Project?
5. What is *population genomics*? Does research in this field pose any special challenges for the ethical objectives articulated in the ELSI Research Program?
6. What are some of the key ethical concerns raised in the case of deCODE Genetics, Inc.? Why are those issues so controversial from an ethical point of view?
7. Which kinds of criticisms have been directed against deCODE? Which arguments have been used to defend it? Do you believe that deCODE's practices are defensible on ethical grounds? Explain.
8. Do ethical issues affecting computational genomics deserve special consideration from a normative—i.e., legal or moral—point of view?
9. What is meant by the expression *genetic exceptionalism*? Which arguments can be used to defend that thesis? Which ("anti-exceptionalist") arguments can be used against it?
10. Can an alternative position to the views advanced by genetic exceptionalists and anti-exceptionalists be framed in a way that is coherent? Explain.
11. What are Onora O'Neill's arguments for why personal genetic information is "distinctive"? Do you agree with O'Neill?
12. What does James Moor mean by the expressions "policy vacuum" and "conceptual muddle"? How can those concepts help us to analyze ethical issues that arise in computational genomics?

Suggested Further Readings
Anderson, James G., and Kenneth W. Goodman, eds. (2002). *Ethics and Information Technology: A Case-Based Approach to a Health Care System in Transition*. New York: Springer.
Backus, Ann, Richard A. Spinello, and Herman T. Tavani, eds. (2004). *Genomics, Ethics, and ICT*. Special Issue of *Ethics and Information Technology* (Vol. 6, No. 2). Dordrecht, The Netherlands: Kluwer.
Cahill, George F. (1996). "A Brief History of the Human Genome Project." In B. Gert, et al., eds. *Morality and the New Genetics: A Guide for Students and Health Care Patients*. Sudbury, MA: Jones and Bartlett, pp. 1–28.
Chadwick, Ruth, and Alison Thompson, eds. (1999). *Genetic Information Acquisition, Access, and Control*. New York: Kluwer/Plenum.

Gert, Bernard, et al., eds. (1996). *Morality and the New Genetics: A Guide for Students and Health Care Patients*. Sudbury, MA: Jones and Bartlett.

Gert, Bernard (2004). "Common Morality and Computing." In R. A. Spinello and H. T. Tavani, eds. *Readings in CyberEthics*. Sudbury, MA: Jones and Bartlett, pp. 96–106.

Goodman, Kenneth W. (1996). "Ethics, Genomics, and Information Retrieval," *Computers in Biology and Medicine*, Vol. 26, No. 3, pp. 223–229.

Goodman, Kenneth W., ed. (1998). *Ethics, Computing, and Medicine: Informatics and the Transformation of Healthcare*. New York: Cambridge University Press.

Goodman, Kenneth W. (2004). *Ethics and Evidence-Based Medicine*. New York: Cambridge University Press.

Moor, James H. (1985). "What Is Computer Ethics?" *Metaphilosophy*, Vol. 16, No. 4, pp. 266–275.

Moor, James H. (2004). "Reason, Relativity, and Responsibility in Computer Ethics." In R. A. Spinello and H. T. Tavani, eds. *Readings in CyberEthics*. 2nd ed. Sudbury, MA: Jones and Bartlett, pp. 40–54.

Murray, Thomas H. (1997). "Genetic Exceptionalism and Future Diaries: Is Genetic Information Different from Other Medical Information?" In M. Rothstein, ed. *Genetic Secrets: Protecting Privacy and Confidentiality*. New Haven, CT: Yale University Press, pp. 61–73.

Murray, Thomas H. (1997). "Ethical Issues in Human Genome Research." In K. Schrader-Frechette, and L. Westra, eds. *Technology and Values*. New York: Rowman and Littlefield, pp. 415–432.

National Academy of Sciences (2002). "Social, Ethical, and Legal Implications of Genetic Testing." In R. Sherlock and J. D. Murray, eds. *Ethical Issues in Biotechnology*. New York: Rowman and Littlefield, pp. 375–418.

O'Neill, Onora (2002). *Autonomy and Trust in Bioethics*. Cambridge: Cambridge University Press.

Schrader-Frechette, Kristin, and Laura Westra, eds. (1997). *Technology and Values*. New York: Rowman and Littlefield.

Sosclone, Colin L. (1997). "Ethical, Legal, and Social Issues Surrounding Studies of Susceptible Populations," *Environmental Health Perspectives*, Vol. 105, pp. 837–841.

Tavani, Herman T. (2004). "Genomic Research and Data-Mining Technology: Implications for Personal Privacy and Informed Consent," *Ethics and Information Technology*, Vol. 6, No. 1, pp. 15–28.

Weinberg, Robert A. (2000). "The Dark Side of the Genome." In A. H. Teich, ed. *Technology and the Future*. 8th ed. New York: St. Martin's, pp. 215–223.

1. Ethics at the Intersection of Computing and Genomics

Herman T. Tavani

In this chapter, we examine a cluster of ethical issues that arise at the intersection of computing and genomics. Before we identify and analyze these issues, however, it is useful to consider some connections that link the fields of computer science and genetics/genomics, both at the theoretical and practical levels.

1. Theoretical and Practical Connections Between Computer Science and Genetics/Genomics

First, consider some theoretical- or conceptual-level connections that link computing and genetics/genomics. James Moor (1999, p. 257) points out that "computing provides a conceptual model for the function and malfunction of our genetic machinery." Noting that researchers sometimes describe the operation of genes as "straightforward computational programs," Moor (p. 258) suggests that the "sequence of nucleotides can be regarded as a kind of computer program for constructing an organism."[1] Antonio Marturano and Ruth Chadwick (2004, p. 43) also describe some theoretical connections between the two fields when they note that genetics is "fully impregnate" with concepts from computer science, including notions such as "coding" and "information."[2] And Manuel Castells (2001, p. 164) remarks on some

---

This chapter, which appears for the first time in *Ethics, Computing and Genomics*, draws from and expands upon material in "Ethical Issues at the Intersection of Population Genomics Research and Information Technology." *Proceedings of the Conference on Privacy, Informed Consent, and Genomic Research*, Harvard School of Public Health, 2003. Copyright © 2005 by Herman T. Tavani. Printed with permission.

[1] Moor draws an analogy between the work of computer science pioneer Alan Turing and biologists James Watson and Francis Crick, noting that the latter were to genetics what the former was to computing. Moor (p. 257) remarks: "Watson and Crick described a simple arrangement of simple components following simple rules that captured a mechanism, a biological computer, that can generate incredible diversity and complexity." Moor's article is included as Chapter 6 of this book.

[2] Marturano and Chadwick (p. 43) note that "information" is a "fundamental concept" that is shared both by computer science and genetics. The authors also point out that the term "genetic information" was used for the first time in 1953 by Watson and Crick.

theoretical connections between genetics and information technologies, pointing out that geneticists focus on "decoding" and eventually "reprogramming" DNA.[3]

Next, consider some practical-level connections between computing and genetics/genomics, where developments in computer science have significantly influenced certain outcomes in genetic and genomic research. Marturano and Chadwick (p. 45) note that a computerized technique referred to as the "shotgun method" accelerated the mapping of the human genome, which contributed significantly to the objective of mapping of the entire human genome by 2003. Other authors have also remarked on ways that computing technology has played an important practical role in the "new genetics" by providing tools and methods critical to genomic research. For example, Kenneth Goodman (1998, p. 16) describes a case in which scientists working in Cambridge, England and St. Louis, Missouri set out to make 900,000 nucleotide bases available to colleagues around the world, who could then join their effort in locating the gene for breast cancer.[4] Goodman notes that in the pre-computer era, these scientists would have had to mail parcels of human tissue to their colleagues in order to accomplish this objective. They also would have had to publish a book or write a paper to describe the results of their research to colleagues. But in this case, the scientists, working at sites geographically distant from one another, were able to make information about the sequenced data available to colleagues simply by placing it on two publicly accessible FTP (File Transfer Protocol) sites on the Internet. More importantly, however, the gene sequences represented in this data had already been identified *computationally*—i.e., the genes had been discovered, sequenced, and analyzed via computers—before the data was shared electronically among an international group of scientists.[5]

The above case illustrates two different kinds of ways in which computers have contributed at the practical level to genetics/genomics research. Computer networks have made possible the communication and sharing of information about genetic data in ways that significantly expedite the data sharing and communication. In this way, computers have significantly *assisted* genetics/genomics researchers. We can refer to this function as the *computer-assisting* role that computing technology has played in genetics/genomics. But computers have also assisted researchers and professionals in other fields as well, helping them to carry out practical tasks associated with a given field or profession. So what, if anything, is unusual or different about the role that computing has played in genetics/genomics research?

There is another, far more significant, contribution that computing technology has made in the case of genetics/genomics research. By providing computational simulations and models that have *enhanced* research in the field, computers have

---

[3] Originally cited in Marturano and Chadwick, p. 44.

[4] The case described by Goodman is originally discussed in Dickson (1995).

[5] Marturano and Chadwick (p. 44) point out that the new genetics relies on computer simulations, computational models, and large data sets. Because of this dependency, Marturano and Chadwick ask whether research on the human genome is more properly described as work in *bioinformatics* than as medical research, given that much of the work in genetics research is now actually done on computers that manipulate *representations* of genetic material (i.e., *information*) rather than on the genetic materials themselves.

contributed significantly to the "new genetics." We can refer to this function as the *computer-enhancing* role that computing technology plays in genetics/genomics.[6]

In the case described by Goodman, we saw that not only was the information about the gene sequences able to be communicated and shared with colleagues via computer networks, but the information itself had become available only as a result of certain computational techniques and methods that were used. In this sense, computing technologies have significantly enhanced genetics/genomics research to the point where the "new genetics" now *depends on* computers—i.e., is dependent on the computational techniques and methodologies made possible by computing technology—to conduct its research and to make new discoveries.[7]

Francis Collins, Director of the National Human Genome Research Institute, clearly articulates this dependency on computational techniques when he asserts:

> *We need* computational *approaches that allow us to take* all the mountain of data *and begin to make sense of it and even to be able to create a model of a cell that captures the most important part of the proteins present there and how they interact. (Emphasis added)*[8]

As the vast amount of genetic/genomic information continues to expand, there is an increased need for computers and computational techniques to manage this data. Goodman (1998, p. 16) points out that the evolution of genetic databases, as well as the tools used for searching them, is proceeding so rapidly that scientists are now building "computational toolkits to hold their computational tools." Thus, it would seem that genetic/genomic research is now inextricably tied to computer/information technology and that the future of genomics is computational.

2. Computational Genomics vs. Bioinformatics, Computational Biology, and Genome Informatics: Some Distinctions

As noted above, our main concern in this book is with identifying and analyzing ethical concerns that arise at the intersection of computing and genomics—an area of inquiry sometimes referred to as *computational genomics*. Yet this field is not clearly defined; nor is it always easily distinguishable from three closely related fields: bioinformatics, computational biology, and genome informatics. Although

---

[6]In the final section of this chapter, we return to this distinction between the computer-enhancing and computer-assisting roles that computing technology plays in genetics/genomics. Our primary concern in the analysis of the moral issues examined in this chapter and book result from the computer-enhancing aspect of genomic research–i.e., issues that arise because of computational techniques and methodologies on which the "new genetics" depends.

[7]We should note that with respect to the "dependency" of genetics/genomics research on computers, Marturano and Chadwick (p. 43) claim that "without computer science there is no new genetics." However, this controversial claim is neither defended nor explored in this chapter.

[8]Francis Collins, interviewed in "Policy Roundup–State of Genomic Research," available at http://www.laskerfoundation.org/homenews/genome/pround_gn.html. Accessed 12/5/2004.

the four terms or expressions are sometimes used interchangeably, and although they often denote fields of inquiry whose content or subject matter is very similar, we propose the following distinctions.[9] *Bioinformatics* can be understood as a field that employs principles and techniques from the broader field of *informatics*,[10] which, in turn, involves the acquisition, storage, manipulation, analyses, transmission, sharing, and visualization of information on a computer (Goodman, 1999, p. 17). Bioinformatics, then, can be understood as the application of the informatics model to the management of biological information; its objective is to make that information more understandable and useful.

Computational biology, like bioinformatics, can be understood as a field that also aims at making complex biological data more understandable and useful. Whereas bioinformatics applies principles of information sciences and technologies to accomplish a set of practical tasks, computational biology goes one step further by using mathematical and computational approaches to address some questions in biology that are more *theoretical and experimental* in nature. As such, computational biology includes the development and application of data-analytical and theoretical methods, mathematical modeling, and computational-simulation techniques to the study of biological systems.[11]

Computational genomics, as we use the term in this book, both differs from and is similar to computational biology and bioinformatics. First, consider some differences. The content or subject matter examined in computational genomics is somewhat narrower in scope than the full range of biological concepts, principles, and systems potentially considered both in bioinformatics and computational biology. Like computational biology, however, computational genomics employs mathematical and computational approaches that address theoretical and experimental questions (in additional to practical ones). And like bioinformatics, computational genomics employs the use of computerized techniques, tools, and approaches to acquire, store, organize, analyze, or visualize data. However, the data involved is limited to genetic and genomic information per se, as opposed to biological data in a broader sense, which can include biological systems as well as medical, behavioral, and health data. So unlike bioinformatics, computational genomics is not concerned with the management and analysis of this broader set of biological data. And unlike computational biology, computational genomics need not be concerned with theoretical and experimental questions in biology that

---

[9]We should note that no universally accepted or even "standard" definitions exist for these fields. So, these definitions are stipulated for purposes of this book.

[10]In using these principles and techniques, bioinformatics shares some common traits with other specialized or sub fields of informatics such as medical informatics and healthcare informatics.

[11]Note that I am basing these distinctions between computational biology and bioinformatics on the NIH (National Institutes of Health) Working Definition of Bioinformatics and Computational Biology, developed by the BISTIC (Biomedical Information Science and Technology Consortium) Definition Committee (July 17, 2000). For additional definitions of "bioinformatics," see the following Web sites: http://brin.uams.edu/bionfodefinition.html and www.geocities.com/bioinformatics web/definition.html.

are beyond the scope of concepts and principles that pertain specifically to genes and genomes.[12]

Genome informatics is another term that is now also used, albeit less frequently than the three expressions we have examined thus far, to describe the field of inquiry involving computing and genomic technologies.[13] We can differentiate genome informatics from bioinformatics and computational genomics in the following ways. Like bioinformatics, genome informatics employs the informatics model in an analysis of practical questions involving genomics. Unlike bioinformatics, however, genome informatics does not address biological questions that are beyond the domain of genomics per se. In this sense, genome informatics is similar to computational genomics. However, the latter field is broader because it examines genomics issues from the perspective of theoretical and experimental questions, in addition to the practical ones considered by genome informatics.

To sum up the similarities and differences between computational genomics and genome informatics and between bioinformatics and computational biology, we can appeal to two key criteria: *content* and *methodology*. Whereas the content or subject matter of both computational genomics and genome informatics is similar, the methodological approaches used in the two fields are different. Computational genomics uses mathematical models to address theoretical and experimental questions in genomics that go beyond the mainly practical questions examined in genome informatics. Insofar as the content (or subject matter) of both bioinformatics and computational biology is "biological" in a broad sense of the term, the two fields are similar. However, they differ with respect to their methodologies, since the techniques used in computational biology address certain kinds of theoretical and experimental questions, in addition to the largely practical questions considered in bionformatics. The methodologies used in computational genomics and computational biology, on the other hand, are similar because of the kinds of computational techniques used in both fields. However, the content area or subject matter of computational genomics, as well as that of genome informatics, is narrower in scope than both bioinformatics and computational biology.

Figure 1 illustrates some ways in which the four fields are both similar and different with respect to the kind of data examined by each, as well as the method used to analyze that data—i.e., distinctions having to do with content and methodology.

Thus far, we have described some features of the *computational* aspect of computational genomics, at least in the context of biological research. But we have said very little about the *genomics* portion of the expression "computational genomics," other than that its content is narrower in scope than the range of topics considered in the broader field of biology. To better understand the relatively new field of genomics, which arguably is emerging as a separate science, it is useful to define, and thus differentiate among, the following key concepts and terms: *gene, genome,*

---

[12] Sometimes the distant fields of genetics, genomics, and proteomics are included together and referred to as a single field of inquiry called GGP.

[13] A series of international conferences on the topic of genome informatics have been held, which suggests that genome informatics may be emerging as a distinct field.

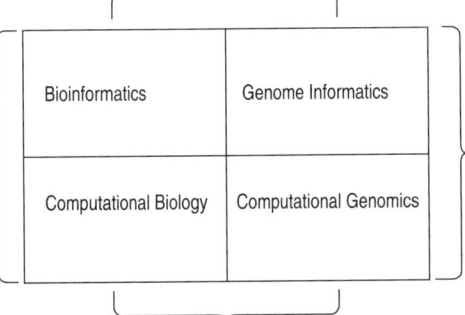

Figure 1 Content and Methodological Distinctions

genomics, and *functional genomics*. Whereas a *gene* is a discrete unit of inherited information that occupies a particular position on the DNA molecule in a chromosome,[14] a *genome* is the total DNA content of an organism. That is, a genome comprises genes as well as other DNA sequences that make up an organism.[15] *Genomics* is the comprehensive analysis of all the genes of an organism. As such, genomics is a field concerned with *acquiring knowledge about genes*, including an understanding of what genes do. In this sense, genomics is said to "yield information" (about genes and DNA segments).[16] Genome projects aim not merely to discover genes but to explore their *functions* in living organisms, populations, and communities.[17]

We have defined the field of computational genomics, and we have identified some theoretical and practical connections that link the fields of computer science and genomics. Given the nature of these links, both at the conceptual and practical levels, we should not be surprised to find that some moral concerns frequently associated with computer/information technology ethics and that some frequently

---

[14] Genes can also exist in the mitochondrial DNA in all eukaryotes, as well as in chromosomes.

[15] It is estimated that only about 4-5% of the human genome consists of gene information. The function of the remaining DNA is unclear; however, it is believed that some of it is likely to have a role in maintaining the stability of the genome.

[16] For example, knowledge about our individual genomes may indicate more precisely which diseases one is susceptible to and which medicines would work best in response.

[17] In this sense, "functional genomics" projects continue to engage researchers, even though the general structure of the human genome has been identified. In some cases, the expressions "functional genomics," "comparative genomics," and "population genomics" have been used to refer to the same field of genomic inquiry.

associated with bioethics/biotechnology ethics—as separate fields of applied ethics—now overlap and converge. However, we have said nothing yet about the ethical issues that arise in the area of computational genomics.

3. Ethical Aspects of Computational Genomics

Before identifying and analyzing particular ethical issues affecting computational genomics as a distinct field of ethical inquiry, we can ask a foundational question: What makes an issue an *ethical* issue? Ethical issues typically arise because of a range of concerns that ultimately have to do with notions such as fairness, autonomy, dignity, and respect for persons. These notions, of course, are very broad categories of ethical concern.[18] But we will see how each is, nonetheless, manifested in particular controversies involving computational genomics. For example, issues having to do with fairness arise because of concerns about the ways that one's personal genetic data can be used in making judgments about individuals, which can result in their being denied employment and health insurance. Autonomy issues arise because of concerns having to do with whether research subjects can make fully informed decisions when consenting to participate in genetics studies that involve the use of computational tools and techniques. And issues involving dignity and respect arise because of the stigmatization that genomics research subjects can experience in being identified with certain groups and subgroups generated by controversial uses of computational tools.

So we can begin to identify, at least in very general terms, some of the ethical concerns that arise in genomics research. However, identifying other, more specific, ethical issues in this area is not always easy because the computational tools and methods used make it difficult to anticipate the kinds of problems that can arise for research subjects. Thus some ethical issues affecting computational genomics research are not as easily recognizable as others, at least not at first glance. Goodman (1996, p. 223) points out that the study of ethical issues at the juncture of information technology and genomics is still mostly "uncharted." He notes that while a mature literature has emerged with respect to ethical, legal, and policy considerations in the distinct fields of information technology and genetics/genomics, the "unification of... ethics *and* [computing/information technology] *and* genomics awaits sustained discussion, rigorous analysis and clear identification of issues" (Italics Goodman).

Among the more easily recognizable ethical issues at the intersection of computing and genomics are concerns about privacy and confidentiality. For example, we noted above that a research subject's genetic data can be used in ways that violate an individual's confidentiality, which can result in some harm to that person.

---

[18] Ethical concerns involving fairness, respect for persons, and autonomy are sometimes considered "core" moral issues because they underlie ethical concerns that are often expressed and articulated at a higher level of discourse in applied ethics. For example, ethical issues identified as "higher-level" categories such as personal privacy and intellectual property express basic (or "core") moral concerns having to do with respect and fairness. Similarly, issues involving informed consent express core moral concerns having to do with autonomy. See Moor (2004, p. 109) for a discussion of how ethical concerns can be identified and understood at these two levels.

But there are many other specific issues that need to be identified and analyzed. For example, closely related to, but also distinct from, privacy and confidentiality issues is a cluster of concerns pertaining to informed-consent policies for research subjects. Consider that individuals who sign consent policies and participate in genetic/genomic studies involving the use of computational methodologies cannot always be told in advance which kinds of risks they may face as a result of their participation in a particular study. (Recall our earlier point that many outcomes made possible by computational methodologies are not easily anticipated.) Additional ethical issues at the intersection of computing and genomics arise because of concerns about property rights. For example, who should have ownership rights to personal genetic information that resides in computer databases? Which parties should be granted access to that information? Should entire genomes, or even DNA sequences and fragments, be eligible for patent protection?[19] Will granting such patents interfere with and possibly impede progress in genomic research?

One way of organizing ethical concerns affecting computational genomics is to include them under specific applied-ethics categories such as privacy, informed consent, and property,[20] as we did in the preceding paragraph. We should also note, however, that other organizing schemes have been suggested. For example, Goodman proposes a somewhat different model in which he articulates categories such as "data sharing," "database accuracy," and "group or subgroup stigma" (in addition to the category of privacy). Goodman's categories can also be helpful in analyzing many of the ethical issues that arise in computational genomics. Although the categories that Goodman articulates do not specifically mention concerns involving informed consent and intellectual property, the topical areas he identifies can be mapped into our three-fold ethical scheme involving privacy, consent, and property. For example, concerns about database accuracy can be included under the broader discussion of privacy; concerns pertaining to group/subgroup stigma can be subsumed within our analysis of controversies surrounding informed consent; and concerns affecting data sharing can be incorporated into in our discussion of intellectual property issues. So alternative schemes of analysis could also work. However, our analysis of ethical issues in this chapter (and in this book) will proceed from the vantage point of the three broad categories described above: personal privacy, informed consent, and intellectual property.

3.1. The ELSI Research Program and its Application to Genomics Projects: A Brief Historical Sketch

It is now widely recognized that information acquired through genetic and genomic research can have significant ethical and social implications for individuals. To address, in a pro-active manner, controversial ethical and social concerns that would likely emerge from the work on the Human Genome Project (HGP), the

---

[19]Of course, patents have already been granted for many genes (e.g., for the gene that codes for insulin). However, the moral question at issue is whether such patents *should* be granted.

[20]Elsewhere (Tavani, 2003), I have argued for this three-fold distinction as an organizing scheme to analyze ethical issues at the intersection of computing and genomics.

National Institutes of Health (NIH) and the U.S. Department of Energy (DOE) jointly proposed the establishment of an ethical, legal, and social implications (ELSI) program to guide genomics researchers.[21] In January 1989, the Program Advisory Committee of the Human Genome Project established a working group on ethics to develop plans for an ELSI component. The working group, which was later named the NIH-DOE Joint Working Group on Ethical, Legal, and Social Implications, held it first meting in 1989. That year, NIH created the National Human Genome Research Institute (NHGRI) to oversee its ELSI research requirements. In 1997, NHGRI's National Advisory Council for Human Genome Research (NACHGR) and DOE's Budget and Environmental Research Advisory Council (BERAC) jointly established the ELSI Research Planning and Evaluation Group (ERPEG). In February 2000, ERPEG issued its final report. Since then, NIH (through its NHGRI office) and DOE have administered their ELSI programs independently of each other. However, the programs are structured in a way to assure coordination between the two groups.[22]

The initial ELSI report, issued jointly by NIH and DOE in 1990, described ELSI's function; it also identified objectives for the ELSI Program, three of which were to: (1) anticipate and address social and ethical implications of mapping and sequencing[23] the human genome, (2) stimulate public discussion of the issues, and (3) develop policy options to ensure that the information would be used to benefit individuals and society.[24] To provide some guidelines for accomplishing these objectives, the ELSI report also identified a number of "focus areas" that included specific activities. The ELSI Research Program was also organized around several "program areas," three of which are important for our purposes: (1) "privacy and fairness" (in the use of genetic information), (2) "clinical issues" (including informed-consent practices in genetic testing), and (3) "commercialization" (of products, including property rights).[25]

The first ELSI program area, privacy and fairness, examines questions concerning who should have access to genetic information. For example, should this information be available to insurers, employers, courts, schools, adoption agencies, and the military? Concerns identified in this program area also focus on ways to prevent the misinterpretation or misuse of genetic information. The second program area, which identifies issues affecting clinical investigation, focuses on concerns relating to the impact of genetic testing on individuals, families, and society.

---

[21] NIH and DOE have devoted 3% to 5% of their annual HGP budgets to funding ELSI projects.

[22] See the Web site for the ELSI Research Program, available at http://www.genome.gov/10001763.

[23] Although "mapping" and "sequencing" are sometimes discussed in ways that suggest they are roughly the same functions, Cahill (1996, p. 10) notes that the following distinctions can be drawn. Whereas "mapping" is the assignment of human genes to locations on various chromosomes, "sequencing" is concerned with finding the precise order of each of the base pairs that comprise a particular gene.

[24] The ELSI Research Program initially listed four distinct functions. For our purposes, however, these functions can be understood in terms of the three objectives described above.

[25] The ELSI Research Program identifies several additional program areas, including "psychological impact and stigmatization," "conceptual and philosophical implications," "genetic testing," and so forth. However, we limit our discussion in this chapter to the three program areas identified above.

Concerns identified in this program area include questions about the design, conduct, participation in and reporting of genetics research. Among the concerns examined in the third program area, commercialization, are those affecting property rights. Questions examined under the category "commercialization" include: Who should have (legal) ownership rights to genes and DNA segments? Will patenting DNA sequences limit their accessibility and development into useful products?

3.2 Applying the ELSI Framework to Computational Genomics Research Involving Populations

The ELSI guidelines were framed at an early stage of HGP, when researchers focused mainly on mapping the human genome to reveal its *structure*. Now that this phase of HGP has been successfully completed, researchers aim at achieving a better understanding of the *functions* of particular genes and DNA segments. Noting that HGP has recently "given birth" to a number of more specialized genomics programs, Eric Juengst (1999) identifies six distinct genomics initiatives: (1) the Human Genome Diversity Project, (2) Complex Trait Genetics, (3) Public Health Genetics, (4) Pharmacogenetics, (5) Functional Genomics, and 6) the Environmental Genome Project (EGP).[26] One question worth considering from the vantage point of ethics is whether these initiatives introduce any new concerns for the ELSI Research Program. In particular, we focus on initiatives such as EGP and Functional Genomics, which can be viewed as programs that are part of, or at least closely associated with, a field that is now sometimes referred to as *population genomics*.

A key objective of the relatively new field of population genomics is to identify the underlying genes responsible for common chronic diseases in large populations. According to Taubes (2001, p. 42), population genomics research uses the following strategy: (1) sift through the DNA of large populations (possibly entire nations) to find identifying causes of diseases; (2) analyze the DNA of a few thousand victims of a disease; (3) compare it to the DNA of healthy individuals; and (4) identify the "salient differences"—i.e., those genetic variations that result in illness on the one hand, and health on the other. Population genomics researchers believe that the study of large families is desirable because members of a family share a common genetic inheritance. Taubes (p. 43) notes that population genomics research proceeds on the following assumption:

> ... *if you can find a few hundred family members with a certain disease and a few hundred without, you can be pretty confident that eventually you will find the mutation that is present in the afflicted members and absent from the DNA of the healthy ones.*

In the past, most human genetics studies focused on specific diseases and used relatively small sample sizes. However, Taubes (p. 43) notes that recent genomic technologies have enabled researchers to test significantly more individuals for sig-

---

[26]The EGP, which was recently launched in the U.S. by the National Institute for Environmental Health Sciences (NIEHS), is examined in detail in Chapter 5.

nificantly more genetic markers, in less time and at lower cost. An early example of this kind of research involved Nancy Wexler's classic study in the 1970s of an extended Venezuelan family to locate the gene for Huntington's disease. Taubes points out that since Wexler's study, faster methods of testing DNA samples have been developed, including computational techniques that compare and contrast DNA variations in larger populations. Because of the advances in computing and genomics technologies, it is now easier to identify individuals and entire populations that may be susceptible to certain kinds of diseases. Also, no hypothesis of disease causation is necessary; researchers can simply compare databases of genetic samples and disease records, employing computerized data-mining operations to find the "causative genes and gene variations at work"(Taubes, p. 42).[27]

Many who support the methods used in population genomics believe that research in this area will dramatically influence healthcare by helping members of populations identify more precisely the diseases to which they are susceptible. However, the data-mining techniques used to make this research possible have been controversial from an ethical perspective.[28] So we can ask whether population genomics research per se raises any special ethical concerns. In particular, we can question whether this research poses any challenges to specific objectives described in the ELSI Research Program, which we briefly examined. Are the ELSI guidelines, framed during the early phase of HGP, sufficiently robust to handle the kinds of ethical concerns that arise in light of the technologies (including data mining), methods, and practices currently used by population genomics researchers? We next consider a case involving deCODE Genetics Inc., a privately owned genetics company in Iceland, whose controversial practices may be in conflict with some objectives stated in the ELSI Program.[29]

3.2.1 The Case of deCODE Genetics, Inc.

Kari Stefansson, the founder of deCODE Genetics, left his academic position as a professor of neurology at Harvard University in 1996 to set up a genetics company in the private sector. While working on a Harvard-sponsored investigation into multiple sclerosis, Stefansson had some concerns about limited resources involving access to data about populations. These concerns influenced his decision to seek private funding to start a genetics company. Stefansson believed that the right kind of resources needed for genetics research could be found in a homogeneous population such as that of Iceland, his native country. Iceland's 280,000 inhabitants are descended from a common set of ancestors (Norse and Celts) who arrived there in the ninth century. Because the country's population was virtually

---

[27]The data-mining technique used in this process is described in detail in Chapters 9 and 10.

[28]Some ethical controversies associated with data mining are examined in Chapters 9 and 10.

[29]Because the ELSI Research Program was established to guide government-funded genomic research and development, but not research conducted in the private sector by genetics firms, it might seem unfair to apply those standards in the deCODE case. However, in assessing the practices of genetics companies in the private sector via the ELSI framework, we have a common structure or "base-line" upon which to articulate ethical concerns involving non-government-sponsored genetic/genomic research. Thus the ELSI Program provides us with a useful framework to assess ethical aspects of cases involving both government-funded and private-sector genomics research.

isolated until World War II, it has remained remarkably homogeneous and thus has provided population genomics researchers like Stefansson with an ideal resource.

Stefansson negotiated with the Icelandic government for access to the nation's health care records, which date back to 1915. He already had access to the Iceland's genealogical data, comprised of records that date back more than 1,000 years. Iceland's Parliament also promised Stefansson access to any government-owned medical information that he would need to conduct his research. With these and other agreements in hand, Stefansson officially formed deCODE Genetics and immediately began to construct a genetics database consisting of information based on DNA samples that would eventually be acquired from 70,000 volunteers. Stefansson's company was then able to link together and cross reference medical/healthcare records, genealogical records, and genetic information included in the three separate databases. With this information, deCODE has thus far successfully identified genes associated with several diseases, including osteoporosis, osteoarthritis, schizophrenia, and psoriasis (Stefansson, 2001, p. 52).

Why, exactly, has the deCODE Genetics case been so controversial from an ethical perspective? First, we should note that deCODE is neither the first nor the most recent private firm to engage in population genomics research. For example, Newfoundland Genomics (Newfoundland, Canada), Autogen (Melbourne, Australia), Uman Genomics (Umea, Sweden), DNA Services (Freemont, California), and Wellcome Trust Medical Research Council (UK) are but a few of the privately owned businesses currently engaged in population genomics research (Taubes, 2001). So deCODE's foray into the field of genomic research as a privately funded commercial enterprise is by no means peculiar. However, deCODE's practices have raised criticisms that extend beyond those leveled against other commercial genetics companies.

3.2.2 ELSI Concerns Raised in the deCODE Case

How do the ethical concerns raised in the deCODE case map into the three ELSI "program areas" that we identified in Section 3.1? The first program area that we identified there focuses on concerns having to do with privacy and fairness. Recall that these concerns include questions about who should have access to genetic information. Some critics believe the Icelandic parliament's decision to grant deCODE exclusive rights (for 12 years) to the information included in the nation's health-records database—in accordance with Iceland's Act on a Health Sector Database (1998)—raises questions having to do with fairness and equal access. As a result of this legislation, researchers who are not affiliated with deCODE are required to pay for access to medical information that was once freely available to them. Other critics have accused deCODE of being unfair because of alleged deception. Members of the Icelandic Medical Association (IMA), for example, believe that because deCODE has exaggerated the results of its research and has grossly underestimated the risks to individual privacy and confidentiality, it seriously misled and deceived the Icelandic population and its government.

As already noted, concerns in the first ELSI program area also include privacy-related issues pertaining to questions about access to and control of personal genetic data that resides in databases. For example, who should (and who should not) have access to that data? In the case of deCODE, personal genetic information resides in company-owned databases. Should deCODE hold exclusive ownership rights to such data? If so, should deCODE hold rights to this data *in perpetuity*, and should deCODE be permitted to do whatever it wishes with that data once it has been acquired? Alternatively, should the Icelandic government be granted custodial rights to the personal genetic information in deCODE's databases, so that it can better protect the interests of Iceland's citizens?

Consider that those who voluntarily provided DNA samples to deCODE had the expectation that their personal genetic data was *confidential* information and that it would be treated and protected accordingly. However, it is difficult to see how those individuals could be assured that their genetic data would remain confidential and continue to be protected once it was aggregated and cross-referenced with information in deCODE's non-genetic (or phenotype) databases.[30] Mannvernd, a physicians-led citizen group of the Association of Icelanders for Ethics in Science and Medicine, has argued that deCODE's practices in such areas violate specific articles of Iceland's Constitution.

The second ELSI program area that we identified, which focuses on clinical issues in genetic testing, addresses concerns pertaining to the principle of informed consent—i.e., consent policies used in getting permissions from research subjects. In the deCODE case, consent problems arise on at least two different levels. For one thing, medical information in the healthcare database that deCODE acquired from the Icelandic government is based on data gained from "presumed" (rather than informed) consent. For another thing, the DNA-specific information allegedly acquired from "informed" volunteers may not meet the required conditions for what is considered "valid informed consent."[31] Even though research subjects who voluntarily provided their genetic data to deCODE agreed to have their data used in a specific context, they had not necessarily consented to have it used in secondary or subsequent contexts as well. That is, they had not explicitly authorized their personal data to be cross-referenced with other personal medical data, such as electronic records residing in the healthcare and the genealogy databases. Nor had they authorized the subsequent "mining" of the cross-referenced information. In this sense, the consent practices used by deCODE would seem to be far more "opaque" than transparent to the research subjects involved.[32]

The third ELSI program area that we identified focuses on concerns involving commercialization, some of which have to do with property rights for genetic data. We already noted some controversies surrounding the ownership of personal genetic data in connection with our discussion of issues involving the first ELSI program area, concerned with privacy and fairness. Other questions involving property rights that arise in the deCODE case have to do with the ownership of

---

[30] As we will see in Chapter 10, normative privacy protection does not extend, at least not explicitly, to personal information once it is part of information in the aggregate.

[31] The conditions required for "valid informed consent" are examined in detail in Chapter 10.

[32] In Chapter 10, we examine some issues involving the lack of transparency in informed-consent policies used by deCODE Genetics, as well as by other organizations that conduct genomics research, to show why they are problematic for research subjects.

non-personal genetic data. For example, should deCODE, and other private genetics companies, be able to patent DNA sequences and thus have exclusive rights to that genetic and genomic data?[33] Also, some critics fear that, as a result of the commercialization of data controlled by deCODE, Iceland's genetic information is being reduced to a marketable commodity.

ELSI-related concerns involving commercialization also include worries related to conflicts of interest. Some critics point to the special relationship that has evolved between deCODE and the Icelandic government. Others have alleged conflict of interest in the commercial sphere by pointing to certain business/financial relationships and arrangements that deCODE has developed with pharmaceutical companies and computer firms. For example, critics note that deCODE signed a contract estimated at $200 million with Roche, a Swiss pharmaceuticals company, and signed a three year contract with IBM to sell jointly an integrated computer system that combines deCODE's proprietary "gene-mining" software with IBM's servers and database software.[34]

We have examined a cluster of ELSI-related concerns in the context of practices involving deCODE Genetics. Some of these practices would clearly seem to challenge ethical guidelines established by the ELSI Research Program. However, we should note that deCODE also has its defenders, including Stefansson, who offer a drastically different perspective on the charges leveled against this genomics company. For one thing, they argue on utilitarian grounds that fellow Icelanders (as well as humans in general) stand to benefit from the genetic discoveries made by deCODE. They also argue that many of the criticisms leveled against deCODE are based on misperceptions about the ways in which deCODE acquires and uses data stored in three very different kinds of databases (i.e., data involving genealogical records, health care information, and DNA). Stefansson (2001) points out that Iceland's genealogical information is "public data" because it is already in the public domain. He also notes that with regard to medical information contained in the health care database, Icelanders have a choice because they can *opt out* of having their medical information used. And he points out that the controversial and highly sensitive information in deCODE's DNA database has been acquired only from those individuals who have expressly given their consent. So, in Stefansson's view, deCODE is not guilty of engaging in any practices that are either illegal or immoral.

4. Do Ethical Issues in Computational Genomics Deserve Special Consideration?

At this point in our discussion, it is useful to recall some key themes examined in this chapter. Thus far, we have defined the field of computational genomics and have identified some ethical concerns that arise in this field. We have also briefly

---

[33] These and similar questions are examined in detail in the chapters that comprise Section IV of this book.

[34] Perhaps deCODE's supporters would not see a conflict of interest here because they could point out that deCODE, as a private firm, has the right (and arguably even an obligation to its stockholders) to join in commercial partnerships with other businesses. Critics, however, worry that such "cozy" business alliances only serve the interests of the commercial entities involved, and not the greater welfare of the public.

examined the ELSI Research Program, which was established to anticipate and address ethical issues in this area. We then considered the deCODE Genetics case in order to illustrate some ethical concerns that arise with respect to the three ELSI program areas we identified. Although some have defended deCODE's practices, it would seem that a number of serious ethical concerns arise in this case and that these issues need further analysis. One question that we have not yet considered, however, is whether those ethical concerns deserve special *normative* protection—i.e., protection in terms of genetic-specific policies or laws. Consider that ethical issues arise in areas and fields other than genetics, as well. So why should ethical concerns involving computational genomics, or for that matter concerns involving the broader field of genetics, deserve any special normative consideration?

Underlying many of the arguments favoring special normative protections for personal genetic information is the assumption that this kind of information is unique and thus is "exceptional" in some sense. In other words, because genetic data is assumed to be sufficiently different from other kinds of personal information, it requires an exception to the rule for how ordinary medical information is protected. Murray (1997), Chadwick and Thompson (1999), and others use the expression *genetic exceptionalism* to describe this position.[35]

4.1 Genetic Exceptionalism

It would seem that an individual's genetic records should enjoy at least the same level of normative protection granted to other kinds of personal health and medical information. But do those records deserve greater protection? We can begin by asking whether the differences between genetic and non-genetic personal data are morally relevant—i.e., whether they make a *moral* difference. In our analysis of this question, it is useful to look more closely at the rationale underlying the thesis of genetic exceptionalism. Onora O'Neill (2002, p. 101) notes that genetic exceptionalists believe that genetic data is "intrinsically unlike other personal, including medical, data" because that data "provides information not only about an individual from whom a sample is taken, but also about related individuals." She points out that in coming to know something about our own genetic make-up (via genetic testing), we may also learn something about a relative's genetic make-up. For example, O'Neill notes that a person with an identical twin may discover a genetic test result as true of her twin as of herself.

To support O'Neill's contention, consider her point that if a grandchild of a grandparent who died of Huntington's disease decides to test for the disease and is determined to have the gene (which is dominant and highly penetrative), then the child's parents can discover (as a result of their child's test) something about themselves even if they elected not to be tested for Huntington's. Facts such as these are often cited to give reasons for regarding genetic data as the *exception* to the rule that medical information is individual (and that all personal medical data pertains

---

[35]Genetic exceptionalism, sometimes also referred to as "genetic essentialism" (see, for example, Marturano (2003), Marturano and Chadwick (2004), and Juengst (2004)), is the view that genetic information is "exceptional" because it is qualitatively different from other kinds of medical data and thus merits special ethical and legal consideration.

only to individuals). The irony here is that an argument used to justify the protection of one's genetic information is based on an exception to the rule that an individual's personal information needs to be protected because it is sensitive data that pertains exclusively to *that individual*.

4.1.1 Evaluating the Thesis of Genetic Exceptionalism

Defenders of the genetic-exceptionalism thesis appeal to the fact that when members of a family suffer from a genetic condition, the problem can bear on many family members—perhaps even on the entire family—and not only on an individual.[36] But is this factor peculiar to knowledge affecting genetic information? Consider that various kinds of medical data can also be highly indicative of risks for others, including risks for one's family members. For example, O'Neill (p. 104) points out that infection, contagion, and exposure to environmental hazards can result in the fact that "cohabitating individuals" share a great deal of medical history, especially if those individuals are related.

In analyzing the arguments for genetic exceptionalism, O'Neill suggests that the very notions of *genetic data* or *genetic information* are themselves problematic because they are "systematically unclear." Initially, one might be inclined to assume that much of their personal genetic information can be revealed only through strict genetic testing. But O'Neill points out that a great deal of what she calls "loosely genetic information" can be revealed in our physical appearances (i.e., in our phenotypes, which can suggest familial relationships). She also notes that some medical or personal information that is "ostensibly not genetic," can nevertheless reflect genetic factors. For example, when insurers ask us to disclose how old our parents or grandparents were at death, they are seeking some "loosely genetic information" about our heredity and life expectancy (O'Neill, p.103).

So, on O'Neill's analysis, it would seem that arguments in support of genetic exceptionalism are based on two assumptions, both of which are problematic: (1) genetic data is the only form of personal medical data that can indicate health risks for members of one's family; and (2) one's genetic information can be revealed only through (strict) genetic testing. As we have seen, non-genetic medical information (as well as genetic information) can indicate health risks for family members, and "loosely genetic information" can be disclosed in non-genetic-specific contexts. So, a person's medical information can reflect a "mix of genetic and non-genetic factors," as O'Neill correctly notes. Thus any claim that genetic information is unique, or is an exception to the rules involving ordinary medical information, cannot be justified on the basis of arguments examined thus far.

4.1.2 Evaluating the Case for and Against Anti-genetic Exceptionalism

Those who reject genetic exceptionalism can be described as "anti-genetic-exceptionalists" or simply as "anti-exceptionalists" (Tavani, 2004). Anti-

---

[36]Goodman (1996, p. 226) points out that genetic information, unlike other kinds of health and medical records, is rarely about a single individual. As he also notes, one person's consent to release genetic information "constitutes a *de facto* release of other individuals, that is, his relatives."

exceptionalists maintain that there is nothing inherent in genetic information itself that makes it unique from a medical perspective. Anti-exceptionalists have attacked the exceptionalist's position not only on the basis of conceptual ambiguities and flaws (similar to those mentioned in O'Neill's criticisms), but also on practical and empirical grounds. Christiani, Sharp, Collman, and Suk (2001, p. 530) believe that most arguments for treating genetic data as a separate category of medical or health-related information tend to be based on "common misperceptions" that many people have about genetic information. The authors concede that special treatment may be appropriate when genetic information is "strongly predictive" of an individual's disease risks—e.g., in the case of Huntington's disease—where: (1) there is a direct cause between a particular gene and the disease, and (2) that link could cause harm to individuals by preventing them from securing employment and health insurance. Christiani et al. (p. 530) also note, however, that it is very important to distinguish between the kinds of protection needed for individuals diagnosed with "disease genes" and "susceptibility genes." The authors point out that, to date, most scholarship on genetic privacy has focused on rare genetic conditions associated with "highly penetrant genes."

Arguably, Christiani et al. provide an important insight into how certain misconceptions about genetic information can foster and reinforce genetic exceptionalism. And they offer a useful distinction between disease genes and susceptibility genes, which could have implications for how genetic privacy laws are framed in the future. But do the distinctions they draw get to the heart of the dispute underlying the debate between exceptionalists and anti-exceptionalists? And, does a "two-tier" distinction, based on disease genes vs. susceptibility genes, offer a plausible rationale for framing special genetics laws? Consider, for example, that the status of some genes originally thought to be either low-penetrant or non-disease genes could later be determined to be either highly penetrant or disease genes, and vice versa. So, any defense of the anti-exceptionalist position that is based on grounds similar to those provided by Christiani et al. can certainly be questioned.

Even though we have seen that there are good reasons to be suspicious of genetic exceptionalism, we can also see that the anti-exceptionalist position is difficult to defend on the basis of the arguments we have examined. Because anti-exceptionalists deny what exceptionalists assert, and vice versa, it would seem that if one thesis is false, the other must be true. However, we can show that both positions proceed from a commonly shared assumption that is false. We can also show how it is possible to frame a coherent alternative position that cuts across both the exceptionalist and anti-exceptionalist views.

4.2. An Alternative Scheme for Analyzing the Exceptionalist versus Anti-exceptionalist Debate about Personal Genetic Data

We have seen that genetic exceptionalists tend to overstate the differences between genetic and non-genetic personal information. Anti-exceptionalists, on the other hand, arguably underestimate some critical differences. Both sides, however, seem to proceed on the basis of an assumption that can be framed in the form of the following conditional statement:

> *If genetic data deserves special normative consideration, then it must be because that data is qualitatively different from other kinds of personal data, including an individual's medical data.*

Genetic exceptionalists affirm the antecedent of this conditional when they claim that personal genetic information deserves special consideration from a normative perspective. However, they then infer from this premise that special consideration is needed *because* personal genetic data is qualitatively different from other kinds of personal information. Conversely, anti-exceptionalists correctly deny the consequent of the above conditional statement, noting that not all genetic data is qualitatively different from other kinds of personal medical information. But they then go on to conclude that personal genetic data does not deserve any special normative consideration. Neither side is entitled to the conclusion it draws, however, because of an erroneous assumption that functions as a premise in the arguments used by both genetic exceptionalists and anti-exceptionalists.

We have also seen that Christiani, Sharp, Collman, and Suk (2001) correctly point out that not all personal genetic information is necessarily problematic from the perspective of personal privacy. However, many anti-exceptionalists also tend to conclude, on the basis of this insight, that no special privacy concerns arise in the use of personal genetic data. Can such an inference be justified? And if not, on which grounds, if any, are distinctions between personal genetic information and non-genetic personal information relevant from the point of view of individual privacy in particular, and ethics in general? O'Neill (2002, p. 105) argues that there are good reasons for regarding certain *uses* of personal genetic data—rather than the data itself—as "distinctive." She believes that this data should be viewed as distinctive, not because of factors cited by genetic exceptionalists but because of the uses to which it can be put by third parties. O'Neill correctly notes that third parties who have "no personal interest"[37] in someone's genetic data are nonetheless sometimes very eager to gain access to this data and use it for ends other than that for which it was intended.

We can agree with O'Neill's suggestion that it is because of the uses to which genetic information can be put, rather than because of anything intrinsic to the information itself, that this information warrants special consideration. However, we do not have to agree with her suggestion that this information is also *distinctive* because of the ways it can be used. In other words, we can decouple the claim that personal genetic data deserves special consideration from any claim that this data itself must also be distinctive (perhaps in some metaphysical sense). But what, then, is it about the use of one's genetic data that is so controversial from an ethical point of view? For example, we have seen that the use of other kinds of personal information, including medical information, can also raise significant ethical concerns for research subjects. So why should the use of personal genetic information deserve any special normative consideration—i.e., any moral or legal consideration in addition to that given to ordinary

---

[37] O'Neill (p. 105) believes that third parties have no "personal interest" in that data in the sense that the data does not bear on their own lives or families and thus is "not relevant to their own health."

medical data? The answer, I believe, is because of the kinds of *unanticipated outcomes* that are made possible by the computational techniques used to manipulate that data.

4.3 Computational Techniques, Unanticipated Outcomes, and the Challenge for Normative Policies

In the first section of this chapter, we saw that because genomic research proceeds computationally, it is *dependent* on computers. We also saw that computers play both computer *assisting* and computer *enhancing* roles in genomics. For example, we saw that computers can assist genomic researchers in carrying out certain kinds of routine, day-to-day tasks in many of the same ways that computers assist other professionals, including medical professionals. But we also saw that computing technologies have enhanced research in genetics/genomics to the point that the "new genetics" would seem to depend on computing technology in ways that other fields do not. Because of its special dependency on computing technology, and primarily because of the computational techniques and analyses involved, the kinds of outcomes made possible by genomics research can be very difficult to anticipate. Furthermore, these outcomes can introduce normative issues for which we have no clear laws or policies. James Moor (1985) refers to the absence of clear normative policies in these situations as *policy vacuums*.

What are policy vacuums, and how do they arise because of the use of computers and computing technologies? Moor has argued that computer technology, unlike other technologies designed to perform one or more specific functions, is "logically malleable" because it can be shaped and molded to perform a variety of tasks. He points out that because of a computer's general-purpose nature, it can generate "new possibilities for human action," which, in turn, generate one or more policy vacuums.[38] These vacuums or voids arise because we have no clear normative rules and policies in place to guide the new choices made possible by computer technology. Initially, the solution to this problem would seem quite simple—first identify the vacuums, and then fill them by framing new policies and laws. However, such a solution is not as simple as it might appear at first glance. Moor notes that sometimes the new possibilities for human action generated by computing technology also introduce *conceptual muddles*. In these cases, we must first clear up the muddles before we can fill the voids in which there is either no policy or no adequate policy to guide our actions.

Consider some of the conceptual muddles that arise in the case of computational genomics, which are illustrated in the following questions. Does the use of computational techniques such as data mining in genomic research necessarily violate the privacy of research subjects who participate in genomics studies? Are categories such as "group confidentiality" and "collective confidentiality"[39] needed to prevent the kind of discrimination and stigmatization that research subjects

---

[38]Moor (1999) points out that policy vacuums are not unique to computing technology. For example, he notes that policy vacuums have been generated in other fields as well, including biotechnology. But Moor also notes the scale on which policy vacuums are generated by computing technology has far exceeded those introduced by alternate technologies.

[39]Goodman (1996, 1998) has used both of these expressions.

can encounter because of their potential identification with one or more "new groups" that can be constructed via computational means? Is the ideal of "valid informed consent" achievable for research subjects when the computational techniques used in genomics studies make it so difficult for the researchers to communicate to participants the kinds of risks they potentially face?[40] Also, what is the legal status of personal genetic data; and who, if anyone, should own it? These are merely a few, but nonetheless representative, examples of the kinds of questions that emerge as a result of computational genomics research.

In responding to these questions and the concerns they raise, we need to frame clear normative policies. However, before we can frame adequate policies for these and similar questions involving genomic research, we first have to elucidate some muddles and conceptual confusions that underlie many of the questions. In the subsequent chapters of this book, many confusions and muddles surrounding moral controversies in computational genomics research are identified and elucidated. In some cases, specific proposals are also put forth with respect to which kinds of normative policies should be adopted. However, our main objective in this book is to identify, elucidate, and analyze these issues.

5. Conclusion

We have identified and examined a cluster of ethical concerns at the intersection of computing and genomics. We began by examining some theoretical and practical connections that link the fields of computer science and genetics. We then defined the field of computational genomics, distinguishing it from related fields such as bioinformatics, computational biology, and genome informatics. Following a brief description of three key ethical concerns affecting computational genomics, we examined the ELSI Research Program that was jointly established by NIH and DOE. Next, we considered a case involving deCODE Genetics, which illustrated how specific principles in the ELSI framework are challenged by computational techniques used in population genomics research. We then questioned whether ethical aspects of computational genomics deserve special normative consideration, and we noted that many arguments for this view are based on the thesis of genetic exceptionalism: the assumption that an individual's genetic information is qualitatively different from other kinds of personal data, including medical data. We concluded by noting that both genetic exceptionalists and anti-exceptionalists proceed from an erroneous assumption; and we argued that a different kind of rationale, based on the kinds of unanticipated outcomes made possible by the computational techniques used in genomic research, can be given to show why the use of personal genetic data warrants special normative consideration.

---

[40]There are controversies surrounding what O'Neill (2002) and others refer to as the "problem of opacity" or lack of transparency in current informed-consent practices. We examine this issue in detail in Chapter 10.

Acknowledgments

I am especially grateful to Mark Bolt, Lloyd Carr, Jeremiah Dolan, Martin Menke, and Lawrence Silbert for some helpful comments and suggestions on an earlier version of this chapter.

References

Anderson, James G., and Kenneth W. Goodman (2002). "The Challenge of Bioinformatics." In J. G. Anderson and K. W. Goodman, eds. *Ethics and Information Technology: A Case-Based Approach to a Health Care System in Transition*. New York: Springer.

Biomedical Information Science and Technology Consortium (BISTIC) Definition Committee (2000). NIH Working Definition of Bioinformatics and Computational Biology, available at http://brin.uams.edu/bionfodefinition.html and www.geocities.com/bioinformaticsweb/definition.html. Accessed on 12/2/04.

Cahill, George F. (1996). "A Brief History of the Human Genome Project." In B. Gert, et al., eds. *Morality and the New Genetics: A Guide for Students and Health Care Patients*. Sudbury, MA: Jones and Bartlett, pp. 1–28.

Castells, Manuel (2001). "Informationalism and the Network Society." Epilogue in Pekka Himanen. *The Hacker Ethic: A Radical Approach to the Philosophy of Business*. New York: Random House, pp. 155–158.

Chadwick, Ruth, and Alison Thompson, eds. (1999). *Genetic Information Acquisition, Access, and Control*. New York: Kluwer/Plenum.

Christiani, David C., Richard R. Sharp, Gwen. W. Collman, and William A. Suk. (2001). "Applying Genomic Technologies in Environmental Health Research: Challenges and Opportunities," *Journal of Occupational and Environmental Medicine*. Vol. 43, No. 6, pp. 526–533.

Collins, Francis S. (2004). Interview (with Bradie Metheny on "Key Facts About Genomic Research") in *Policy Roundup–State of Genomic Research*, available at http://www.laskerfoundation.org/homenews/genome/pround_gn.html. Accessed 12/5/2004.

Dickson, David (1995). "Open Access to Sequenced Data Will Boost Hunt for Breast Cancer Gene," *Nature*, Vol. 328, November, p. 425.

ELSI Research Program. National Human Genome Research Institute. Available at: http://www.genome.gov. Accessed 12/5/2004.

Goodman, Kenneth W. (1996). "Ethics, Genomics, and Information Retrieval," *Computers in Biology and Medicine*, Vol. 26, No. 3, pp. 223–229.

Goodman, Kenneth W. (1998) "Bioethics and Health Informatics: An Introduction." In K. W. Goodman, ed. *Ethics, Computing, and Medicine: Informatics and the Transformation of Healthcare*. New York: Cambridge University Press, pp. 1–31.

Goodman, Kenneth W. (1999). "Bioinformatics: Challenges Revisited," *MD Computing,* May/June, pp. 17–20.

Juengst, Eric (1999). "Anthropology, Genetic Diversity, and Ethics." Presented at the Workshop for Twentieth Century Studies, University of Wisconsin - Milwaukee, Feb. 12.

Juengst, Eric (2004). "Face Facts: Why Human Genetics Will Always Provoke Bioethics," *Journal of Law, Medicine, and Ethics*, Vol. 32.

Marturano, Antonio (2003). "Molecular Biologists as Hackers of Human Data: Rethinking IPR for Bioinformatics Research," *Journal of Information, Communication, and Ethics in Society*, Vol. 1, No. 4, pp. 207–215.

Marturano, Antonio and Ruth Chadwick (2004). "The Role of Computing is Driving New Genetics' Public Policy," *Ethics and Information Technology*, Vol. 6, No. 2, pp. 43–53.

Moor, James H. (1985). "What Is Computer Ethics?" *Metaphilosophy*, Vol. 16, No. 4, pp. 266–275.

Moor, James H. (1999). "Using Genetic Information While Protecting the Privacy of the Soul," *Ethics and Information Technology*, Vol. 1, No. 4, pp. 257–263.

Moor, James. H. (2004). "Just Consequentialism and Computing." In R. A. Spinello and H. T. Tavani, eds. *Readings in CyberEthics*. 2nd ed. Sudbury, MA: Jones and Bartlett, pp. 107–113.

Murray, Thomas (1997). "Genetic Exceptionalism and Future Diaries: Is Genetic Information Different from Other Medical Information?" In M. Rothstein, ed. *Genetic Secrets: Protecting Privacy and Confidentiality*. New Haven, CT: Yale University Press, pp. 61–73.

O'Neill, Onora (2002). *Autonomy and Trust in Bioethics*. Cambridge: Cambridge University Press.

Stefansson, Kari (2001). "Population, Inc." Interviewed in *Technology Review*, Vol. 104, No. 3, pp. 50–55.

Taubes, Gary (2001). "Your Genetic Destiny for Sale," *Technology Review*, Vol. 104, No. 3, pp. 40–46.

Tavani, Herman T. (2003). "Ethical Issues at the Intersection of Population Genomics Research and Information Technology." In A. Backus and H. T. Tavani, eds. *Proceedings of the Conference on Privacy, Informed Consent, and Genomic Research*, Harvard School of Public Health, pp. 1–12.

Tavani, Herman T. (2004). "Genomic Research and Data-Mining Technology: Implications for Personal Privacy and Informed Consent," *Ethics and Information Technology*, Vol. 6, No. 1, pp. 15–28.

II: MORAL, LEGAL, POLICY, AND (SCIENTIFIC) RESEARCH PERSPECTIVES

The chapters that comprise Section II provide four distinct methodological perspectives—moral/philosophical, legal, public policy, and scientific research, respectively—from which the specific controversies and issues in computational genomics examined in Sections III-V can be analyzed. In Chapter 2, Bernard Gert examines aspects of the Human Genome Project (HGP) from the point of view of moral theory. Gert shows how his account of "common morality," which is developed more fully in his books *Common Morality* (2004) and *Morality* (2005), can be applied to specific issues affecting HGP. To appreciate Gert's contribution, it is useful to contrast his notion of common morality with some traditional ethical theories.

Ethical Theory and "Common Morality"

In evaluating the moral status of social policies, philosophers tend to appeal to one or more "standard" ethical theories, which are typically based on criteria such as *consequences* or *duties*.[1] Utilitarianism, one form of consequentialist ethical theory, proceeds on the notion that we can determine whether or not a particular social policy is morally acceptable simply by assessing the consequences that would likely result from implementing that policy. Generally speaking, utilitarians are interested in advancing only those social policies that produce the greatest good (social utility) for the greatest number of individuals. Deontological (or duty-based) ethical theories, on the other hand, reject the view that consequences themselves can be used as the appropriate criterion in determining whether a particular social policy is morally acceptable or unacceptable. Deontologists point out that it is possible for a policy to yield desirable consequences for the greatest number of people and still be a morally unacceptable policy.

For deontologists, a social policy is morally acceptable only when everyone affected by that policy is respected as an individual and is given equal consideration. In this scheme, it is not morally permissible for some individuals to be used as a means to some further end (e.g., an end in which the majority of individuals are affected favorably at the expense of the minority). Rather, each individual is

---

[1] In addition to consequence-based and duty-based theories, other kinds of ethical theories that have received serious consideration in the philosophical literature are character-based and contract-based. See Tavani (2004) for a comparative analysis of these four types of ethical theories.

considered to be an *end-in-him/her-self*, and deontologists argue that we have a moral duty to ensure that each individual is treated accordingly.

Critics of utilitarianism often note that because utilitarians are so concerned with promoting happiness for the majority, they ignore the importance of justice and fairness for each individual. Conversely, deontologists are often accused of ignoring the importance of happiness and overall social utility, because they focus exclusively on the primacy of notions such as duty, autonomy, rights, and respect for each individual. Presumably, an ideal solution would be to frame a comprehensive ethical theory that combined the strengths of both utilitarian and deontological theories while avoiding the respective weaknesses of each. However, philosophers have not had an easy time in reconciling these theories in a way that is both coherent and logically consistent.

Gert argues that standard ethical theories, including utilitarian and deontological theories, are both artificial and simplistic and thus are inadequate models for understanding and analyzing everyday moral issues. He proposes that our system of ordinary moral rules, which Gert calls *common morality*, should be used as an alternative to those theories. He defines "common morality" as "the moral system that most people use, usually not consciously, in deciding how to act when confronting moral problems and in making moral judgments." Gert believes that the common moral system, which he also describes as one that is "public and informal," has rules that are already known, at least implicitly, by everyone.[2]

Gert suggests that because there is far more agreement than disagreement on most moral matters, we should try to better understand the reasons *why* we agree in these cases. He also notes that the fact there is disagreement on some controversial issues should not necessarily deter us from striving to find some common positions on those moral issues. Gert's account of morality acknowledges that reasonable people can and will disagree on what the "best" answer is to a particular moral issue. However, he also points out that even if there is no uniquely right answer to a particular moral question, there are still plenty of *wrong* answers. In this way, Gert avoids the charge of ethical relativism[3]—i.e., the view that any position is acceptable so long as one group or even one individual holds it.[4]

Legal, Policy, and (Scientific) Research Perspectives

Chapters 3-5 analyze controversies involving computational genomics from the methodological perspectives of law, public policy, and scientific research, respectively. Chapter 3 offers a *legal* perspective. Here, law professor Dan Burk notes that recent advances in genetic engineering make possible the design of "pro-

---

[2] He elaborates on these and other aspects of his notion of moral system in Gert (2004, 2005). For an excellent summary and analysis of Gert's moral system, as well as an account of how Gert's system can be applied to issues involving computing, see Triplett (2002). Some of the ways that Gert's moral system can be applied to issues in computing are also examined in Tavani (2004).

[3] Triplett (2002) argues that Gert's account of morality avoids the kind of relativism rampant in our popular culture as well as the dogmatism found in many absolutist theories.

[4] Interested readers may wish to apply Gert's notion of common morality to cases and scenarios examined in the chapters included in Sections III-V, as well as the deCODE Genetics case examined in Chapter 1.

grammable biological artifacts." Burk points out that such programming may include "usage constraints" that will alter the balance of ownership and control for biotechnology products. As a result of these concerns, Burk believes that we are presented with new conceptual problems that require a more comprehensive evaluation of the interplay between law and "technologically embedded values."[5] In particular, Burk's chapter highlights the problem of distinguishing "coded constraints" that we might view as "equivalent to law" from other types of technologically embedded values. In this sense, Burk also addresses some of the broader questions involving what he describes as "long standing discussions about contract laws' effective application to technological restraint."

In Chapter 4, Ruth Chadwick and Antonio Marturano examine some public-policy implications of genomics research, which they argue have increasing relevance to policy issues involving computer/information technology. The authors note that some of the most pressing policy issues have "the potential to increase or decrease inequalities." Chadwick and Marturano begin by asking whether genomics should be considered a "global public good" and thus be viewed as something that is "globally relevant." But the authors note that even if genomic data and genomic databases are globally relevant, the genomic information contained in those databases may not be "globally available." Thus the authors illustrate how an ethical aspect of genomics intersects with a concern that is central to computer/information technology ethics: the *digital divide*. Chadwick and Marturano point out that because the field of genomics depends on the "power of computing," access to genomic information requires technology. Thus technology is a "limiting factor in facilitating access" to genomic data at all levels, including global access. The authors argue that because of this, public-policy considerations should include attention to the kind of infrastructure investment that is required for the "potential benefits of genomics to accrue" and for a "proper 'benefits-sharing' to be feasible" at the global level.[6] Chadwick and Marturano also note, however, that the infrastructure questions they examine are "just the beginning" of the public policy issues at the intersection of genomics and computer/information technology that need to be addressed.

In Chapter 5, David Christiani, Richard Sharp, Gwen Collman, and William Suk examine some controversies surrounding genomic technologies from the perspective of scientific research, which can arise at the intersection of genomics research and environmental health. Noting that recent discoveries in molecular biology and genetics have made it possible for environmental health researchers to examine how genetic characteristics affect response to environmental exposures, the authors point out that understanding such gene-environment interactions offers exciting possibilities for the prevention and control of environmentally induced diseases. Despite the potential benefits, however, the authors worry that

---

[5]Burk's chapter also addresses some controversies involving intellectual property rights and genetics/genomics, which are examined in detail in the chapters comprising Section IV. So his chapter can also be read in conjunction with the chapters in that section, including Chapter 14, where Burk examines some legal implications in applying the "open source" model in computer software to property–rights issues involving genomics.

[6]Marturano also addresses some concerns that arise because of access to genomic information in Chapter 13, where he proposes the application of a model based on the "open source movement" in computer software to the regulation of genomic data.

the collection and analysis of genetic information in environmental health present many of the same kinds of challenges involving ethical, legal, and social implications (ELSI)[7] that apply in other types of genetic research.

Christiani et al. describe some specific ELSI challenges that relate to research involving the Environmental Genome Project (EGP). For example, the authors show how some individuals participating in EGP-related studies could be at increased risk for stigmatization and discrimination because of the sensitive nature of the genetic information collected and used in those studies. However, the authors also worry that "stringent constraints" affecting recent informed-consent policies, which have been advocated by Institutional Review Boards (IRBs) on behalf of research subjects in genetic testing, may threaten research in this area.[8] They also note that this could have negative implications for future research involving public health. Christiani et al. conclude their essay with some specific recommendations for how EGP research should move forward with respect to ELSI challenges.

Review Questions for Chapters in Section II

1. Identify some of the key differences between consequence-based (utilitarian) and duty-based (deontological) ethical theories.
2. What does Bernard Gert mean by "common morality"?
3. Why does Gert believe that morality is a *system* comprised of rules that are both "public" and "informal"?
4. How does Gert propose that his moral system be applied to ethical issues involving the Human Genome Project?
5. How could Gert's account of common morality be applied to the case involving deCODE Genetics (examined in Chapter 1)?
6. What does Dan Burk mean when he says that recent advances in genetic engineering "now allow the design of programmable biological artifacts"?
7. What does Burk mean when he says that we are presented with new conceptual problems that require a more comprehensive evaluation of the interplay between law and technologically embedded values?
8. What do Ruth Chadwick and Antonio Marturano mean by the claim that some of the most pressing public-policy issues involving genomics and information technology have "the potential to increase or decrease inequalities"? How do they relate this concern to ongoing issues affecting the digital divide?
9. Assess Chadwick and Marturano's claim that public-policy considerations need to attend to the infrastructure investments required for the potential benefits of genomics to accrue and to be shared at the global level. Do you agree with the authors? Explain.
10. What is the Environmental Genome Project (EGP)? What kinds of challenges do Christiani, Sharp, Collman, and Suk believe that EGP research poses for the ELSI Research Program?

---

[7]For a discussion of the ELSI Research Program, see Chapter 1.

[8]Concerns raised about informed-consent policies for research subjects participating in environmental genomics studies, and the role that IRBs play in this process, are examined in detail in Chapter 10.

11. Why do Christiani et al. believe that research subjects participating in EGP studies are at increased risk for discrimination and stigmatization? Why do the authors also believe that stringent informed-consent policies, such as those advocated by Institutional Review Boards (IRBs) on behalf of research subjects, are "inappropriate"? And why do they believe that these policies could have negative implications for public health research?
12. Do you agree with Christiani et al. that the existing ELSI framework needs to be expanded in light of challenges posed by EGP research? Do you believe that the authors' distinction between "susceptibility genes" and "disease genes" can be useful in framing more "appropriate" privacy and informed-consent policies for research subjects? Explain.

References

Gert, Bernard. (2004). *Common Morality: Deciding What to Do*. New York: Oxford University Press.

Gert, Bernard. (2005) *Morality: Its Nature and Justification*. New York: Oxford University Press.

Tavani, Herman T. (2004). *Ethics and Technology: Ethical Issues in Information Technology*. Hoboken, NJ: John Wiley and Sons.

Triplett, Timm (2002). "Bernard Gert's *Morality* and Its Application to Computer Ethics," *Ethics and Information Technology*, Vol. 4, No. 1, pp. 79–92.

Suggested Further Readings

American Association for the Advancement of Science, Mark S. Frankel, and Audrey R. Chapman (2002). "Human Inheritable Genetic Modifications: Assessing Scientific, Ethical, Religious, and Policy Issues." In. R. Sherlock and J. D. Murray, eds. *Ethical Issues in Biotechnology*. New York: Rowman and Littlefield, pp. 495–502.

Anderson, James G., and Kenneth W. Goodman, eds. (2002). *Ethics and Information Technology: A Case-Based Approach to a Health Care System in Transition*. New York: Springer.

Annas, George J. (1998). *Some Choice: Law, Medicine, and the Market*. New York: Oxford University Press.

Burk, Dan L. (2002). "Open Source Genomics," *Boston University Journal of Science and Technology Law*, Vol. 8, No. 1, Winter, pp. 254–271.

Cahill, George F. (1996). "A Brief History of the Human Genome Project." In B. Gert, et al., eds. *Morality and the New Genetics: A Guide for Students and Health Care Patients*. Sudbury, MA: Jones and Bartlett, pp. 1–28.

Castells, Manuel (2001). "Informationalism and the Network Society." Epilogue in Pekka Himanen. *The Hacker Ethic: A Radical Approach to the Philosophy of Business*. New York. Random House, pp. 155–158.

Chadwick, Ruth, and Alison Thompson, eds. (1999). *Genetic Information Acquisition, Access, and Control*. New York: Kluwer/Plenum.

DeCew, Judith Wagner (2004). "Privacy and Policy for Genetic Research," *Ethics and Information Technology*, Vol. 6, No. 1, pp. 5–14.

Foung, Mira (2002). "Genetic Trespassing and Environmental Ethics." In. R. Sherlock and J. D. Murray, eds. *Ethical Issues in Biotechnology*. New York: Rowman and Littlefield, pp. 89–96.

Gert, Bernard, et al., eds. (1996). *Morality and the New Genetics: A Guide for Students and Health Care Patients*. Sudbury, MA: Jones and Bartlett.

Gert, Bernard (2004). "Common Morality and Computing." In R. A. Spinello and H. T. Tavani, eds. *Readings in CyberEthics*. 2nd ed. Sudbury, MA: Jones and Bartlett, pp. 96–106.

Goodman, Kenneth W., ed. (1998). *Ethics, Computing, and Medicine: Informatics and the Transformation of Healthcare*. New York: Cambridge University Press.

Goodman, Kenneth W. (2004). *Ethics and Evidence-Based Medicine*. New York: Cambridge University Press.

Gostin, Lawrence, et al. (1996). "The Public Health Information Infrastructure: A National Review of the Law on Health Information Privacy," *Journal of the American Medical Association*, Vol. 275, pp. 1921–1927.

Juengst, Eric (1991). "Priorities in Professional Genetics and Social Policy for Human Genetics," *Journal of the American Medical Association*, Vol. 266, pp. 1835–1836.

Juengst, Eric (2004). "Face Facts: Why Human Genetics Will Always Provoke Bioethics," *Journal of Law, Medicine, and Ethics*, Vol. 32.

Marturano, Antonio and Ruth Chadwick (2004). "The Role of Computing is Driving New Genetics' Public Policy," *Ethics and Information Technology*, Vol. 6, No. 2, pp. 43–53.

Mills, Claudia, ed. (1993). *Values and Public Policy*. New York: Harcourt, Brace, and Jovanovich.

O'Neill, Onora (2002). *Autonomy and Trust in Bioethics*. New York: Cambridge University Press.

Samet, J. M., and L. A. Bailey (1997). "Environmental Population Screening." In M. Rothstein, ed. *Genetic Secrets: Protecting Privacy and Confidentiality*. New Haven, CT: Yale University Press, pp. 197–211.

Schrader-Frechette, Kristin (1984). *Science Policy, Ethics, and Economic Methodology: Some Problems With Technology Assessment and Environmental-Impact Analysis*. Dordrecht, The Netherlands: Kluwer.

Schrader-Frechette, Kristin (1994). *The Ethics of Scientific Research*. Lanham, MD: Rowman and Littlefield.

Schrader-Frechette, Kristin, and Laura Westra, eds. (1997). *Technology and Values*. New York: Rowman and Littlefield.

Sharp, Richard R., and J. Carl Barrett (1999). "The Environmental Genome Project and Bioethics," *Kennedy Institute of Ethics Journal*, Vol. 9, pp. 175–188.

Sharp, Richard R., and J. Carl Barrett (2000). "The Environmental Genome Project: Ethical, Legal, and Social Implications," *Environmental Health Perspectives*, Vol. 108, No. 4, pp. 279–281.

Tavani, Herman T. (2000). "Technology, Policy, Ethics, and the Public Health: A Select Bibliography," *IEEE Technology and Society*, Vol. 19, No. 3, pp. 26–34.

Teich, Albert H., ed. (2000). *Technology and the Future*. 8th ed. New York: St. Martin's Press.

Weiss, Marcia J. (2004). "Beware! Uncle Sam Has Your DNA: Legal Fallout From Its Use and Misuse in the U.S." *Ethics and Information Technology*, Vol. 6, No. 1, pp. 55–63.

2. Moral Theory and the Human Genome Project

Bernard Gert

Any useful attempt to resolve the moral problems that may arise because of the new information that is generated by the Human Genome Project requires an explicit, clear, and comprehensive account of morality. Since some of the problems that will be generated by the Human Genome Project seem to be so different from the kinds of moral problems we normally confront, it is likely that many people will find it difficult to apply their intuitive understanding of morality to these problems. This chapter is an attempt to provide a clear and explicit description of our common morality; it is not an attempt to revise it. Common morality does not provide a unique solution to every moral problem, but it always provides a way of distinguishing between morally acceptable answers and morally unacceptable answers, i.e., it places significant limits on legitimate moral disagreement.

One reason for the widely held belief that there is no common morality is that the amount of disagreement in moral judgments is vastly exaggerated. Most people, including most moral philosophers, tend to be interested more in what is unusual than in what is ordinary. It is routine to start with a very prominent example of unresolvable moral disagreement—e.g., abortion—and then treat it as if it were typical of the kinds of issues on which one must make moral judgments. It may, in fact, be typical of the kinds of issues on which one makes moral judgments, but this says more about the word "issues" than it does about the phrase "moral judgments." Generally, the word "issues" is used when talking about controversial matters. More particularly, the phrase "moral issues" is always used to refer to matters of great controversy. Moral judgments, however, are not usually made on moral issues; we condemn murderers and praise heroic rescuers, we reprimand our children or our neighbor's children for taking away the toys of smaller children, we condemn cheating and praise giving to those in need. None of these are "moral issues," yet they constitute the subject matter of the vast majority of our moral judgments. These moral judgments, usually neglected by both philosophers and others, show how extensive our moral agreement is.

This essay originally appeared in *Morality and the New Genetics*. (eds. B. Gert et al.) Jones and Bartlett, 1996, pp. 29–55. Copyright © 1996 by Bernard Gert. Reprinted by permission.

Areas of Moral Agreement

There is general agreement that such actions as killing, causing pain or disability, and depriving of freedom or pleasure are immoral unless one has an adequate justification. Similarly, there is general agreement that deceiving, breaking a promise, cheating, breaking the law, and neglecting one's duties also need justification in order not to be immoral. There are no real doubts about this. There is some disagreement about what counts as an adequate moral justification for any particular act of killing or deceiving, but there is overwhelming agreement on some features of an adequate justification. There is general agreement that what counts as an adequate justification for one person must be an adequate justification for anyone else in the same situation, i.e., when all of the morally relevant features of the two situations are the same. This is part of what is meant by saying that common morality requires impartiality.

There is also general agreement that everyone knows what kinds of behavior common morality prohibits, requires, encourages, and allows. Although it is difficult even for philosophers to provide an explicit, clear, and comprehensive account of common morality, most cases are clear enough that almost everyone knows whether or not some particular piece of behavior is morally acceptable. No one engages in a moral discussion of questions like "Is it morally acceptable to deceive patients in order to get them to participate in an experimental treatment that one wants to test?" because everyone knows that such deception is not justified. The prevalence of hypocrisy shows that people do not always behave in the way that common morality requires or encourages, but it also shows that everyone knows what kind of behavior common morality does require and encourage. This is part of what is meant by saying that common morality is a public system.

Finally, there is general agreement that the world would be a better place if everyone acted morally, and that it gets worse as more people act immorally more often. This explains why it makes sense to try to teach everyone to act morally even though we know that this effort will not be completely successful. Although in particular cases a person might benefit personally from acting immorally, e.g., providing false information in order to get government medicare payments when there is almost no chance of being found out, even in these cases it would not be irrational to act morally, viz., not to provide this kind of information even though it means one will not get those payments. We know that the providing of such false information is one of the causes of the problems with the health care system, which results in many people suffering. Common morality is the kind of public system that every rational person can support. This is part of what is meant by saying that common morality is rational.

A Moral Theory

A moral theory is an attempt to make explicit, explain, and, if possible, justify common morality, i.e., the moral system that people use in making their moral judgments and in deciding how to act when confronting moral problems. It attempts to provide a usable account of our common morality; an account of the moral system that can actually be used by people when they are confronted with new or difficult moral decisions.[1] It must include an accurate account of the concepts of rationality, impartiality and a public system, not only because they are necessary for providing a justification of common morality, but also because they are essential to providing an adequate account of it. Indeed, a moral theory can be thought of as an analysis of the concepts of rationality, impartiality, a public system, and common morality itself, showing how these concepts are related to each other. In this chapter we hope to use the clear account of common morality or the moral system presented by the moral theory to clarify and resolve some of the moral problems that have arisen and will arise from the new information that has been and will be gained from The Human Genome Project.

Rationality is the fundamental normative concept. A person seeking to convince people to act in a certain way, must try to show that this way of acting is rational, i.e., either rationally required or rationally allowed. I use the term "irrational" in such a way that everyone would admit that if a certain way of acting has been shown to be irrational, i.e., not even rationally allowed, no one ought to act in that way.[2] But that a way of acting is rationally allowed, does not mean that everyone agrees that one ought to act in that way. On the contrary, given that it is often not irrational, i.e., rationally allowed, to act immorally, it is clear that many hold that one should not act in some ways that are rationally allowed. However, there is universal agreement that any action that is not rationally allowed ought not be done, i.e., no one ever ought to act irrationally. If rationality is to have this kind of force, the account of rationality must make it clear why everyone immediately agrees that no one ever ought to act irrationally.

To say that everyone agrees that they ought never act irrationally is not to say that people never do act irrationally. People sometimes act without considering the harmful consequences of their actions on themselves; and although they do not generally do so, strong emotions sometimes lead people to act irrationally. But regardless of how they actually act, people acknowledge that they should not act irrationally. A moral theory must provide an account of rationality such that, even though people do sometimes act irrationally, no one thinks that he ought to act irrationally. It must also relate this account of rationality to common morality.

Impartiality is universally recognized as an essential feature of common morality. A moral theory must make clear why common morality requires impartiality only when one acts in a kind of way that harms people or increases their probability of suffering harm, and does not require impartiality when deciding which people to help—e.g., which charity to give to. Most philosophical accounts of common morality are correctly regarded as having so little practical value because of their failure to consider the limits on the moral requirement of impartiality. That an adequate account of impartiality requires relating impartiality to some group, e.g., as a father is impartial with regard to his children, explains why abortion and the treatment of animals are such difficult problems. People may differ concerning the size of the group with regard to which common morality requires impartiality; some holding that this group is limited to actual moral agents, some holding that it should include potential moral agents—e.g., fetuses, and still others claiming that it includes all sentient beings, e.g., most mammals. I do not think there are conclusive arguments for any of these views.

Most moral theories, unfortunately, present an oversimplified account of common morality. Philosophers seem to value simplicity more than adequacy as a feature of their theories. Partly, this is because they do not usually think that their theories have any practical use. Many are more likely to accept theories that lead to obviously counter-intuitive moral judgments than to make their theories complex enough to account for many of our actual considered moral judgments. This has led many in applied ethics to claim to be anti moral theory. They quite rightly regard these very simple kinds of theories as worse than useless. Unfortunately, they seem to accept the false claim of the theorists that all ethical theories must be very simple. Thus they become anti theory and are forced into accepting the incorrect view that moral reasoning is ad hoc or completely relative to the situation.

The correct Aristotelian middle ground is that moral reasoning is neither ad hoc nor is there any simple account of common morality that is adequate to account for our considered moral judgments. Any adequate moral theory must recognize that neither consequences nor moral rules, nor any combination of the two, are the only matters that are relevant when one is deciding how to act in a morally acceptable way or in making moral judgments. Other morally relevant features—e.g., the relationship between the parties involved—were almost universally ignored until feminist ethical theory emphasized them. When these other features change, they change the kind of action involved and thus may change the moral acceptability of the action under consideration even though the consequences and the moral rules remain the same.

Another reason for the current low esteem in which philosophical accounts of common morality are held is that most of these accounts present common morality as if it were primarily a personal matter. It is as if each person decides for herself not only whether or not she will act morally, but also what counts as acting morally. But everyone agrees that the moral system must be known to everyone who is judged by it, and moral judgments are made on almost all adults. This means that common morality must be a public system, one that is known to all responsible adults; all of these people must know what common morality requires of them. In order to justify common morality a moral theory must show that common morality is the kind of public system that all impartial rational persons support.

Rationality as Avoiding Harms

Rationality is very intimately related to harms and benefits. Everyone agrees that unless one has an adequate reason for doing so, it would be irrational to avoid any benefit or not to avoid any harm. The present account of rationality, although it accurately describes the way in which the concept of rationality is ordinarily used, differs radically from the accounts normally provided by philosophers in two important ways. First, it starts with irrationality rather than rationality, and second, it defines irrationality by means of a list rather than a formula. The basic definition is as follows: *An action is irrational when it significantly increases the probability* that the agent, or those for whom (s)he cares, will suffer (avoidable) death, pain, disability, loss of freedom or loss of pleasure; and there is not an adequate reason for so acting.

The close relationship between irrationality and harm is made explicit by this definition, for this list also defines what counts as a harm or an evil. Everything that anyone counts as a harm or an evil—e.g., thwarted desires, diseases or maladies, and punishment—is related to at least one of the items on this list. All of these items are broad categories, so that nothing is ruled out as a harm or evil that is normally regarded as a harm. That everyone agrees on what the harms are does not mean that they all agree on the ranking of these harms. Further, pain and disability have degrees, and death occurs at very different ages, so that there is no universal agreement that one of these harms is always worse than the others. Some people rank dying several months earlier as worse than a specified amount of pain and suffering while other people rank that same amount of pain and suffering as worse. Thus, for most terminally ill patients, it is rationally allowed either to refuse death delaying treatments or to consent to them.

Most actual moral disagreements—e.g., whether or not to discontinue treatment of an incompetent patient—are based on a disagreement on the facts of the case, e.g., how painful would the treatment be and how long would it relieve the painful symptoms of the patient's disease? Differences in the rankings of the harms account for most of the rest, e.g., how much pain and suffering is it worth to cure some disability? Often the factual disagreements about prognoses are so closely combined with different rankings of the harms involved that they cannot be distinguished. Further complicating the matter, the probability of suffering any of the harms can vary from insignificant to almost certain, and people can differ in the way that they rank a given probability of one harm against different probabilities of different harms. Disagreement about involuntary commitment of people with mental disorders that make them dangerous to themselves, involves a disagreement about both what percent of these people would die if not committed and whether a significant probability (say, 10 %) of death within one week, compensates for a 100% percent probability of three to five days of a very serious loss of freedom and a significant probability (say, 30%) of long term mental suffering. Actual cases usually involve much more uncertainty about outcomes as well as the rankings of many more harms. Thus complete agreement on what counts as a harm or evil is compatible with considerable disagreement on what counts as the lesser evil or greater harm in any particular case.

If a person knowingly makes a decision that involves an increase in the probability of her suffering some harm, her decision will be personally irrational unless she has an adequate reason for that decision. Thus, not only what counts as a reason, but also what makes a reason adequate must be clarified. *A personal reason is a belief that one's action will help anyone, not merely oneself or those one cares about, avoid one of the harms, or gain some good—viz., consciousness, ability, freedom, or pleasure—and this belief is not seen to be inconsistent with one's other beliefs by almost everyone with similar knowledge and intelligence.*[3] What was said about evils or harms in the last paragraph also holds for the goods or benefits mentioned in this definition of a reason. Everything that people count as a benefit or a good, e.g., health, love, and friends, is related to one or more of the items on this list or to the absence of one or more of the items on the list of harms. Complete agreement on what counts as a good is compatible with considerable disagreement on whether one good is better than another, or whether gaining a given good or benefit adequately compensates for suffering a given harm or evil.

A reason is adequate if any significant group of otherwise rational people regard the harm avoided or benefit gained as at least as important as the harm suffered. People are otherwise rational if they do not knowingly suffer any avoidable harm without some reason. No rankings that are held by any significant religious, national, or cultural group count as irrational, e.g., the ranking by Jehovah's Witnesses of the harms that would be suffered in an afterlife as worse than dying decades earlier than one would if one accepted a transfusion, is not an irrational ranking. Similarly, psychiatrists do not regard any beliefs held by any significant religious, national, or cultural group as delusions or irrational beliefs, e.g., the belief by Jehovah's Witnesses that accepting blood transfusions will have bad consequences for one's afterlife is not regarded as an irrational belief or delusion. The intent is to not rule out as an adequate reason any relevant belief that has any plausibility; the goal is to count as irrational actions only those actions on which there is close to universal agreement that they should not be done.

Any action that is not irrational is rational. This results in two categories of rational actions, those that are rationally required and those that are merely rationally allowed. Since no action will be irrational if one has a relevant religious or cultural reason for doing it and that reason is taken as adequate by a significant group of people, in what follows I shall assume that the persons involved have no beliefs that are not commonly held. Given this assumption, an example of a rationally required action—i.e., an action that it would be irrational not to do—would be taking a proven and safe antibiotic for a life threatening infection. On the same assumption, refusing a death delaying treatment for a painful terminal disease will be a rationally allowed action—i.e., an action which it is neither irrational to do or not to do. These two categories share no common feature except that they are both not irrational. This account of rationality has the desired result that everyone who is regarded as rational always wants himself and his friends to act rationally. Certainly, on this account of rationality, no one would ever want themselves or anyone for whom they are concerned to act irrationally.

Although this account of rationality may sound obvious, it is in conflict with the most common account of rationality, where rationality is limited to an instrumental role. A rational action is often defined as one that maximizes the satisfaction of all of one's desires, but without putting any limit on the content of those desires. This results in an irrational action being defined as any action that is inconsistent with such maximization. But unless desires for any of the harms on the list are ruled out, it turns out that people would not always want those for whom they are concerned to act rationally. If a genetic counselor has a young patient who, on finding out that he has the gene for Huntington's disease, becomes extremely depressed and desires to kill himself now, more than twenty years before he will become symptomatic, no one would encourage him to satisfy that desire even if doing so would maximize the satisfaction of his present desires. Rather, everyone concerned with him would encourage him to seek counseling. They would all hope that he would be cured of his depression and then come to see that he has no adequate reason to deprive himself of twenty good years of life.[4] That rationality has a definite content and is not limited to a purely instrumental role, e.g., acting so as to maximize the satisfaction of all one's desires, conflicts with most philosophical accounts of rational actions.[5]

Scientists may claim that both of these accounts of rationality are misconceived. They may claim that on the basic account of rationality, it is not primarily related to actions at all, but rather rationality is reasoning correctly. Scientific rationality consists of using those scientific methods best suited for discovering truth. Although I do not object to this account of rationality, I think that it cannot be taken as the fundamental sense of rationality. The account of rationality as avoiding harms is more basic than that of reasoning correctly, or scientific rationality. Scientific rationality cannot explain why it is irrational not to avoid suffering avoidable harms when no one benefits in any way. The avoiding-harm account of rationality does explain why it is rational to reason correctly and to discover new truth, viz., because doing so helps people to avoid harms and to gain benefits.

Rationality, Common Morality, and Self-Interest

Although common morality and self-interest do not usually conflict, the preceding account of rationality makes clear that when they do conflict, it is not irrational to act in either way. Although this means that it is never irrational to act contrary to one's own best interests in order to act morally, it also means that it is never irrational to act in one's own best interest even though this is immoral. Further, it may even be rationally allowed to act contrary to both self-interest and common morality, if e.g., friends, family, or colleagues benefit. This is often not realized, and some physicians and scientists believe that they cannot be acting immorally if they act to benefit their colleagues when this is contrary to their own self-interest. This leads some to immorally cover up the mistakes of their colleagues, believing that they are acting morally, because they, themselves, have nothing to gain and are even putting themselves at risk.

Although some philosophers have tried to show that it is irrational to act immorally, this conflicts with the ordinary understanding of the matter. There is general agreement, for example, that it may be rational for someone to deceive a client about a mistake that one's genetic counseling facility has made, even if this is acting immorally. [In this chapter we do not attempt to provide the motivation for one to act morally.] That motivation primarily comes from one's concern for others, together with a realization that it would be arrogant to think that common morality does not apply to oneself and one's colleagues in the same way that it applies to everyone else. Our attempt to provide a useful guide for determining what ways of behaving are morally acceptable presupposes that the readers of this chapter want to act morally.

Impartiality

Impartiality, like simultaneity, is usually taken to be a simpler concept than it really is. Einstein showed that one cannot simply ask whether A and B occurred simultaneously, one must ask whether A and B occurred simultaneously with regard to some particular observer, C. Similarly, one cannot simply ask if A is impartial, one must ask whether A is impartial with regard to some group in a certain respect. The following analysis of the basic concept of impartiality shows that to fully understand what it means to say that a person is impartial involves knowing both the group with regard to which her impartiality is being judged and the respect in which her actions are supposed to be impartial with regard to that group. *A is impartial in respect R with regard to group G if and only if A's actions in respect R are not influenced at all by which members of G benefit or are harmed by these actions.*

The minimal group toward which common morality requires impartiality consists of all moral agents (those who are held morally responsible for their actions), including oneself, and former moral agents who are still persons (incompetent but not permanently unconscious patients). This group is the minimal group because everyone agrees that the moral rules—e.g., Do not kill and Do not deceive—require acting impartially with regard to a group including at least all of these people. Further, in the United States and the rest of the industrialized world, almost everyone would include in the group toward whom the moral rules require impartiality, infants and older children who are not yet moral agents. However, the claim that moral rules require impartiality with regard to any more inclusive group is more controversial. Many hold that this group should not be any more inclusive while many others hold that this group should include all potential moral agents, whether sentient or not, e.g. a fetus from the time of conception. Still others hold that this group should include all sentient beings, i.e., all beings who can feel pleasure or pain, whether potential moral agents or not—e.g., all mammals.

The debates about abortion and animal rights are best understood as debates about who should be included in the group toward which the moral rules require impartiality. Since fully informed rational persons can disagree about who is included in the group toward which common morality requires impartiality, there is no way to resolve the issue philosophically. This is why discussions of abortion and animal rights are so emotionally charged and often involve violence. Common morality, however, does set limits to the morally allowable ways of settling unresolvable moral disagreements. These ways cannot involve violence or other unjustified violations of the moral rules, but must be settled peacefully. Indeed, one of the proper functions of a democratic government is to settle unresolvable moral disagreements by peaceful means.

The respect in which common morality requires impartiality toward the minimal group (or some larger group) is when considering violating a moral rule—e.g., killing or deceiving. Persons are not required to be impartial in following the moral ideals, e.g., relieving pain and suffering. The failure to distinguish between moral rules, which can and should be obeyed impartially with respect to the minimal group, and moral ideals, which cannot be obeyed impartially even with regard to this group, is the cause of much confusion in discussing the relationship of impartiality to common morality. The kind of impartiality required by the moral rules involves allowing a violation of a moral rule with regard to one member of the group, e.g., a stranger, only when such a violation would be allowed with regard to everyone else in the group, e.g., friends or relatives. It also involves allowing a violation of a moral rule by one member of the group, e.g., oneself, only when everyone else in the group, e.g., strangers, would be allowed such a violation.

Acting in an impartial manner with regard to the moral rules is analogous to a referee impartially officiating a basketball game, except that the referee is not part of the group toward which he is supposed to be impartial. The referee judges all participants impartially if he makes the same decision regardless of which player or team is benefited or harmed by that decision. All impartial referees need not prefer the same style of basketball; one referee might prefer a game with less bodily contact, hence calling more fouls, while another may prefer a more physical game, hence calling fewer fouls. Impartiality allows these differences as long as the referee does not favor any particular team or player over any other. In the same way, moral impartiality allows for differences in the ranking of various harms and benefits as long as one would be willing to make these rankings part of the moral system and one does not favor any particular person in the group, including oneself or a friend, over any others when one decides to violate a moral rule or judges whether a violation is justified.

A Public System

A public system is a system that has the following two characteristics. In normal circumstances, (1) All persons to whom it applies—i.e., those whose behavior is to be guided and judged by that system—understand it, i.e., know what behavior the system prohibits, requires, encourages, and allows. (2) It is not irrational for any of these persons to accept being guided and judged by that system. The clearest example of a public system is a game. A game has an inherent goal and a set of rules that form a system that is understood by all of the players—i.e., they all know what kind of behavior is prohibited, required, encouraged, and allowed by the game; and it is not irrational for all players to use the goal and the rules of the game to guide their own behavior and to judge the behavior of other players by them. Although a game is a public system, it applies only to those playing the game. Common morality is a public system that applies to all moral agents; all people are subject to common morality simply by virtue of being rational persons who are responsible for their actions.

In order for common morality to be known by all rational persons, it cannot be based on any beliefs that are not shared by all rational persons. Those beliefs that are held by all rational persons (rationally required beliefs) include general factual beliefs such as: people are mortal, can suffer pain, can be disabled, and can be deprived of freedom or pleasure; also people have limited knowledge—i.e., people know some things about the world, but no one knows everything. On the other hand, not all rational people share the same scientific and religious beliefs, so that no scientific or religious beliefs can form part of the basis of common morality itself, although, of course, such beliefs are often relevant to making particular moral judgments. Parallel to the rationally required general beliefs, only personal beliefs that all rational persons have about themselves, e.g., beliefs that they themselves can be killed and suffer pain, etc. can be included as part of the foundation for common morality. Excluded as part of a foundation for common morality are all personal beliefs about one's race, sex, religion, etc., because not all rational persons share these same beliefs about themselves.

Although common morality itself can be based only on those factual beliefs that are shared by all rational persons, particular moral decisions and judgments obviously depend not only on the moral system, but also on factual beliefs about the situation. Most actual moral disagreements are based on a disagreement about the facts of the case, but particular moral decisions and judgments may also depend on the rankings of the harms and benefits. A decision about whether to withhold a proband's genetic information from him involves a belief about the magnitude of the risk—e.g., what the probability is of the information leading him to kill himself—and the ranking of that degree of risk of death against the certain loss of freedom to act on the information that would result from withholding that information. Equally informed impartial rational persons may differ not only in their beliefs about the degree of risk, but also in their rankings of the harms involved, and either of these differences may result in their disagreeing on what morally ought to be done.

Common Morality

Although common morality is a public system that is known by all those who are held responsible for their actions (all moral agents), it is not a simple system. A useful analogy is the grammatical system used by all competent speakers of a language. Almost no competent speaker can explicitly describe this system, yet they all know it in the sense that they use it when speaking and in interpreting the speech of others. If presented with an explicit account of the grammatical system, competent speakers have the final word on its accuracy. They should not accept any description of the grammatical system if it rules out speaking in a way that they regard as acceptable or allows speaking in way that they regard as completely unacceptable.

In a similar fashion, a description of morality or the common moral system that conflicts with one's own considered moral judgments normally should not be accepted. However, an explicit account of the systematic character of common morality may make apparent some inconsistencies in one's own moral judgments. Moral problems cannot be adequately discussed as if they were isolated problems whose solution does not have implications for all other moral problems. Fortunately, everyone has a sufficient number of moral judgments that they know to be both correct and consistent so that they are able to judge whether a proposed moral theory provides an accurate account of common morality. Although few, if any, people consciously hold the moral system described in this chapter, I believe that this moral system is used by most people when they think seriously about how to act when confronting a moral problem themselves, or in making moral judgment on others.

Providing an explicit account of common morality may reveal that some of one's moral judgments are inconsistent with the vast majority of one's other judgments. Thus one may come to see that what was accepted by oneself as a correct moral judgment is in fact mistaken. Even without challenging the main body of accepted moral judgments, particular moral judgments, even of competent people, may sometimes be shown to be mistaken, especially when long accepted ways of thinking are being challenged. In these situations, one may come to see that one was misled by superficial similarities and differences and so was led into acting or making judgments that are inconsistent with the vast majority of one's other moral judgments. For example, today most doctors in the United States regard the moral judgments that were made by most doctors in the United States in the 1950's about the moral acceptability of withholding information from their patients as inconsistent with the vast majority of their other moral judgments. However, before concluding that some particular moral judgment is mistaken, it is necessary to show how this particular judgment is inconsistent with most of one's more basic moral judgments. These basic moral judgments are not personal idiosyncratic judgments, but are shared by all who accept any of the variations of our common moral system—e.g., that it is wrong to kill and cause pain to others simply because one feels like doing so.

Common morality has the inherent goal of lessening the amount of harm suffered by those included in the protected group, either the minimal group or some larger group; it has rules which prohibit some kinds of actions, e.g., killing, and which require others, e.g., keeping promises; and [it has] moral ideals which encourage certain kinds of actions, e.g., relieving pain. It also contains a procedure for determining when it is justified to violate a moral rule—e.g., when a moral rule and a moral ideal conflict. *Common morality does not provide unique answers to every question, rather it sets the limits to genuine moral disagreement.* One of the tasks of a moral theory is to explain why, even when there is complete agreement on the facts, genuine moral disagreement cannot be eliminated, but it must also explain why this disagreement has legitimate limits. It is very important to realize that unresolvable moral disagreement on some important issues, e.g., abortion, is compatible with total agreement in the overwhelming number of cases on which moral judgments are made.

One of the proper functions of a democratic government is to choose among the morally acceptable alternatives when faced with an unresolvable moral issue. One important task of this book is to show how to determine those morally acceptable alternatives, in order to make clear the limits of acceptable moral disagreement. Within these limits, it may also be important to show that different rankings of harms and benefits have implications for choosing among alternatives. If one justifies refusing to allow job discrimination on the basis of race or gender because one ranks the loss of the opportunity to work as more significant than the loss of the freedom to choose whom one will employ, impartiality may require one to refuse to allow job discrimination against those suffering disabilities because of their genetic condition.

Moral disagreement not only results from factual disagreement and different rankings of the harms and benefits, but also from disagreement about the scope of common morality, i.e., who is protected by common morality. This disagreement is closely related to the disagreement about who should be included in the group toward which common morality requires impartiality. Some maintain that common morality is only, or primarily, concerned with the suffering of harm by moral agents, while others maintain that the death and pain of those who are not moral agents is as important, or almost so, as the harms suffered by moral agents. Abortion and the treatment of animals are currently among the most controversial topics that result from this unresolvable disagreement concerning the scope of common morality. Some interpret the moral rule, "Do not kill" as prohibiting killing fetuses and some do not. Some interpret the moral rule, "Do not kill" as prohibiting killing animals and some do not. But even if one regards fetuses and animals as not included in the group impartially protected by common morality, this does not mean that one need hold they should receive no protection. There is a wide range of morally acceptable options concerning the amount of protection that should be provided to those who are not included in the group toward which common morality requires impartiality.

Disagreement about the scope of common morality is only one of the factors that affect the interpretation of the rules. Another factor is disagreement on what counts as breaking the rule—e.g., what counts as killing or deceiving, even when it is clear that the person killed or deceived is included in the group impartially protected by common morality. People sometimes disagree on when not feeding counts as killing, or when not telling counts as deceiving. But although there is some disagreement in interpretation, most cases are clear and there is complete agreement on the moral rules and ideals to be interpreted. All impartial rational persons agree on the kinds of actions that need justification, e.g., killing and deceiving, and the kinds that are praiseworthy, e.g., relieving pain and suffering. Thus all agree on what moral rules and ideals they would include in a public system that applies to all moral agents. These rules and ideals are part of our conception of common morality, for it is our view that a moral theory must explain, and if possible, justify our conception of common morality, it should not, as most moral theories do, put forward some substitute for it.

With regard to (at least) the minimal group, there are certain kinds of actions that everyone regards as being immoral unless one has an adequate justification for doing them. Among these kinds of actions are killing, causing pain, deceiving, and breaking promises. Anyone who kills people, causes them pain, deceives them, or breaks a promise, and does so without an adequate justification, is universally regarded as acting immorally. Saying that there is a moral rule prohibiting a kind of act is simply another way of saying that a certain kind of act is immoral unless it is justified. Saying that breaking a moral rule is justified in a particular situation—e.g., breaking a promise in order to save a life—is another way of saying that a kind of act that would be immoral if not justified, is justified in this kind of situation. When no moral rule is being violated, saying that someone is following a moral ideal—e.g., relieving pain—is another way of saying that he is doing a kind of action regarded as morally good. Using the terminology of moral rules and moral ideals, and justified and unjustified violations, allows us to formulate a precise account of common morality, showing how its various component parts are related. I believe such an account may be helpful to those who must confront the problems raised by the information that has been and will be gained from the new genetics.

A Justified Moral System

A moral system that all impartial rational persons could accept as a public system that applies to all rational persons is a justified moral system. Like all justified moral systems, the goal of our common morality is to lessen the amount of harm suffered by those protected by it; it is constrained by the limited knowledge of people and by the need for the system to be understood by everyone to whom it applies. It includes rules prohibiting causing each of the five harms that all rational persons want to avoid and ideals encouraging the prevention of each of these harms.

The Moral Rules

Each of the first five rules prohibits directly causing one of the five harms or evils:

- Do not kill. (equivalent to causing permanent loss of consciousness.);
- Do not cause pain. (includes mental suffering, e.g., sadness and anxiety.);
- Do not disable. (includes loss of physical, mental and volitional abilities.);
- Do not deprive of freedom. (includes freedom to act and from being acted on.);
- Do not deprive of pleasure. (includes sources of pleasure.)

The second set of five rules includes those rules that, when not followed in particular cases, usually cause harm, and general disobedience always results in more harm being suffered:

- Do not deceive. (includes more than lying.);
- Keep your promise. (equivalent to Do not break your promise.);
- Do not cheat. (primarily violating rules of a voluntary activity.);
- Obey the law. (equivalent to Do not break the law.);
- Do your duty. (equivalent to Do not neglect your duty.) The term "duty" is being used in its everyday sense to refer to what is required by one's role in society, primarily one's job, not as philosophers customarily use it, which is to say, simply as a synonym for "what one morally ought to do."

The Moral Ideals

In contrast with the moral rules, which prohibit doing those kinds of actions which cause people to suffer some harm, or increase the risk of their suffering some harm, the moral ideals encourage one to do those kinds of actions which lessen the amount of harm suffered (including providing goods for those who are deprived), or decrease the risk of people suffering harm. As long as one avoids violating a moral rule, following any moral ideal is encouraged. In particular circumstances, it may be worthwhile to talk of specific moral ideals, e.g., one can claim that there are five specific moral ideals involved in preventing harm, one for each of the five kinds of harms. Physicians seem primarily devoted to the ideals of preventing death, pain, and disability. Genetic counselors may have as their primary ideal, preventing the loss of freedom of their clients. One can also specify particular moral ideals that involve preventing unjustified violations of each of the moral rules. Insofar as a misunderstanding of common morality may lead to unjustified violations of the moral rules, providing a proper understanding of common morality may also be following a moral ideal.

Although it is not important to decide how specific to make the moral ideals, it is important to distinguish moral ideals from other ideals. Utilitarian ideals involve promoting goods, e.g., abilities and pleasure, for those who are not deprived. Such ideals are followed by those who train athletes or who create delicious new recipies. Religious ideals involve promoting activities, traits of character, etc., which are idiosyncratic to a particular religion or group of religions. Personal ideals involve promoting some activities, traits of character, etc., which are idiosyncratic to particular persons, e.g., ambition, about which there is not universal agreement. Except in very special circumstances, only moral ideals can justify violating a moral rule with regard to someone without her consent.

It is the possibility of being impartially obeyed all of the time that distinguishes the moral rules from the moral ideals. Impartial rational persons favor people following both the moral rules and the moral ideals, but it is only failure to obey a moral rule that requires an excuse or a justification. This account of moral rules and ideals should not be surprising at all. All that is being claimed is that everyone counts certain kinds of actions as immoral—e.g., killing, causing pain, deceiving, and breaking promises—unless one can justify doing that kind of act; and that no one doubts that acting to relieve pain and suffering is encouraged by common morality. That two moral rules can conflict—e.g., doing one's duty may require causing pain—makes it clear that it would be a mistake to conclude that one should always avoid breaking a moral rule. Sometimes breaking one of these rules is so strongly justified that, not only is there nothing immoral about breaking it, it would be immoral not to break the rule. A physician who, with the rational informed consent of a competent patient, performs some painful procedure in order to prevent much more serious pain or death, breaks the moral rule against causing pain, but is not doing anything that is immoral in the slightest. In fact, refusing to do the necessary painful procedure, given the conditions specified, would itself be a violation of one's duty as a doctor and thus would need justification in order not to be immoral. It is clear, therefore, to say that someone has broken a moral rule is not, by itself, to say that anything wrong has been done, it is only to say that some justification is needed.

What Counts as a Violation of a Moral Rule?

As mentioned earlier, there is often a difference in interpretation about what counts as breaking the rule. Sometimes people will disagree whether to consider an action a justified violation of a moral rule, as described above, or an action that is not even a violation of a rule. Not every action that results in someone suffering a harm or an evil counts as breaking one of the first five rules. A scientist who discovers that another scientist's important new discovery is, in fact, false, may know that publishing this will result in the second scientist feeling bad. But publishing her findings is not a violation of the rule against causing pain. Almost no one would say that it was, but determining whether or not it was depends upon the practices and conventions of the society. Often these are not clear—e.g., if a genetic counselor responds to a couple's question and informs them that their fetus has some serious genetic problem (e.g., trisomy 18), she may know that this will result in their suffering considerable grief. However, if she has verified the information and told them in the appropriately considerate way, then many would say that she did not break the rule against causing pain and her action requires no justification. Indeed, not responding truthfully to their question would be an unjustified violation of the rule against deception. This interpretation is taking the counselor to be acting like the scientist reporting a mistake by another scientist. Others might take the genetic counselor to be acting like the doctor justifiably breaking the rule against causing pain because she is doing so with the consent of the couple and for their benefit. In either case, it is at least a moral ideal to be as kind and gentle in telling that truth as one can. Indeed, many would claim it is a duty of genetic counselors to minimize the suffering caused by providing information about serious genetic problems.

It is quite clear that lying, making a false statement with the intent to deceive, counts as a violation of the rule prohibiting deception, as does any other action which is intentionally done in order to deceive others. But it is not always clear when withholding information counts as deception. Thus it is not always clear that one needs a justification for withholding some information—e.g., that the husband of the woman whose fetus is being tested did not father that fetus. In scientific research, what counts as deceptive is determined in large part by the conventions and practices of the field or area of research. If it is a standard scientific practice not to report unsuccessful experiments or to smooth the curves, then doing so is not deceptive, even if some people are deceived. However, a practice that results in a significant number of people being deceived is a deceptive practice even if it is a common practice within the field or area, e.g., releasing to the public press a premature and overly optimistic account of some genetic discovery, thereby creating false hope for those suffering from the related genetic malady. Recognition that one's action is deceptive is important, for then one realizes that one needs a justification for it or else one is acting immorally.

Justifying Violations of the Moral Rules

Almost everyone agrees that the moral rules are not absolute, that they have justified exceptions; most agree that even killing is justified in self-defense. Further, there is widespread agreement on several features that all justified exceptions have. The first of these involves impartiality. There is general agreement that all justified violations of the rules are such that if they are justified for any person, they are justified for every person when all of the morally relevant features are the same. The major, and probably only, value of simple slogans like the Golden Rule, "Do unto others as you would have them do unto you" and Kant's Categorical Imperative, "Act only on that maxim that you could will to be a universal law" are as devices to persuade people to act impartially when they are contemplating violating a moral rule. However, given that these slogans are often misleading, it would be better to consider whether an impartial rational person could publicly allow that kind of violation, when trying to decide what to do in difficult cases.

There is almost complete agreement that it has to be rational to favor everyone being allowed to violate the rule in these circumstances. Suppose that someone suffering from a mental disorder both wants to inflict pain on others and wants pain inflicted on him. He favors allowing any person who wants others to cause him pain, to cause pain to others, whether or not they want pain inflicted on them. Whether or not this person is acting in accord with the Golden Rule or the Categorical Imperative, it is not sufficient to justify that kind of violation. No impartial rational person would favor allowing those who want pain caused to them to cause pain to everyone else, whether or not they want pain caused to them. The result of allowing that kind of violation would be an increase in the amount of pain suffered with almost no compensating benefit, which is clearly irrational.

Finally, there is general agreement a violation is justified only if it is rational to favor that violation even if everyone knows that this kind of violation is allowed, i.e., the violation must be publicly allowed. A violation is not justified simply if it would be rational to favor allowing everyone to violate the rule in the same circumstances, but only if almost no one knows that it is allowable to violate the rule in those circumstances. For example, it might be rational for one to favor allowing a physician to deceive a patient about his diagnosis if that patient were likely to be upset by knowing the truth, when almost no one knows that such deception is allowed. But that would not make deception in these circumstances justified. It has to be rational to favor allowing this kind of deception when everyone knows that one is allowed to deceive in these circumstances. One must be prepared to publicly defend this kind of deception, if it were discovered. Only the requirement that the violation be publicly allowed guarantees the kind of impartially required by common morality.

Not everyone agrees on which violations satisfy these three conditions, but there is general agreement that no violation is justified unless it satisfies all three of these conditions. Allowing for some disagreement while acknowledging the significant agreement concerning justified violations of the moral rules, results in the following formulation of the appropriate moral attitude toward violations of the moral rules: *Everyone is always to obey the rule unless an impartial rational person can advocate that violating it be publicly allowed. Anyone who violates the rule when no impartial rational person can advocate that such a violation be publicly allowed may be punished.* (The 'unless clause' only means that when an impartial rational person can advocate that such a violation be publicly allowed, impartial rational persons may disagree on whether or not one should obey the rule. It does not mean that they agree one should not obey the rule.)

The Morally Relevant Features

When deciding whether or not an impartial rational person can advocate that a violation of a moral rule be publicly allowed, the kind of violation must be described using only morally relevant features. Since the morally relevant features are part of the moral system, they must be such that they can be understood by all moral agents. This means that any description of the violation that one offers as appropriate to determine whether or not an impartial rational person could publicly allow it, must be such that it can be reformulated in a way that all moral agents could understand it. Limiting the way in which a violation can be described makes it easier for people to discover that their decision or judgment is biased by some consideration that is not morally relevant. All of the morally relevant features that I have discovered so far are answers to the following questions. It is quite likely that other morally relevant features will be discovered, but I think that I have discovered the major features. Of course, in any actual situation, the particular facts of the situation determine the answers to these questions, but all of these particular facts can be redescribed in a way that can be understood by all moral agents.

1. What moral rules are being violated?
2. What harms are being (a) avoided, (b) prevented, (c) caused?
3. What are the relevant beliefs and desires of the people toward whom the rule is being violated? (This explains why it is important to provide the patient with adequate information and to find out what they want.)
4. Does one have a relationship with the person(s) toward whom the rule is being violated such that one has a duty to violate moral rules with regard to the person(s)? (This explains why a parent or guardian is allowed to make decisions about treatment that cannot be made by the health care team.)
5. What benefits are being promoted?
6. Is an unjustified or weakly justified violation of a moral rule being prevented?
7. Is an unjustified or weakly justified violation of a moral rule being punished?
8. Are there any alternative actions that would be preferable?[6]
9. Is the violation being done intentionally or only knowingly?[7]
10. Is the situation an emergency that no person is likely to plan to be in?[8]

When considering the harms being avoided (not caused), prevented, or caused, and the benefits being promoted, one must consider not only the kind of benefit or harm involved, one must also consider their seriousness, duration, and probability. If more than one person is affected, one must consider not only how many people will be affected, but also the distribution of the harms and benefits. If two violations are the same in all of their morally relevant features then they count as the same kind of violation. Anyone who claims to be acting or judging as an impartial rational person who holds that one of these violations be publicly allowed must hold that the other also be publicly allowed. This follows from the account of impartiality. However, this does not mean that two people, both impartial and rational, who agree that two actions count as the same kind of violation, must always agree on whether or not to advocate that this kind of violation be publicly allowed, for they may differ in their estimate of the consequences of publicly allowing that kind of violation or they may rank the benefits and harms involved differently.

To act or judge as an impartial rational person one decides whether or not to advocate that a violation be publicly allowed by estimating what effect this kind of violation, if publicly allowed, would have. If all informed impartial rational persons would estimate that less harm would be suffered if this kind of violation were publicly allowed, then all impartial rational persons would advocate that this kind of violation be publicly allowed and the violation is strongly justified; if all informed impartial rational persons would estimate that more harm would be suffered, then no impartial rational person would advocate that this kind of violation be publicly allowed and the violation is unjustified. However, impartial rational persons, even if equally informed, may disagree in their estimate of whether more or less harm will result from this kind of violation being publicly allowed. When this happens, even if they are impartial, they will disagree on whether or not to advocate that this kind of violation be publicly allowed and the violation counts as weakly justified. Sometimes, primarily when considering the actions of governments, it is also appropriate to consider not only the harms but also the benefits that would result from this kind of violation being publicly allowed.

Disagreements in the estimates of whether a given kind of violation being publicly allowed will result in more or less harm may stem from two distinct sources. The first is a difference in the rankings of the various kinds of harms. If someone ranks a specified amount of pain and suffering as worse than a specified amount of loss of freedom, and someone else ranks them in the opposite way, then although they agree that a given action is the same kind of violation, they may disagree on whether or not to advocate that this kind of violation be publicly allowed. The second is a difference in estimates of how much harm would result from publicly allowing a given kind of violation, even when there seems to be no difference in the rankings of the different kinds of harms. These differences may stem from differences in beliefs about human nature or about the nature of human societies. In so far as these differences cannot be settled by any universally agreed upon empirical method, such differences are best regarded as ideological. The disagreement about the acceptability of voluntary active euthanasia of patients with terminal illnesses is an example of such a dispute. People disagree on whether publicly allowing voluntary active euthanasia will result in various bad consequences, including significantly more people dying sooner than they really want to. However, it is quite likely that most ideological differences also involve differences in the rankings of different kinds of harms, e.g., does the suffering prevented by voluntary active euthanasia rank higher or lower than the earlier deaths that might be caused? But sometimes there seems to be an unresolvable difference when a careful examination of the issue shows that there is actually a correct answer.

Applying Common Morality to a Particular Case

For example, one genetic counselor may claim that deception about a diagnosis, e.g. of Huntington's Disease in a young adult, to avoid causing a specified degree of anxiety and other mental suffering is justified. He may claim that withholding unpleasant findings in these circumstances will result in less overall harm being suffered than if such deception were not practiced. He may hold that patients are often not able to deal with bad news and are very unlikely to find out about the deception. Thus he may claim that this kind of deception actually results in patients suffering less harm than if they were told the truth. However, another genetic counselor may claim that deception, no matter how difficult it will be for the client to accept the facts or how confident the counselor is that the deception will not be discovered, is not justified. The latter may hold that deception of this kind will actually increase the amount of harm suffered because patients will be deprived of the opportunity to make decisions based upon the facts and that if they do find out about the deception they will not only have less faith in statements made by the counselor, they will also have less faith in statements made by other health care providers, thus increasing the amount of anxiety and suffering. Thus there is a genuine empirical dispute about whether withholding bad news from patients is likely to increase or decrease the amount of harm suffered. Which of these hypotheses about the actual effects of deception in the particular circumstances is correct, I do not know, but if one is concerned with the moral justifiability of such deception it does not matter.

The morally decisive question is not "What are the consequences of this particular act?" but rather "What would be the consequences if this kind of deception were publicly allowed?" Neither counselor has taken into account that a justifiable violation against deception must be one that is publicly allowed, i.e., one that everyone knows is allowed. Once one realizes that in making a moral decision one must consider the consequences if everyone knows that it is allowable to deceive in certain circumstances, e.g., to withhold bad news in order to avoid anxiety and other mental suffering, then the loss of trust involved will obviously have worse consequences than if everyone knew that such deception was not allowed. It is only by concentrating on the results of one's own deception, without recognizing that morally allowed violations for oneself must be such that everyone knows that they are morally allowed for everyone, that one could be led to think that such deception was justified. Consciously holding that it is morally allowable for oneself to deceive others in this way although, of course, one would not want everyone to know that everyone is morally allowed to deceive others in the same circumstances, is exactly what is meant by arrogance, viz., the arrogating of exceptions to the moral rules for oneself which one would not want everyone to know are allowed for all. This arrogance is clearly incompatible with the kind of impartiality that common morality requires with regard to obeying the moral rules.

Contrasting Common Morality with Other Systems for Guiding Conduct

For those who are concerned with the philosophical foundations of bioethics, it may clarify our account of the moral system to compare it with the views put forward by many contemporary followers of Immanuel Kant (1724-1804) and John Stuart Mill (1806-1873). The Kantian Categorical Imperative, "Act only on that maxim whereby you can at the same time will that it be a universal law of nature," and Mill's Utilitarian Greatest Happiness Principle, "Act so as bring about the greatest happiness for the greatest number," are two of the most popular and influential moral philosophical slogans. But these slogans, though often cited, are inadequate, by themselves, to provide a useful moral guide to conduct. It is not fair to Kant and Mill or their contemporary followers, to compare these slogans with our account of the moral system sketched in this chapter, for Kant and Mill and their contemporary followers have far more to say than simply working out the consequences of these slogans. However, popular use of these slogans, especially in medical contexts, is often as simple as I shall characterize it. Further, neither Kant nor Mill nor their contemporary followers provide a list of morally relevant features, i.e., there is little effort devoted to providing plausible accounts of how one determines whether two violations count as violations of the same kind for the purpose of moral evaluation.

On a popular interpretation of a Kantian deontological system, one should never act in any way that one cannot will to be a universal law. If it would be impossible for everyone always to do a specific kind of action, then everyone is prohibited from doing that kind of action. For example, that it is impossible for everyone always to make lying promises (for then there could be no practice of promising), is what makes it morally prohibited to make lying promises. On the system of common morality, one is prohibited from doing a kind of action only if, given the morally relevant facts, no impartial rational person would publicly allow that kind of action. A Kantian system seems to rule out ever making lying promises, whereas our common morality allows the making of lying promises in some circumstances, e.g., when it is necessary to make the lying promise to prevent a harm sufficiently great that less overall harm would be suffered even if everyone knew such lying promises were allowed.

On a popular interpretation of a Utilitarian or consequentialist system (Bentham and Mill), one not only may, but should, violate any rule if the foreseeable consequences of that particular violation, including the effects on future obedience to the rule, are better than the consequences of not violating the rule. A consequentialist system is concerned only with the foreseeable consequences of the particular violation, not with the foreseeable consequences of that kind of violation being publicly allowed. But on our moral system, it is precisely the foreseeable consequences of that kind of violation being publicly allowed that are decisive in determining whether or not it is morally allowed. The consequences of the particular act are important only in determining the kind of violation under consideration. A consequentialist system favors cheating on an exam if one were certain that one would not get caught and no harm would result from that particular violation of the rule against cheating. Assuming that the exams serve a useful function, our moral system would not allow this kind of violation of the rule against cheating, for if this kind of violation were publicly allowed, it would make it pointless to have exams.

According to consequentialism, the only morally relevant features of an act are its consequences. It is, paradoxically, the kind of moral theory usually held by people who claim that they have no moral theory. Their view is often expressed in phrases like the following: "It is all right to do anything as long as no one gets hurt," "It is the actual consequences that count, not some silly rules," or "What is important is that things turn out for the best, not how one goes about making that happen." According to Classical Utilitarianism (Bentham and Mill), the only relevant consequences are pleasure and pain. That act is considered morally best which produces the greatest balance of pleasure over pain. On our moral system, pleasure and pain are not the only consequences that count, and it is not the consequences of the particular violation that are decisive in determining its justifiability, but rather the consequences of publicly allowing such a violation.

Common morality differs from a Kantian system and resembles a consequentialist system in that it has a purpose, and consequences are explicitly taken into consideration. It resembles a Kantian system and differs from a consequentialist system in that common morality must be a public system in which rules are essential. The role of impartiality also differs. The Kantian system requires all of one's actions to be impartial, and consequentialist systems require one to regard the interests of everyone impartially. Common morality does not require impartiality with regard to all of one's actions; it requires impartiality only with respect to obeying the moral rules. Nor does common morality require one to regard the interests of everyone impartially; it only requires that one act impartially when violating a moral rule. Indeed, it is humanly impossible to regard the interests of everyone impartially, when concerned with all those in the minimal group. Impartiality with respect to the moral ideals (Kant would call these imperfect duties) is also humanly impossible. That all of the moral rules are or can be taken as prohibitions, is what makes it humanly possible for them to be followed impartially. The public nature of common morality and the limited knowledge of rational persons help to explain why impartial obedience to the moral rules is required to achieve the point of common morality, lessening the suffering of harm. Common morality also differs from both systems in that it does not require all moral questions to have unique answers, but explicitly allows for a limited area of disagreement among equally informed, impartial, rational persons.

Endnotes

1. A more extended account of common morality, and of the moral theory that justifies it is contained in *Morality: Its Nature and Justification*, revised edition, (Oxford University Press, 2005) 411 pp. A shorter account is contained in *Common Morality: Deciding What To Do*, (Oxford University Press, 2004) 179 pp. Both are by Bernard Gert.
2. I am aware that the terms "rational" and "irrational" are used in many different ways, e.g., "irrational" means spontaneous. However, I think that there is a basic normative concept of rationality and that is the one that I am attempting to describe. I call this the objective sense of an irrational action. However, there is a corresponding personal sense, which involves the beliefs of the agent and whether he is motivated by them. Although this is a more commonly used sense, it is not the basic normative sense. For a more detailed account of these senses of irrationality, see the books cited in the previous note and *Brute Rationality: Normativity and Human Action* by Joshua Gert (Cambridge University Press, 2004).
3. An objective reason is a corresponding fact about one's action helping anyone, not merely oneself or those one cares about, avoid one of the harms, or gain some good, viz., consciousness, ability, freedom, or pleasure.
4. See "Irrationality and the *DSM-III-R* Definition of Mental Disorder," by Bernard Gert, *Analyze & Kritik*, Jahrgang 12, Heft 1, July 1990, pp 34-46
5. See the books cited in notes 1 and 2
6. This involves trying to find out if there any alternative actions such that they would either not involve a violation of a moral rule, or that the violations would differ in some morally relevant features especially, but not limited to, the amount of evil caused, avoided, or prevented.
7. "Free Will as the Ability to Will," *Nous*, Vol. 13, No. 2, May 1979, pp. 197-217. Bernard Gert and Timothy Duggan. Reprinted in *Moral Responsibility*, edited by John Martin Fisher, 1986. (b) freely or because of coercion? (c) knowingly or without knowledge of what is being done? (d) is the lack of knowledge excusable or the result of negligence? whose answers will affect the

moral judgment that some people will make. The primary reason for not including answers to these questions as morally relevant features is that our goal in listing morally relevant features is to help those who are deciding whether or not to commit a given kind of violation; so we did not want to include those features that are solely of value in judging violations that had already been committed, and cannot be used in deciding how to act. For questions (a), (b), (c), and (d) one cannot decide whether or not to commit one rather than another of these kinds of violations, hence they are not useful in deciding how to act.

Although one does not usually decide whether or not to commit a violation intentionally or only knowingly, sometimes that is possible. For violations that are alike in all of their other morally relevant features, a person might not publicly allow a violation that was done intentionally, but might publicly allow a violation that was not done intentionally, even though it was done knowingly. For example, many people would publicly allow nurses to administer morphine to terminally ill patients in order to relieve pain even though everyone knows it will hasten the death of the patient but, with no other morally relevant changes in the situation, they would not allow nurses to administer morphine in order to hasten the death of the patient. This distinction explains what seems correct in the views of those who endorse the doctrine of double effect. I think that such a distinction may also account for what many regard as a morally significant difference between lying and other forms of deception, especially withholding information. Nonetheless, it is important to remember that many, perhaps most, violations that are morally unacceptable when done intentionally are also morally unacceptable when done only knowingly.

8. We are talking about the kind of emergency situation that is sufficiently rare that no person is likely to plan or prepare for being in it. This is a feature that is necessary to account for the fact that certain kinds of emergency situations seem to change the moral judgments that many would make even when all of the other morally relevant features are the same. For example, in an emergency when large numbers of people have been seriously injured, doctors are morally allowed to abandon patients who have a very small chance of survival in order to take care of those with a better chance, in order that more people will survive. However, in the ordinary practice of medicine they are not morally allowed to abandon patients with poor prognoses in order to treat those with better prognoses, even if doing so will result in more people surviving.

3. Lex Genetica: The Law and Ethics of Programming Biological Code

Dan L. Burk

Introduction

Human artifacts embody human values, but more than any other human artifact, information technology embodies within its design rules for its use. Biotechnology has now arrived as true information technology, permitting technological constraints to be purposefully programmed into genetic code. The emerging ability to program genetic code in this fashion blurs the line between law and artifact, and promises to challenge long-held assumptions in the legal regime of ownership and control over such biological artifacts.

This essay illuminates some of these emerging problems by drawing upon insights that have been developed in parallel discussions regarding digital technology, recognizing that those parallel discussions are themselves in their infancy, and may need to be enhanced or extended for my purposes here. In particular, this discussion highlights the problem of distinguishing coded constraints that we might treat as equivalent to law from other types of technologically embedded values. In doing so, the essay touches upon the broader questions involving long-standing discussions about contract law's effective application to technological constraints.

I shall begin by describing the recent advances in genetic design that allow constraints on the use of plants or other transgenic organisms to be programmed into the organism itself. I then show that this development parallels that in other programmable information technologies, and that current trends in the analysis electronic digital technologies may be properly applied to biological technologies. In particular, the development of programmable biological code raises a series of difficult policy questions regarding the market power of commodity producers, the

---

This essay originally appeared in *Ethics and Information Technology*, 2002; **4**:109–121. Copyright © 2002 by Dan L. Burk and Kluwer Academic Publishers. Reprinted by permission.

autonomous choices of commodity users, and the proper role of the state in regulating programmed constraints.

Coded Constraints

Gene splicing techniques have enabled the creation of many types of sexually reproducing plants with commercially attractive characteristics: increased nutritional value, resistance to drought and pests, herbicide resistance, and medicinal properties, to name only a few.[1] The economic challenge to development of such plant varieties is that plants reproduce, as living organisms are wont to do. As a consequence, these new varieties are relatively expensive to create, but are trivially inexpensive to propagate once they are in existence – and, indeed, may propagate unintentionally. This 'public goods' problem of distribution at a marginal cost close to zero is common in other areas of innovation, even where the subject matter does not reproduce itself.[2] Legal prohibitions have been the typical solution to this problem, although technological solutions have also been employed. Both these strategies have now emerged in the case of genetically engineered plants, though with important and unusual characteristics not seen in previous incarnations of the problem.

Anti-germination technology

Society has primarily relied upon written rules to govern the use of biological inventions and in particular to address the public goods problem inherent in their creation. For example, in the United States, a form of intellectual property called Plant Variety Protection encourages development of new varieties of sexually reproducing plants by granting the developer broad control over the growth, use, importation, and sale of a new plant.[3] This American statute is a version of an international plant variety protection treaty, UPOV. As required under UPOV, the Plant Variety Protection Act includes some important exceptions to a seed developer's control, such as provisions allowing farmers to save seed from a proprietary crop, or permitting agricultural research involving the plant.[4]

Plant variety owners might prefer that their control over the variety were not subject to such exceptions, and so as a condition of access to their seeds, routinely require that farmers contractually waive their rights to save seed or engage in other legally permissible uses.[5] Often the terms of this contract are printed on or attached to the bag of seed; by using the seed, the contractual 'fine print' purports that the farmer has agreed to the terms. However, it is difficult to police the use

---

[1] United States Congress Office of Technology Assessment. A New Technological Era for American Agriculture. U.S. Government Printing Office, Washington D.C., 1992.

[2] William M. Landes and Richard A. Posner. An Economic Analysis of Copyright Law. *Journal of Legal Studies*, 18: 325–363, 1989.

[3] 7 U.S.C. § 2402.

[4] International Convention for the Protection of New Varieties of Plants, Dec. 2. 1961, as revised 33 U.S.T. 2703, 815 U.N.T.S. 89.

[5] Neil D. Hamilton. Legal Issues Shaping Society's Acceptance of Biotechnology and Genetically Modified Organisms. *Drake Journal of Agricultural Law*, 6: 81–117, 2001, pp. 90–91.

of seed and to enforce the terms of such 'seedwrap' licenses. To do so, seed developers must send agents out into farmers' fields to sample crops, looking for unlicensed users of proprietary seed. When such uses are found, costly legal procedures may be necessary to halt the use, force acceptance of a license, or recover unpaid royalties.

The problems of detection and enforcement might be lessened if seed could be designed to be 'selfpolicing,' that is, unsuitable for use without the developer's permission. Newly available transgenic technology allows for the creation of such 'selfpolicing' seed.[6] Genetic elements that produce a toxin late in seed development may be introduced into the plant variety.[7] The toxin kills the seeds after the plant has matured, producing a viable crop for the farmer, but forcing him to return to the seed producer for new seed each year. Even in the absence of a contractual obligation not to save seed, the technology makes saving seed impossible. Thus, the genetically altered seed in essence carries within its own makeup a prohibition on unlicensed use.

Indeed, the 'terms' of usage embedded in such genetic code may be quite sophisticated. In one embodiment of the technology, it is possible to introduce into the seed a genetic 'switch' that will repress, or turn off, the toxin production when the seed is exposed to a particular chemical. This in effect supplies a chemical 'password' to activate germination, and which can be used to control the terms of seed usage from year to year. Yearly application of the control chemical, obtained from the seed owner for payment, would allow the owner to activate or deactivate seeds in return for prescribed payment. One can easily envision other types of switches, sensitive to temperature, precipitation, soil alkalinity, or other environmental factors, that could be used to limit use of the seed to certain geographical regions or seasonal applications. Indeed, plants could be engineered for various desirable properties—pest resistance, drought resistance, superior yield, and so on—and particular attributes activated or deactivated depending on the price paid by the purchaser.

Although the patent on this technology is directed to control of plant development, similar genetic control elements are known in other organisms, and there is no particular reason that such technology need be confined to plants. Since the advent of genetically engineered animals, beginning with the 'Harvard Oncomouse,' the ability of the animal to reproduce has posed a challenge to the owners of proprietary rights in the organism: does the purchase of a patented animal confer the right to breed or use subsequent generations of the animal, and if not, how can the patent holder control subsequent generations?[8] Much as in the case of genetically altered seed, this problem has been largely handled via licenses that either

---

[6] Keith Aoki. Neocolonialism, Anti-Commons Property, and Biopiracy in the (Not-So-Brave) New World Order of International Intellectual Property Protection. *Indiana Journal of Global Legal Studies* 6: 11–58, 1998, p. 54.

[7] M.L. Crouch. *How the Terminator Terminates: An Explanation for the Non-scientist of a Remarkable Patent for Killing Second Generation Seeds of Crop Plants*, Edmonds Institute, 1998. (http://www.bio.indiana.edu/people/terminator.html)

[8] Rebecca Dresser. Ethical and Legal Issues in Patenting New Animal Life. *Jurimetrics Journal* 28: 399–435, 1988; United States Congress Office of Technology Assessment. *New Developments in Biotechnology: Patenting Life*. U.S. Government Printing Office, Washington D.C., 1989, p. 121.

include or exclude the right to breed the animal.[9] But once again, due to policing and enforcement problems, as well as the opportunity for price discrimination, the availability of a genetic system to activate or deactivate a genetically engineered trait might be highly attractive to the creators of such animals.

The prospect of germ-line alteration of human subjects has been even more controversial although for somewhat different reasons, primarily relating to the ethical controversy of altering traits in future generations who have had no opportunity to consent to such alterations.[10] A full exploration of this issue is impossible within the scope of this essay, but I will note that some of the ethical objections to germ-line therapy might be addressed by a control system that could deactivate germ-line therapies in future generations, unless perhaps they requested activation of the trait. A more likely, but no less troublesome application of the technology, might be found in somatic cell therapies. Genetic regulatory elements analogous to those in plant applications could equally well be added to the transgenic DNA cassettes contemplated for human gene therapy, placing recombinant genes in human cells under similar proprietary control.

One can easily envision genetic therapies for certain diseases, such as diabetes or hemophilia, which are caused by the failure of a particular gene in the body to produce a particular protein. A recombinant genetic 'cassette' containing a healthy copy of the defective gene could be introduced into the patient's cells in order to supply the missing protein.[11] The cassette could include regulatory elements allowing the gene to be activated or deactivated by administration of a proprietary pharmaceutical; so long as the patient were supplied with the pharmaceutical, the gene would continue to prevent the disease. Such a system might perhaps to allow the recipient to pay for the therapy over an extended period of time, rather than all at once. The supplier of the treatment could exercise self-help if payment were not forthcoming. Of course, under the current system, the supplier would presumably have legal recourse for non-payment, but for the reasons described above, self-help might be a more attractive form of recourse.

Content management technology

The description of seed licensing offered above bears an uncanny resemblance to the history of content licensing in digital media.[12] Digital technology offers inexpensive and widespread access to the means of reproducing and distributing copyrighted materials. As PVPA provides legal protection for seeds, copyright law affords the owners of digital content some recourse against many unauthorized uses of their material, but copyright is subject to a host of consumer uses that require no authorization from the copyright holder. Owners of digital content, much like seed

---

[9] Robert P. Merges, Intellectual Property in Higher Life Forms: The Patent System and Controversial Technologies. *Maryland Law Review* 47: 1051–1075, 1988.

[10] LeRoy Walters and Julie Gage Palmer. *The Ethics of Human Gene Therapy*. Oxford University Press, New York, 1997.

[11] P.D. Robbins. Retroviral Vectors. In Thomas Blankenstein editor, *Gene Therapy, Principles and Applications*. Birkhauser, Basel, 1999, p. 18.

[12] Charles R. McManis, The Privatization (or 'Shrink-Wrapping') of American Copyright Law. *California Law Review*, 87: 173–190, 1999.

owners, have long wished to escape the consumer privileges afforded by copyright law. They have done so through the fiction of the 'shrinkwrap' license, which purports to restrict a purchaser's use of the accompanying product.[13] The license takes its name from the legal fiction that the purchaser demonstrates agreement to the license terms by breaking the 'shrinkwrap' cellophane on the product package. More recently, such 'clickwrap' using the mouse to click on a graphic labeled 'I agree.'[14]

However, courts in the United States have in many cases been reluctant to enforce such agreements because the purchaser may have no opportunity to review the license prior to opening the package.[15] Proponents of mass-market licenses for software have complained that such agreements have long since been accepted in most other areas of commerce.[16] This observation is true, so far as it goes, but the consumer of, say, a car rental agreement has at least a nominal opportunity to read the agreement before the rental occurs; in the case of shrinkwrapped licenses, even the fiction of a pre-transaction opportunity to review is absent. 'Clickwrap' agreements similarly often involve after-market agreement to use software pre-installed on a computer the consumer has already purchased. The situation has not changed appreciably with the advent of electronic commerce; proposed rules for information licensing would permit a merchant to change the terms of the agreement by posting the new terms somewhere on the Internet, or by sending the purchaser an e-mail message that would be considered effective even if the purchaser never actually received the message. In the face of uncertain enforcement by the courts, software vendors have sought to legitimate such practices by promulgation of the Uniform Computer Information Transaction Act, or UCITA, which has been adopted in two states.[17]

Yet even if such licenses become more frequently enforceable, it is still extremely difficult for copyright holders to police such agreements. Consequently, copyright owners have begun deploying sophisticated software 'lock-out' systems that prevent access to digitized content except on the terms dictated by the owner.[18] Such content management software may govern the number of uses, or their duration, or the payment schedule for additional access.[19] For example, access to technologically controlled content may be provisioned on agreement to a clickwrap-type

---

[13]David W. Maher. The Shrink-Wrap License: Old Problems in a New Wrapper. *Journal of the Copyright Society*, 34: 292–312, 1987; Deborah Kemp. Mass Marketed Software: The Legality of the Form License Agreement. *Louisiana Law Review*, 48: 87–128, 1987.

[14]Mark Lemley. Shrinkwraps in Cyberspace. *Jurimetrics Journal*, 35: 311–323, 1995.

[15]Mark Lemley. Beyond Preemption: The Law and Policy of Intellectual Property Licensing. *California Law Review*, 87: 111–172, 1999; Mark Lemley. Intellectual Property and Shrinkwrap Licenses. *Southern California Law Review*, 68(5): 1239–1294, 1995.

[16]Robert W. Gomulkiewicz and Mary L. Williamson. A Brief Defense of Mass Market Software License Agreements. *Rutgers Computer and Technology Law Journal*, 22: 335–367, 1996.

[17]Niva Elkin Koren. A Public-Regarding Approach to Contracting Over Copyrights. In Rochelle Cooper Dreyfuss, Diane Leenheer Zimmerman, and Harry First, editors, *Expanding the Boundaries of Intellectual Property: Innovation Policy for the Knowledge Society*, pp. 191–221. Oxford University Press, Oxford, 2001.

[18]Julie E. Cohen. Reverse Engineering and the Rise of Electronic Vigilantism: Intellectual Property Implications of 'Lock-Out' Programs. *Southern California Law Review*, 68(5): 1091–1202, 1995; Julie E. Cohen. Some Reflections on Copyright Management Systems and Laws Designed to Protect Them. *Berkeley Technology Law Journal*, 12(1): 161–187, 1997.

[19]Mark Stefik. Shifting the Possible: How Trusted Systems and Digital Property Rights Challenge Us to Rethink Digital Publishing. *Berkeley Technology Law Journal*, 12(1): 137–160, 1997.

license.[20] Similarly, the content management system may permit the owner to shut off the software remotely if the user fails to make the required payment in a timely manner; a controversial provision of the UCITA statute makes agreement to such 'self-help' a valid term of computer information licenses.[21]

In this environment where technology provides the first line of defense against unauthorized uses of content, the legal protection preferred by content owners may be not so much a deterrent against violation of copyright or similar proprietary rights, but legal deterrents against circumvention of technological protections.[22] In the United States, they have gained such protection in the form of the Digital Millennium Copyright Act, or DMCA, which prohibits circumvention of technical protection measures, and trafficking in technology that would facilitate such circumvention.[23] This statute effectively provides content owners a new right of technological access, independent of any intellectual property right. Language promulgating similar legal measures has appeared in a recent European Union copyright directive.[24]

The implications of this development are striking: By implementing technical constraints on access to and use of digital information, a copyright owner can effectively supersede the rules of intellectual property law. For example, as described above, the copyright owner may decide that the technological controls will not permit any copying of the controlled content, whether or not the copying would be fair use. If the integrity of the controls is backed by the state, as it is under the DMCA's anti-circumvention provisions, the result is to shift enforcement of the rights-holder's interest from penalties for unauthorized infringement to penalties for unauthorized access. When combined with UCITA provisions favoring the licensing terms promulgated by information producers, these developments dramatically alter the balance of ownership and control of new technologies.[25]

Toward Lex Genetica

It is important to underscore how this insight shapes the unusual nature of the genetic information issues that I have detailed above. There exists a large and rapidly growing literature addressing the legal and the ethical issues related to genetic information, and entire research programs devoted to expanding that literature. Contributions to that literature have typically focused on issues raised by genotechnology as a means of tampering with human genetics, either as a matter of med-

---

[20] Michael J. Madison. Legal-Ware: Contract and Copyright in the Digital Age. *Fordham Law Review*, 67(3): 1025–1143, 1998.

[21] Eric Schlachter. The Intellectual Property Reniassance in Cyberspace: Why Copyright Law Could Be Unimportant on the Internet. *Berkeley Technology Law Journal*, 12(1): 15–52, 1997.

[22] Kenneth W. Dam. Self-Help in the Digital Jungle. *The Journal of Legal Studies*, 28: 393–412, 1999.

[23] Digital Millennium Copyright Act, Pub. L. No. 105–304, 112 Stat. 2860 (1998).

[24] Directive 2001/29/EC of the European Parliament and of the Council of 22 May 2001 on the harmonization of certain aspects of copyright and related rights in the information society, 2001 O.J. (L. 167) 10.

[25] Niva Elkin Koren. A Public-Regarding Approach to Contracting Over Copyrights. In Rochelle Cooper Dreyfuss, Diane Leenheer Zimmerman, and Harry First, editors, *Expanding the Boundaries of Intellectual Property: Innovation Policy for the Knowledge Society*, pp. 191–221. Oxford University Press, Oxford, 2001.

ical treatment or of eugenics. But those issues, as important as they may be, are not our focus here. Rather, the question here relates to design of genetic products, the constraints or values embedded in those designs, and the ability of consumers to exercise choice regarding the use of those products.

Where technological constraints substitute for legal constraints, control over the design of information rights is shifted into the hands of private parties, who may or may not honor the public policies that animate public access doctrines such as fair use or a 'farmer's exemption.' Rights-holders can effectively write their own intellectual property statute in either software or DNA. This shift in control challenges the traditional role of the state in determining the limits of property and contract, as well as the accepted ethical assumptions underlying these legal institutions.

Lex informatica

The development of digital content management systems has been recognized as a graphic demonstration of the power of technology to regulate behavior. As both Larry Lessig and Joel Reidenberg have pointed out, technical standards are within the control of the designer, and so confer upon the designer the power to govern behavior with regard to that system.[26] Once constraints on behavior are built into the technical standards governing a technology, the technical standards effectively become a new method for governing use of that technology—in essence, the technical standards become a type of law. Such technical rule sets may supplement or even supplant the legal rule sets designed to govern the same behavior.

Consider, for example, an example suggested by Latour, in which the state wishes to enforce safety standards by requiring all automobile drivers to use seat belts.[27] One method to produce the desired behavior is to pass laws penalizing the failure to use such harnesses. However, an alternative method to produce the desired behavior is to fit automobiles with seat-belt interlocks that prevent the car's ignition from functioning unless the seatbelt is fastened to complete an electronic circuit. Thus, government may choose to employ or enforce technical standards to achieve goals that might otherwise be achieved by legal rule-making. Such use of technological rules to govern behavior has been dubbed by Joel Reidenberg as 'lex informatica.'[28]

Reidenberg in particular has examined in detail the complex set of interactions through which governmental action can shape technological standards into a substitute for legal controls. For example, the state may implement the technological alternative through a variety of regulatory mechanisms, from a variety of sources. Most directly, the state might simply require automobile manufacturers to install

---

[26]Joel Reidenberg. Lex Informatica: The Formulation of Information Policy Rules Through Technology. *Texas Law Review*, 76: 553–593, 1998; Lawrence Lessig. *Code and Other Laws of Cyberspace*, Basic Books, 1999.

[27]Bruno Latour. Where are the Missing Masses? The Sociology of a Few Mundane Artifacts. In Weibe E. Bijker and John Law, editors, *Shaping Technology/Building Society: Studies in Sociotechnical Change* pp. 225–258. The MIT Press, Cambridge, Massachusetts, 1992.

[28]Joel Reidenberg. Lex Informatica: The Formulation of Information Policy Rules Through Technology. *Texas Law Review*, 76: 553–593, 1998.

seatbelt interlocks on all cars produced.[29] Alternatively, courts or legislatures acting through courts could impose liability for deaths or injuries on manufacturers who fail to install seatbelt interlocks, creating an incentive to include the feature in cars. Similar liability could be imposed on car drivers or owners, creating a consumer demand for manufacturers to install the devices.

Private lawmaking
The design of technological rule sets, however, is not the sole provenance of the state; indeed, it is more often left to private parties. In the case of digital content management systems, copyright owners determine the rules that are embedded into the technological controls. Moreover, to the extent that the DMCA appears to legitimate technological controls over copyrighted works, without regard to their effect on public policy, the statute effectively grants rubber-stamp approval to such private legislation.[30] Although there exists, at present, no similar anti-circumvention statute for genotechnology, other private property statutes might be impressed into service to produce the same result. For example, the anti-germination technology described here is patented, so that attempts to tamper with it or reverse engineer it could constitute patent infringement.[31]

The development of such technological use controls, whether in either software or transgenic corn, has raised concern because it substitutes private technological rules for the public statutory rules declared by Congress in either the Copyright Act or the Plant Variety Protection Act. Producers who employ such lock-out technology may in essence become private legislatures, imposing rules of usage without regard to the broader public interest that informs democratic rule-making.[32] This problem has been well explored with regard to digital technology; the instantiation of a proprietary rule in genetic code, which following Reidenberg we might call 'lex genetica,' is the first example of the regulation by means of genetic code, but is unlikely to be the last.

Of course, the promulgation of technologically embedded rule sets is not the first situation in which private allocation of rights in information has been encouraged and enforced by public institutions. Most notably, the coercive power of the state is routinely brought to bear in the case of contractual agreements, such as confidentiality agreements and intellectual property licenses. Since technical controls can impose conditions that formerly might have been the subject of a detailed license agreement, such controls might be viewed as equivalent to a sort of licensing regime. Then, extending the analogy, penalties for circumvention of the technological constraints simply stand in for the private law of contract, which penalizes breach of license.

---

[29] Jerry L. Mashaw and David L. Harfst. *The Struggle for Auto Safety*. Harvard University Press, 1990.

[30] Niva Elkin-Koren. The Privatization of Information Policy. *Ethics and Information Technology*, 2: 201–209, 2000.

[31] U.S. Patent No. 5,723,765 (Mar. 3 1998).

[32] J.H. Reichman and Jonathan Franklin. Privately Legislated Intellectual Property Rights: Reconciling Freedom of Contract With Public Good Uses of Information. *University of Pennsylvania Law Review*, 147(4): 875–970, 1999.

But such a comparison to contract law by no means justifies employment of technical controls that contravene established public policy. Where traditional contracts are at issue, carte blanche enforcement of private agreements has never been the rule in Anglo-American law. When such agreements are found illegal, unconscionable, or simply in violation of public policy, they are deemed unenforceable.[33] Because contract law is state law, a similar result also may be reached on grounds of federalism: where enforcement of a state law contract would violate the public policy inherent in the federal intellectual property scheme or embedded in the United States Constitution itself, such contractual provisions are preempted. An attempt to leverage the federal statutory right beyond the limits set by federal policy constitutes grounds for voiding the contract.

To the extent that 'code' confronts us with behavioral constraints that are somehow analogous to legally enforceable contractual provisions, we presumably face much the same dilemma with regard to hardwired constraints that we have previously faced when dealing with contractual constraints. This point has perhaps been argued most forcefully by Julie Cohen, although not in precisely these terms, when she opines on the potential for constitutional preemption of certain technological content management constraints.[34] Cohen suggests that the coercive power of the state should be extended in support of technological constraints no farther than it may be extended to enforce statutory or contractual constraints. This conjecture, which Lessig has dubbed the 'Cohen Theorem,' might be applied in either private or public law settings to restrain the implementation of technological constraints by either individuals or the state.

Under the 'Cohen Theorem' analysis, there is no reason to suppose that technological analogs to contracts should be privileged over the legal instruments themselves. Where rights management systems attempt to impose restrictions on access to or use of informational content that would be improper in a contractual agreement, the restrictions should be viewed as equally repugnant to public policy and equally void. Stated differently, where the Constitution imposes limits on the government's creation and recognition of property rights in intellectual goods, those limits apply equally to both legally and technologically delineated property. In some instances of overreaching via technological controls, the Constitution may even demand a limited self-help right, or 'right to hack,' to surmount privately erected technological barriers to information that the Constitution requires be publicly accessible.

It is less clear what might form the jurisprudential basis for such a right outside the context of digital technology. The tension between free speech and copyright is well-defined and well-documented, and the limits upon Congressional power have been the subject of long scrutiny; technological controls over creative works are only the most recent chapter in that policy discussion. Biological controls lack any similar policy precedent. Unlike content management systems, antigermination systems do not implicate a fundamental human right to receive

---

[33]Restatement (Second) of Contracts, § 178.

[34]Julie E. Cohen. Copyright and the Jurisprudence of Self-Help. *Berkeley Technology Law Journal,* 13(3): 1090–1143, 1998.

information. No court has ever recognized a constitutional right to save seed, or to engage in agricultural research. Some commentators have argued in favor of a general First Amendment right to engage in scientific research,[35] but the legitimacy of such arguments is unsettled, and their application to proprietary organisms uncertain.[36]

Where lex genetica is applied to the human body, the established jurisprudence of rights may prove somewhat more fruitful. Certain Supreme Court holdings suggest a constitutional right to bodily integrity,[37] as for example where state-sponsored invasive procedures would 'shock the conscience.'[38] Other cases establish a constitutional prohibition against intrusive state intervention into personal or medical decision-making, especially in reproductive matters, although the exact parameters of this right tend to shift from year to year.[39] Such constitutional guarantees might override contractual or patent prohibitions against tampering with biological controls, but it is difficult to know what type of genetic programming might be considered sufficiently 'shocking' or intrusive to invoke such rights. Moreover, even if a sound legal basis for overriding legal protections can be found, the practical implementation of a 'right to hack' may be problematic outside the context of digital technology. There appears to be no comparable community of biological 'hackers' who might either personally have the skill to circumvent biological lock-out coding, or to supply users with the tools to circumvent such code.

Technological Scripts

Technical controls on digital or biological systems therefore challenge the existing order of control and ownership for technology. However, the concept of technological constraints predates programmable artifacts. The idea that technology embodies rules is not new. Bruno Latour identified the 'scripted' nature of different artifacts, pointing out for example that automobile seatbelts with ignition interlocks embody a type of 'script' requiring a driver to take the particular action of fastening the seatbelt before driving.[40] Similarly, a locked door effectively embodies a rule against unauthorized entry. These artifacts are not programmable in the

---

[35]Harold P. Green, Constitutional Implications of Federal Restrictions on Scientific Research and Communication. *UMKC Law Review*, 60: 619–643 (1992); Richard Delgado & David R. Millen. God, Galileo, and Government: Toward Constitutional Protection for Scientific Inquiry. *Washington Law Review*, 53: 349–404, 1978; John A. Robertson, The Scientist's Right to Research: A Constitutional Analysis. *California Law Review*, 51: 1203–1281, 1977.

[36]Roy G. Spece, Jr. & Jennifer Weinziel, First Amendment Protection of Experimentation: A Critical Review and Tentative Synthesis/Reconstruction of the Literature. *Southern California Interdisciplinary Law Review*, 8: 185–228, 1998; Gary L. Francione, Experimentation and the Marketplace Theory of the First Amendment, *University of Pennsylvania Law Review*, 136: 417–512, 1987.

[37]*Cruzan v. Director*, Missouri Department of Health, 497 U.S. 261 (1990); Stenberg v. Carhart, 120 S.Ct. 2597 (2000).

[38]*Rochin v. California*, 342 U.S. 165 (1952).

[39]*Skinner v. Oklahoma*, 319 U.S. 535 (1942); *Griswold v. Connecticut*, 381 U.S. 479 (1965); *Eisenstadt v. Baird*, 405 U.S 438 (1972); *Roe v. Wade*, 410 U.S. 113 (1973); *Washington v. Glucksberg*, 521 U.S. 707 (1997).

[40]Bruno Latour. Where are the Missing Masses? The Sociology of a Few Mundane Artifacts. In Weibe E. Bijker and John Law, editors, *Shaping Technology/Building Society: Studies in Sociotechnical Change*, pp. 225–258. The MIT Press, Cambridge, Massachusetts, 1992.

sense that software or DNA may be programmed with a wide range of attributes, but nonetheless the physical construction of the door enforces its particular prohibition, just as the electromechanical 'script' of the ignition interlock enforces its particular prohibition.

Thus, although programmability certainly increases the range and complexity of artifactual 'scripts,' this may represent a difference of degree, rather than a difference of kind. Myriad user constraints are routinely built into all kinds of artifacts, and all of them will entail some set of values: the hinge design causes the door to swing in a particular direction, the doorknob is set at a particular height and requires a certain degree of manipulation to open, and so on. Many of these constraints go unnoticed as part of the artifactual backdrop of society, while other constraints implicate important social values, either supporting or frustrating such values. Such effects may be intentional or unintentional; the door may be unintentionally difficult for physically disabled persons to open, or may be intentionally difficult for small children to open, or may even unintentionally frustrate use by the physically disabled precisely because it was designed to retard use by small children.

The creation of such artifactual 'scripts' may be influenced by state action. As suggested by the seatbelt example above, technological design may be either directly or indirectly determined by a range of regulatory interventions.[41] But in market based economies, such intervention is typically limited to design features that have a noticeable effect on public health or safety, or to extraordinary regulation, such as removal of architectural barriers to the disabled. The vast majority of technological choices go largely unregulated, as we primarily entrust to market forces the task of weeding out over time the most inefficient or unusable designs. Although it is understood that such markets may be subject to network effects, incomplete information, and a wide range of market failures that could in fact hamper the efficient development of such designs, the market approach is assumed on the whole to operate more ably than command and control intervention by the state. At the same time, this market approach itself undoubtedly imbues the resulting artifacts with particular embedded values.

At the same time, users of any given technology will for the most part be unaware of the values embedded in a given technological system. Indeed, this is one of Reidenberg's key objections to a wholly 'free market' approach to information technology development: that all unknown to the general populace, it cedes to technologists choices that may later dictate the freedom or constraints upon users.[42] Reidenberg's preferred solution appears to be one of governmental oversight or involvement, at least in democratic states. Governmental bodies may exercise such oversight through a variety of channels, including direct regulation, standard-setting, procurement, criminal or civil penalties, and so on. Reidenberg reasons that involvement by elected officials, or at least by bureaucrats answering

---

[41] Jerry L. Mashaw and David L. Harfst. *The Struggle for Auto Safety*. Harvard University Press, 1990.

[42] Joel Reidenberg. Lex Informatica: The Formulation of Information Policy Rules Through Technology. *Texas Law Review*, 76: 553–593, 1998.

to elected officials, presumably better reflects democratic values than leaving the choice to technologists.

But as detailed above, explicit legal or regulatory intervention into technological design is relatively rare. Unless we are willing to countenance wholesale state oversight of every routine design decision, we must somehow separate out those design constraints that implicate public policy from those that we have previously treated as innocuous, or at least as routine. This separation has long been taken for granted; in a conventional transaction involving the use or exchange of an artifact, we have typically separated the values embedded in the artifact from the disembodied values instantiated in the law governing the transaction. For example, when a consumer purchases an automobile featuring seatbelt interlocks, we could conceptualize as a term of the transaction, embedded in the artifact, 'the purchaser will be required to fasten her seatbelt prior to driving.' We have not done so, however, and in fact tend to separate even a public legal requirement to use seatbelts from the terms of the private transaction; no promise to use seatbelts is written into automobile sales contracts or leases, despite laws requiring seatbelt use.

To be sure, some regulatory intervention may occur at the point of legal transaction if the nexus between the two seems sufficiently close. The licensing of the vehicle, or transfer of the license, may be incorporated into the transaction, if for no other reason than it provides a convenient control point for the state to ensure that such licensing occurs. But conceptual nexus for such incorporation has been relatively rare. Returning to the case of the automobile, other regulatory interventions, such as the requirement that the driver be licensed, or carry proper accident insurance, appear to have an insufficient nexus with the sale of the vehicle.

In much the same way, explicit legal or regulatory intervention into 'private lawmaking' via contract is relatively rare. If our past experience with law in fact maps onto the territory of technological constraints, we would expect only a subset of such constraints to trigger legal safeguards, such as the Cohen Theorem—the vast majority of both private and public lawmaking goes relatively unremarked, routinely functioning without the application of extraordinary judicial or constitutional remedies. Only a small number of contracts are struck down as unconscionable or void for public policy, just as few statutes are struck down as unconstitutional. Yet the current literature analyzing technological constraints gives no clear guidance on where routine or garden variety design choices may begin to shade over into legally cognizable constraints, or which legally cognizable constraints should be the abrogated as contrary to existing public policy.

Taking Code Seriously

Summing up to this point: I have argued that the advent of programmable technical constraints creates two intertwined difficulties: first, determining where legally cognizable technology choices leave off and routine, if sometimes troubling technology choices begin; and second, once legally relevant technology choices have been identified, determining how social policy choices that have been implemented in law will be implemented in its technological analogs. Moreover, the lines drawn in each case may differ according to the technology involved, as biological 'lock-out' systems arise in a different milieu than analogous digital control mechanisms. To illustrate these issues, I turn now to the specific example of translating to programmed artifacts the values of autonomy as they have been instantiated in the substantive law of contract, as well as in relevant principles of informed consent.

Law and autonomy

Modern contractual theory incorporates concepts of autonomy under two broad categories. The first of these categories focuses directly on the importance of contract as a means of promoting individual choice or autonomy, or on autonomy as an animating principle to justify a theory of contract.[43] The second broad category of contractual theory focuses on efficiency as the primary purpose of contract. These latter 'law and economics' formulations of contract owe much to the utilitarian tradition, but focus on maximization of wealth as a proxy or substitute for ensuring the greatest happiness to the greatest number of people.[44] Under such theories, individual choice still plays a central role in implementing the decentralized allocation of resources; by encouraging self-interested activity with minimal outside interference, resources are moved to their optimal use. Thus, in this second set of theories, autonomy functions within this framework as a means to an end, rather than as an end in itself. At the same time, some apologists for an economic approach have melded the two theories, turning the relationship between autonomy and efficiency around to argue that a market-based approach to contract is desirable because it promotes autonomy.[45]

Under either set of theories, excessive governmental intervention into the bargain may be decried as 'paternalism' or interference with the autonomy of the parties. But the focus on autonomy in private bargaining creates a potential paradox regarding state intervention, or paternalism. State intervention into the transaction may be decried as an imposition on the autonomy or contractual freedom of the parties. At the same time, state intervention may be necessary to preserve the autonomy or contractual freedom of certain parties, particularly where one party stands in a position of overwhelming power or influence. Typically such asymmetrical bargaining positions are perceived to occur where one party has far more information than the other, or where one party's range of choices are highly constrained due to lack of competitive alternatives. In such situations, the terms of the agreement may be perceived as imposed by the stronger party, without the free consent of the other. The classic case for such asymmetrical bargains are mass-market consumer transactions, where a large corporate entity may have access to far more information about a product than the typical consumer, or where the consumer's bargaining choices may be limited to few or even one vendor. Either situation may be conceived in an economic framework as a form of market failure; were

---

[43]Randy E. Barnett. A Consent Theory of Contract. *Columbia Law Review*, 86: 269–321, 1986.

[44]Jeffrie Murphy and Jules Coleman. *Philosophy of Law: An Introduction to Jurisprudence*, 2nd ed. Westview Press, 1990.

[45]Richard Posner. The Ethical and Political Basis of the Efficiency Norm in Common Law Adjudication. *Hofstra Law Review*, 8: 487–507, 1980; Richard Posner. Utilitarianism, Economics, and Legal Theory. *Journal of Legal Studies*, 8: 103–140, 1979.

the market to operate perfectly, market forces would act to discipline contractual overreaching.

Such market failure situations may in fact be very common, but where the social system puts its faith in markets, the law assumes that they will be rare. The tradition in Anglo-American law has been that for the most part, the state avoids intervention into particular terms of the contract. Courts typically refuse to inquire, for example, into the adequacy of consideration.[46] The state may withhold its coercive power in those rare cases where a party falls into a category clearly classified as lacking legally cognizable autonomy, such as that of minors or the mentally incompetent.[47] Equally rarely, a court may invoke a doctrine such as unconscionability to protect otherwise competent parties, and most especially individual consumers, from exploitation by more powerful or better informed parties.[48] Autonomy may also be husbanded in unusual situations by other doctrines, such as recission,[49] misrepresentation,[50] or mistake,[51] that might be viewed as designed to nullify agreements a party has entered into without full information, which may be to say without full autonomy.

However, such doctrines are invoked rarely and with some reluctance because of their potential to supersede 'freedom of contract.' Judges are reluctant to override terms that may have been the preference of the contracting parties. Libertarian analysts denounce the doctrines for introducing the heavy hand of the state into private bargaining. Economic analysts decry the potential for inefficiency. Even analysts outside the free-market economics tradition may be wary of such doctrines because they are highly interventionist—assuming, for example, that certain classes of individuals cannot understand contractual terms or cannot formulate a legally recognizable desire to be bound by contractual terms. The historical inclusion of women together with children and mentally handicapped individuals as legal incompetents amply illustrates the objection that imposition of judicial preferences may be dangerous to individual autonomy.

Consequently, although the state may forbid or invalidate certain contractual terms, it will more often intervene by mandating disclosure of terms. For example, certain key terms to a mass-market contract must be 'conspicuous,' which typically means printed in a larger, bolder, or more prominent typeface than terms considered less important or less potentially troublesome.[52] Similarly, under conditions requiring medical consent, physicians may be required to be especially forthcoming regarding particularly troublesome risks or outcomes attending a treatment. Such 'paternalism light' is intended to secure autonomous decision

---

[46]Restatement (Second) of Contracts, § 79; E. Allen Farnsworth. *Farnsworth on Contracts* § 2.11, 2000.

[47]Restatement (Second) of Contracts, § 12.

[48]E. Allen Farnsworth. *Farnsworth on Contracts*. § 4.28, 2000.

[49]Restatement (Second) of Contracts § 283.

[50]Restatement (Second) of Contracts § 164.

[51]Restatement (Second) of Contracts § 153.

[52]Uniform Commercial Code § 2-316(2); E. Allen Farnsworth. *Farnsworth on Contracts*. § 4.29a, 2000.

making by ensuring that information deemed important to a decision is available, without dictating the decision itself. This approach is, of course, laden with important underlying assumptions that the recipient of the information both understands the information provided and has the circumstantial latitude to act freely on it, and the more interventionist doctrines may be invoked in those unusual occasions where the law may believe such latitude is lacking.

In the case involving human application of genetic programming, a second source of autonomous consent comes into play, that of informed medical consent. In the context of medical treatment, the question of autonomy had played a somewhat different role, as the focus is on assent and authorization, rather than upon contractual consideration. The issue here is typically not framed as one of contract, perhaps because the problem is seldom conceived in terms of a bargained-for exchange. Patients whose medical treatment proves sub-standard are seldom interested in a contractual remedy, such as getting their money back; neither are human research participants who are subjected to unconsented research procedures, interested in demanding performance of the experiments to which they thought they had agreed. Under the Anglo-American legal system, such a claim sounds in tort rather than contract, in large part due to the development of informed consent out of the waiver doctrine in law of battery.[53]

Consequently, in matters of informed consent, the stigma of paternalism is not directed to the imposition of governmental restraints on the parties bargaining, but to the imposition of the physician's preferences or decisionmaking upon the patient. The autonomy question is less an issue of governmental intervention than one of medical intervention: the assumption of a decisionmaking role by the physician.[54] The issue of governmental intervention is of course lurking in the background. Legal duties may be imposed on a physician, perhaps mandating a certain level of disclosure, discouraging or prohibiting certain interventions without proper consent, or in rare cases, requiring intervention regardless of consent. Yet governmental paternalism has received relatively little attention in this context, perhaps because it is assumed that one party to the transaction—the physician—is routinely and uniformly in possession of asymmetrically greater information and situational control. Thus, unlike the contract situation in which arm's length bargaining is routinely assumed, in the informed consent setting, market failure is routinely assumed.

Code and autonomy

Bargained-for contract and informed consent may thus be viewed as polar opposites in the approach to disclosure and the preservation of a legal regime of autonomy. Their commonality lies in the underlying assumption that first line of defense in preserving autonomy, or at least the semblance of autonomy, should be the requirement of disclosure; for both market contracting or medical consent, autonomous consent given in either setting requires information upon which to

---

[53] Ruth Faden and Thomas Beauchamp. *A History and Theory of Informed Consent*. Oxford University Press, New York, 1986.

[54] Carl Schneider. *The Practice of Autonomy*. Oxford University Press, New York, 1998.

decide. But in past situations, even where both sets of obligations might be present, they could be treated under entirely different assumptions: one for the bargained-for transaction, and one for treatment, each requiring a different duty of disclosure. While the process, effects, and outcomes related to the treatment might in some sense have been considered terms of the business transaction, the persistent asymmetry of information and control, favoring the physician, gave rise to disclosure requirements never seen under the rubric of contract.

But this compartmentalization of assumptions begins to blur when the features of the technology coincide with the terms of the transaction. Programmable biological elements, when used for human treatment, make the terms of the bargain a characteristic of the treatment. And even where human treatment is not involved, the same persistent asymmetry of information will exist. Courts have shown some reluctance to enforce written shrinkwrap licenses where information material to the transaction is disclosed subsequent to the transaction. Adherence to 'freedom of contract' in such situations may be little more than a sham, and the resistance to state intervention little more than an excuse to give the more powerful party in the transaction the maximum latitude to impose unrestrained oppressive or overreaching terms. The potential for abuse is far greater when the information material to the transaction is never disclosed, but remains embedded in the artifact—the consumer and producer of the artifact stand in a relationship of persistent informational asymmetry, much like the relation of the physician and patient under informed consent.

This suggests a pressing need to equalize the informational disparity, but if disclosure is to be the mechanism for equalization, the precise contours of the needed disclosure remain problematic. In the contractual setting, disclosure requirements have typically been limited to terms considered 'material' to the transaction—terms such as warranties, disclaimers, and remedies. Design choices or embedded technological values have simply not been part of that constellation of terms. The rare instances where design choices are the subject of disclosure tend to arise in the area of products liability, where an industrial product is found to have dangerous characteristics not apparent upon consumer examination.[55] Much as in the case of informed medical consent, disclosure of the potential danger allows the manufacturer of the product to avoid liability for injury by virtue of the consumers' voluntary acceptance of the danger. But non-dangerous design choices are typically mandated by neither contract nor tort theories. Courts do not require that an automobile seller reveal, for example, that a car was designed on the assumption that exhaust manifolds would need replacement every 10 years, or that gasoline prices would remain at $1.35 per gallon, or that automobile factory workers' wages would remain stable, or that state law on 'plug-molds' would continue to provide a cheap source of replacement automobile body parts, or that Americans in the next decade would value mobility over ecology.

Indeed, the law has been somewhat hostile to mandating disclosure when technologies render products that are not materially different, but morally different. In those rare instances where consumers have displayed an interest in knowing,

---

[55] Restatement (Third) of Torts: Products Liability § 2(c).

for example, where particular meats originated, or whether recombinant gene products were used in the production of milk or vegetable produce, both courts and legislatures have been resistant to imposing a legal disclosure requirement.[56] In some of these cases, market demand has prompted producers to provide products carrying the desired disclosures, obviating the need for legal or political intervention.[57] But where a market for the information has not developed, there has to date been little state intervention to solve the market failure, or to force disclosure for the sake of a broader conception of informed product consent.

Thus, if disclosure is to remain the first line of defense in protecting contractual autonomy, reevaluation of our previous approach to disclosure seems in order. Although the purchaser of seed may have the opportunity to read the agreement on the side of the bag, he has no ability to examine the programming of a seed, and cannot determine its constraints by examining the product, any more than the patient has the ability to divine the likely outcome of a medical procedure. Human applications of 'lex genetica' offer the clearest case for an increased duty of disclosure, but the same considerations will remain in other applications. Preservation of the value of autonomy in the face of embedded terms requires the creation of criteria to determine when an embedded term is legally relevant, and then determination of the level of state intervention that is appropriate. At a minimum, this likely means mandating disclosure of biological product characteristics that are material to the use of the product.

Conclusion

This essay closes having likely raised more questions than it has answered. That was in part the intent—to indicate how sparse is our current understanding of technological constraints as a matter of policy, and in particular the need for some criterion to distinguish relatively routine technological constraints that might deserve a social response from extraordinary constraints that deserve a legal response. But at the same time, this discussion moves us closer to answering such questions by demonstrating how current analysis of technological constraints may be extended. The development of biologically programmable artifacts indicates that the issue is a general question of technology policy, and not idiosyncratic to digital technology such as the Internet.

Acknowledgements

I am grateful for comments on previous drafts of this article offered by Julie Cohen, Joel Reidenberg, Michael Madison, Helen Nissenbaum, and Lucas Introna.

---

[56] Dan L. Burk. The Milk-Free Zone: Federal and Local Interests in Regulating Recombinant BST. *Columbia Environmental Law Review*, 22(2): 227–317, 1997.

[57] This has occurred, for example in the case of milk from cows treated with recombinant bovine somatotropin (rBST), to which some consumers may have social or moral objections, in response to which the producers have supplied milk from untreated herds at a higher price.

References

Keith Aoki. Neocolonialism, Anti-Commons Property, and Biopiracy in the (Not-So-Brave) New World Order of International Intellectual Property Protection. *Indiana Journal of Global Legal Studies*, 6: 11–58, 1998, p. 54.

Randy E. Barnett. A Consent Theory of Contract. *Columbia Law Review*, 86: 269–321, 1986.

Dan L. Burk. The Milk-Free Zone: Federal and Local Interests in Regulating Rebombinant BST. *Columbia Environmental Law Review*, 22(2): 227–317, 1997.

Julie E. Cohen. Copyright and the Jurisprudence of Self-Help. *Berkeley Technology Law Journal*, 13(3): 1090–1143, 1998.

Julie E. Cohen. Some Reflections on Copyright Management Systems and Laws Designed to Protect Them. *Berkeley Technology Law Journal*, 12(1): 161–187, 1997.

Julie E. Cohen. Reverse Engineering and the Rise of Electronic Vigilantism: Intellectual Property Implications of 'Lock-Out' Programs. *Southern California Law Review*, 68(5): 1091–1202, 1995.

M.L. Crouch. *How the Terminator Terminates: An Explanation for the Non-scientist of a Remarkable Patent for Killing Second Generation Seeds of Crop Plants*. Edmonds Institute, 1998. (http://www.bio.indiana.edu/people/terminator.html)

Cruzan v. Director, Missouri Department of Health, 497 U.S. 261 (1990).

Kenneth W. Dam. Self-Help in the Digital Jungle. *The Journal of Legal Studies*, 28: 393–412, 1999.

Richard Delgado and David R. Millen. God, Galileo, and Government: Toward Constitutional Protection for Scientific Inquiry. *Washington Law Review*, 53: 349–404, 1978.

Directive 2001/29/EC of the European Parliament and of the Council of 22 May 2001 on the harmonization of certain aspects of copyright and related rights in the information society, 2001 O.J. (L. 167) 10.

Rebecca Dresser. Ethical and Legal Issues in Patenting New Animal Life. *Jurimetrics Journal*, 28: 399–435, 1988. United States Congress Office of Technology Assessment. *New Developments in Biotechnology: Patenting Life*.U.S.Government Printing Office, Washington D.C., 1989, p. 121.

Eisenstadt v. Baird, 405 U.S. 438 (1972).

Niva Elkin-Koren. The Privatization of Information Policy. *Ethics and Information Technology*, 2: 201–209, 2000.

Niva Elkin-Koren. A Public-Regarding Approach to Contracting Over Copyrights. In Rochelle Cooper Dreyfuss, Diane Leenheer Zimmerman, and Harry First, editors, *Expanding the Boundaries of Intellectual Property: Innovation Policy for the Knowledge Society*, pp. 191–221. Oxford University Press, Oxford, 2001.

Ruth Faden and Thomas Beauchamp. *A History and Theory of Informed Consent*. Oxford University Press, New York, 1986.

E. Allen Farnsworth. *Farnsworth on Contracts*. §§ 2.11, 4.28, 4.29a, 2000.

Gary L. Francione, Experimentation and the Marketplace Theory of the First Amendment. *University of Pennsylvania Law Review*, 136: 417–512, 1987.

Harold P. Green, Constitutional Implications of Federal Restrictions on Scientific Research and Communication. *UMKC Law Review*, 60: 619–643 (1992).

Griswold v. Connecticut, 381 U.S. 479 (1965).

Robert W. Gomulkiewicz and Mary L. Williamson. A Brief Defense of Mass Market Software License Agreements. *Rutgers Computer and Technology Law Journal*, 22: 335–367, 1996.

Neil D. Hamilton. Legal Issues Shaping Society's Acceptance of Biotechnology and Genetically Modified Organisms. *Drake Journal of Agricultural Law*, 6: 81–117, 2001. pp. 90–91.

International Convention for the Protection of New Varieties of Plants, Dec. 2. 1961, as revised 33 U.S.T. 2703, 815 U.N.T.S. 89.

Deborah Kemp. Mass Marketed Software: The Legality of the Form License Agreement. *Louisiana Law Review*, 48: 87–128, 1987.

William M. Landes and Richard A. Posner. An Economic Analysis of Copyright Law. *Journal of Legal Studies*, 18: 325–363, 1989.

Bruno Latour. Where are the Missing Masses? The Sociology of a Few Mundane Artifacts. In Weibe E. Bijker and John Law, editors, *Shaping Technology/Building Society: Studies in Sociotechnical Change* pages 225–258. The MIT Press, Cambridge, Massachusetts, 1992.

Mark Lemley. Beyond Preemption: The Law and Policy of Intellectual Property Licensing. *California Law Review*, 87: 111–172, 1999.

Mark Lemley. Shrinkwraps in Cyberspace. *Jurimetrics Journal*, 35: 311–323, 1995.

Mark Lemley. Intellectual Property and Shrinkwrap Licenses. *Southern California Law Review*, 68(5): 1239–1294, 1995.

Lawrence Lessig. *Code and Other Laws of Cyberspace*, Basic Books, 1999.

Charles R. McManis. The Privatization (or 'Shrink-Wrapping') of American Copyright Law. *California Law Review*, 87: 173–190, 1999.

Michael J. Madison. Legal-Ware: Contract and Copyright in the Digital Age. *Fordham Law Review*, 67(3): 1025–1143, 1998.

Jerry L. Mashaw and David L. Harfst. *The Struggle for Auto Safety*. Harvard University Press, 1990.

Robert P. Merges, Intellectual Property in Higher Life Forms: The Patent System and Controversial Technologies. *Maryland Law Review* 47: 1051–1075, 1988.

David W. Maher. The Shrink-Wrap License: Old Problems in a New Wrapper. *Journal of the Copyright Society*, 34: 292–312, 1987.

Jeffrie Murphy and Jules Coleman. *Philosophy of Law: An Introduction to Jurisprudence*, 2nd ed. Westview Press, 1990.

Richard Posner. The Ethical and Political Basis of the Efficiency Norm in Common Law Adjudication. *Hofstra Law Review*, 8: 487–507, 1980.

Richard Posner. Utilitarianism, Economics, and Legal Theory. *Journal of Legal Studies*, 8: 103–140, 1979.

J.H. Reichman and Jonathan Franklin, Privately Legislated Intellectual Property Rights: Reconciling Freedom of Contract With Public Good Uses of Information. *University of Pennsylvania Law Review*, 147(4): 875–970, 1999.

Joel Reidenberg. Lex Informatica: The Formulation of Information Policy Rules Through Technology. *Texas Law Review*, 76: 553–593, 1998.

P.D. Robbins. Retroviral Vectors. In Thomas Blankenstein editor, *Gene Therapy, Principles and Applications*, page 18. Birkhauser, Basel, 1999.

John A. Robertson, The Scientist's Right to Research: A Constitutional Analysis. *California Law Review*, 51: 1203–1281, 1977.

Rochin v. California, 342 U.S. 165 (1952).

Roe v. Wade, 410 U.S. 113 (1973).

Skinner v. Oklahoma, 319 U.S. 535 (1942).

Eric Schlachter. The Intellectual Property Renaissance in Cyberspace: Why Copyright Law Could Be Unimportant on the Internet. *Berkeley Technology Law Journal*, 12(1): 15–52, 1997.

Carl Schneider. *The Practice of Autonomy*. Oxford University Press, New York, 1998.

Roy G. Spece, Jr. and Jennifer Weinziel. First Amendment Protection of Experimentation: A Critical Review and Tentative Synthesis/Reconstruction of the Literature. *Southern California Interdisciplinary Law Review*, 8: 185–228, 1998.

Mark Stefik. Shifting the Possible: How Trusted Systems and Digital Property Rights Challenge Us to Rethink Digital Publishing. *Berkeley Technology Law Journal*, 12(1): 137–160, 1997.

Stenberg v. Carhart, 120 S.Ct. 2597 (2000).

Restatement (Second) of Contracts, §§ 12, 79, 153, 164, 283.

Restatement (Third) of Torts: Products Liability § 2(c).

Uniform Commercial Code § 2-316(2).

United States Congress Office of Technology Assessment. *A New Technological Era for American Agriculture*. U.S. Government Printing Office, Washington D.C., 1992.

United States Congress Office of Technology Assessment. *New Developments in Biotechnology: Patenting Life*. U.S. Government Printing Office, Washington D.C., 1989, p. 121.

U.S. Patent No. 5,723,765 (Mar. 3 1998).

LeRoy Walters and Julie Gage Palmer. *The Ethics of Human Gene Therapy*. Oxford University Press, New York, 1997.

Washington v. Glucksberg, 521 U.S. 707 (1997).

4. Computing, Genetics, and Policy: Theoretical and Practical Considerations

Ruth Chadwick and Antonio Marturano

Introduction

It is currently fashionable to speak of the convergence of technologies: nanotechnology, biotechnology, information technology and cognitive science. Convergence itself is a subject of study regarding both its meaning and its social and ethical implications. Nanotechnology is arguably the key to convergence in so far as it enables the bringing together of all these. Computers and information technology, however, have for some time converged in their role in the development of genomics and postgenomics. What this paper will argue is that this is the case not only in the practical sense but also theoretically, and that the interface between information technology is central to the assessment of and realisation of the purported benefits of genomics.

The extent to which there is the potential for benefit at all remains a contested issue. The hopes and fears surrounding genomics are multiple: on the one hand, there have been fears that genomics might be used to increase discrimination and health inequalities; on the other hand, there are promises that genomics will lead to improved and indeed individualised health care and that it will provide new cures for hitherto intractable conditions. There are those who see the potential for genomics to be a global public good, on the one hand; while on the other some commentators have pointed to the potential for, at worst, biopiracy and exploitation of genetic resources of population groups. The issues in the debates are complex. We aim to explore the extent to which the connections between genetics and computing have a bearing on the issues. It will be argued that policy concerning IT has increasing relevance to some of the most pressing issues in genetics, in relation to, for example, the potential to increase or decrease inequalities.

This essay appears for the first time in *Ethics, Computing, and Genomics: Moral Controversies in Computational Genomics*. Copyright © 2005 by Ruth Chadwick and Antonio Marturano. Printed by permission.

Genomics as a Public Good

It has been argued that genomics should be considered a global public good (Thorsteindottir et al., 2003). In its central form this argument depends on regarding genomics as a form of knowledge, where knowledge is construed as the archetypal global public good. Public goods are defined as those which are non-rivalrous and non-excludable. Knowledge, when enjoyed by one, does not preclude someone else from knowing it: one person's knowledge does not deplete this good for others. It therefore meets the criteria, apparently. In these respects it is different from a piece of fruit which, when eaten by one, is no longer available for anyone else. Chadwick and Wilson (2004) have reconstructed the argument as follows:

1. Public goods are goods which are non-rivalrous and non-excludable;
2. Global public goods are public goods the enjoyment of which is not limited to any specific geographical area;
3. Knowledge is the archetypal global public good;
4. Genomics is a form of knowledge;
5. Genomics knowledge is a global public good.

Taking up this point, the HUGO Ethics Committee in its Statement on Human Genomic Databases (2002) applied the concept of global public goods to human genomic databases, saying that such databases *are* global public goods. The Committee defined a genomic database as "a collection of data arranged in a systematic way so as to be searchable. Genomic data can include *inter alia*, nucleic acid and protein sequence variations (including neutral polymorphisms, susceptibility alleles to various phenotypes, pathogenic mutations), and polymorphic haplotypes". As Chadwick and Wilson have shown, if it is indeed accepted that genomics *per se* is a public good, on the grounds that it is a form of knowledge, by a further step in the argument it seems to follow that

6. *a fortiori*, genomic databases, in so far as they contain genomic knowledge, are a global public good.

There are several criticisms that may be made of the argument. As far as step (6) is concerned, as Chadwick and Wilson also argued, there are queries about the 'global' nature of genomic databases: proposed national biobank initiatives such as those in Estonia or the UK may be more accurately construed as a local good, benefiting a specific population. On the other hand it may be argued that the knowledge acquired from a specific biobank initiative has global relevance to the whole of humankind, in so far as the human genome is considered to be the common heritage of humanity. In addition to this, however, steps (1) to (5) are not unproblematic. It is our aim to examine the ways in which the interface between information technology and genomics has a bearing on the issues, starting with a further point about (6).

Access and the Digital Divide

Even if genomics and genomic databases are globally *relevant*, the information contained therein may not be globally *available*. Indeed there is a difference of opinion about the extent to which it *ought* to be globally available. One important aspect of this relates to intellectual property regimes and the ways in which they can act to restrict access. There have certainly been calls for the data produced by human genome research to be freely available. As John Sulston has said, "The future of biology is strongly tied to that of bioinformatics, a field of research that collects all sorts of biological data, tries to make sense of living organisms in their entirety and then make predictions... If we wish to move forward with this fascinating endeavour, which will undoubtedly translate into medical advances, the basic data must be freely available for everyone to interpret, change and share, as in the open-source software movement" (Sulston, 2002). Debates about the desirability of patenting in the context of genomics are however by no means resolved. On the one hand there are arguments about the applicability of principles that have been traditional in the patenting debate, including the arguments about the desirability of rewarding creativity and the need to provide incentives; the distinction between discovery and invention and how that applies in the genomics context. On the other hand there are calls for rethinking the terms of the debate. Much is at stake here in terms of the potential impact on health inequalities worldwide, in so far as the upshot of patenting leads to restricted access to genetic tests and control of genetic resources.

Implementation and health service delivery

Beyond this principled debate, however, there are very practical issues about infrastructure. In the postgenome era attention is currently turning to implementation of genomics in health service delivery and the in the United Kingdom, for example, the Department of Health White Paper, *Our Inheritance, Our Future* (2003) acknowledged the need for considerable investment in information systems to cope with new demands and facilitate the effective utilisation of 'genetic knowledge' (sic) . The needs for information systems in this context are said to include access to sources of clinical information and decision support; co-ordination of electronic storage and handling of patients' clinical records; protection of confidentiality.

Not all societies are in a position to make this kind of investment. Minakshi Bhardwaj, writing about the genomics divide in relation to developing countries, has drawn attention to accessibility in the face of another divide – the digital. One consequence of the fact that genomics depends on the power of computing is that access to the information requires technology, and this is a limiting factor in facilitating access. Issues of access to the information thus depend on a number of variables. These include necessary IT to access what is contained in databases. A database is defined by HUGO as 'searchable' – but how? It is not just a matter of IT, however, but of other features of the social context, including governance arrangements, which control who has access to the information contained in the databases. Where such factors can affect access, it is problematic to say the least to say that information or indeed knowledge is a global public good. It is for good reason that knowledge is described as power – but it conveys power especially in relation to those who do not have it. Bhardwaj writes:

> Global inequity also exists in the form of information inequity, which is fundamentally related to the issue of accessibility. ....The lack of basic infrastructure facilities and virtual connectivity is a handicap for development... Figures show that even the free access to databases in public domain, the number of hits from developing countries is very low (Bhardwaj, 2004).

Thus public policy considerations include attention to the required infrastructure investment for the potential benefits of genomics to accrue, and for a proper 'benefit-sharing' (cf HUGO, 2000) to be feasible.

Information Technology: Solving and Creating Problems of Access

Information *sometimes* and to some *extent* needs to be protected by restricting access. This is particularly the case when it is, or could be, linked to particular individuals, as in human genomic databases. One of the principal concerns about the establishment of genetic databases has been the potential for third parties to acquire information about individuals that could be used to their disadvantage.

The power of computers here might be appealed to in order to 'solve', practically rather than theoretically, ethical problems. The debates about the principles of procedures of codification and anonymisation presuppose mechanisms for bringing about situations where information can be protected in such ways. But the very mechanisms which can assist in the protection of privacy and confidentiality can also undermine it. For even if there are adequate anonymisation procedures in place, there are increasing concerns about the possibilities of data mining. Anton Vedder has drawn attention to the possibilities, for example, of profiling through data mining, where data mining refers to a set of techniques called 'knowledge discovery in databases', or KDD. Vedder says:

> The basic aims and greatest opportunities of data mining are description and prediction through the discovery of significant patterns in the relationships of whatever kind between data. By 'profiling through data mining', I refer to those data-mining processes that result in profiles of groups of persons, i.e. characterisations of groups that can be assigned to those groups and to the members of those groups. The process of profiling through data mining aims at tracking down significant relationships between characteristics that define and identify a group of persons on the one hand and whatever properties or characteristics on the other (Vedder, 2000).

In this article Vedder is primarily concerned with data mining for the purposes of epidemiology and prediction of diseases. Profiling of this type could in principle operate to the disadvantage of certain defined subgroups of the population. The general point, then, is that information technology is both required to protect the data and can be used to mine it.

We have looked at the ways in which information technology is relevant to the availability and protection of information arising out of genomics. We will now move to look at the implications of information technology for the assessment of the status of the information itself. In so far as the global public good argument depends on an analogy with knowledge, it is pertinent to ask about what sort of information is at stake, and the answer to this question is not unconnected to the relationship between genomics and information technology.

Knowledge or Information?

To what extent is it legitimate, as in step (4) of the above argument, to speak of 'knowledge' in this context? What may be at issue, as indicated by the definition of genomic databases included in the HUGO Ethics Committee Statement, is massive amounts of *information*. But what is meant by this?

Acquiring the 'information': the Human Genome Project

The Human Genome Project, in producing a map and sequence of the location of genes on the chromosomes, has given rise to vast amounts of information. The production of this information itself depended on the interrelatedness between computer technology and molecular biology—for example, through the so-called 'shotgun' sequencing method. The shotgun method is a mathematical algorithm – that is, basically computer software. The shotgun method involves randomly sequencing the cloned sections of the genome, with no foreknowledge of where on a chromosome the sequence originally came from. The partial sequences are then reassembled to a complete sequence by use of computers. The advantage of this method is that it eliminates the need for time-consuming mapping because of increased computer speeds to solve such complicated algorithms (Trivedi, 2000). In fact, other methods of sequencing such as BAC to BAC sequencing need to create a crude physical map of the whole genome before sequencing the DNA. In the BAC to BAC sequencing computers became important in the final step of the protocol with the sequences collected in the so-called M13 libraries being fed into a computer program called PHRAP that looks for common sequences that join two fragments together (Trivedi, 2000).

In the case of the shotgun method we cannot see geneticists as working on living material. Rather they are working on representations of DNA sequences. Moreover, the shotgun method is able to shift two steps in the sequencing, because of its abstractness. The work of genome scientists therefore becomes similar to the software developers who are working on complicated mathematical functions in order to create useful software.

This is one context in which the relationship between computing and genomics becomes particularly pertinent, and it also forces us to reflect on the nature of the 'information' produced. The concept of 'information' is fundamental to both IT and molecular biology, although it is not clear that the concept has the same meaning in both fields. The notion of information developed by Shannon (1948) was less one concerned with meaning than a statistical one. Information reduces uncertainty in the receiver. The expression 'genetic information', on the other hand, used for the first time by Watson and Crick (1953), denoted the specification of the amino acid sequence in the proteins. Specificity and information were based on the concept of uniqueness of the sequence as a condition for replication. This is very different from the notion of information in computing, as it had developed by the end of the 50s, which did not attach any importance to *meaning* attached to transmitted information. Bosnack, however (1961) proposed a model in which information was interpreted as 'specificity' that is the difference between two entropies (that one before choosing a particular message and that after such a choice) within a repertoire.

During the 1960s, despite these characteristics of the nature of information in terms of specificity, a metaphorical notion of information became fully absorbed into the vocabulary of molecular biology. But it was not until 1977 that robust and generally applicable sequencing methods were developed and even then the modern bioinformatics techniques of gene discovery were still years away. The result has been that although the development of information/processing by computers proceeded contemporaneously with progress in research into biological and biochemical information processing, their trajectories, though they sometimes touched each other, were never unified.

According to Manuel Castells, however, such a theoretical convergence between information and genetics is now realised (2001, p. 164). This is due to the fact that, according to Gezelter (1999), modern science relies to a very large extent on computer simulations, computational models and computational analyses of large data sets. Genomics, indeed, while it may be considered a process entirely independent from microelectronics, is not really so independent. First, Castells argues that analytically these technologies are obviously information technologies, since they are focused on the decoding and eventually the reprogramming of DNA, the information code of living matter. And, more importantly, "Without massive computing power and the simulation capacity provided by advanced software, the Human Genome Project would not have been completed – nor would scientists be able to identify specific functions and the locations of specific genes" (ibid).

Beyond this, as Holdsworth (1999) suggests, it is not just that computer tools are rather convenient for doing genomics and protein sequencing. Rather, these disciplines have reorganised themselves around the bioinformatics paradigm and have absorbed Shannon's notion of information.

Thus genomics turns into an information science. Says Karp, "Many biologists consider the acquisition of sequencing to be boring. But from a computer science point of view, these are first rate and challenging algorithmic questions" (cited in Kaku, op. cit. p. 157). According to Lewontin, "Many of the founders of molecular biology began as physicists, steeped in the lore of the quantum mechanical revolution of the 1920s. The Rousseau of molecular biology was Erwin Schrodinger, the inventor of the quantum wave equation, whose *What is Life?* (Schrodinger, 1944) was the ideological manifesto of the new biology" (Lewontin, 2000, p. 136). Today "the intrusion of computers into molecular biology shifted power into the hands of those with mathematical aptitudes and computer savvy" (Kaku, p. 158).

The upshot of this for the global public goods argument is as follows. While information, in the form for example of raw sequence data, may be a *good*, it is not the *same* good as 'knowledge'. This suggests that it is too simplistic to try to argue, for example, that because genomic databases contain genomic information they are a global public good. Not only is it the case that most people would be unable to understand the 'information' therein, but the concept of 'information' itself may be one that does not include meaningfulness.

The Postgenome Era

The postgenome era might be characterised in terms of the search for meaning in the light of the acquisition of the information. The shotgun and other methods of genome sequencing have facilitated the completion of the Human Genome Project, but what most of the genes look like is still unknown, as is their expression of proteins. The genome by itself does not provide answers to any of these questions. What is necessary is to determine their significance, their location and how their control signals work (Sulston, 2002).

The next step, which is under way, is thus the understanding of the meaning and function of genes (Trivedi, 2000). According to Lewontin (1999) the same 'words' (that is the amino acid codon) have different meanings in different contexts and multiple functions in a given context, as in any complex language. Lewontin also claims that we do not know how the cell divides among possible interpretations. In working out the interpetative rules (that is, rules which determine whether code sequences are meaningful), it certainly helps to have very large numbers of gene sequences; and Lewontin himself suspected that the claimed significance of the genome sequencing project for human health is an elaborate cover story for the computerised hermeneutics of biological scripture.

A Right Not to Know?

Beyond the ascription of meaning to particular sequences, however, a further step is the determination of the significance of this information for particular individuals and population groups. It is here that debate has raged over whether individuals then have the right either to know or not to know facts about their own genetic information. The term 'know' tends to be used here even though the individual-related information, having meaning, remains in the overwhelming majority of cases probabilistic rather than certain.

For present purposes however the significant point is that even if the information were believed to be accurate and certain, the idea that the 'knowledge' would be unequivocally a good is by no means in all cases accepted, as is demonstrated by the increasing popularity of 'right not to know' arguments (See, e.g., Chadwick et al., 1997). The reasons for supporting a right not to know may be based on the distress and unhappiness that it may cause, especially if no ameliorative action can be taken in the light of an unwelcome result of a genetic test. Apart from the psychological consequences for those undergoing tests, however, there are wider concerns about the increasing reliance on genetic forms of explanation of health and other social phenomena, in so far as such explanations might have associated opportunity costs, leading us to play down the potential of other forms of explanation and intervention. As has already been noted, moreover, scepticism about benefits includes concerns about potential abuse of information – thus realisation of any benefit requires attention to the possibility of prevention of such abuse by protection of the information by bringing it about that *no one* knows it.

In so far as the claim that genomics is a global public good relies on genomics as knowledge, therefore, there are several possible objections. We have seen that there are various reasons for doubting point (4), first because of uncertainty over whether genomics is knowledge; and second because even where it is appropriate to speak in terms of knowledge it is not an unquestioned good. Even if this were not the case, that is if it were accepted that genomics is knowledge and that knowledge is a good, the move from (4) to (5) is not a simple matter. That is because, as we have seen, social and political realities can affect access to the good.

Conclusion

The future of molecular biology, according to Sulston (2002) is strongly tied to that of bioinformatics, a field of research that involves management, processing and modification of genetic information by means of computers, of all sorts of biological information, trying to make sense of living organisms in their entirety and then makes predictions. Possession of bioinformatics capabilities is a *sine qua non* of being an information rich country (Marturano 2003). There are clear policy implications for harnessing genomics to reduce health inequalities.

The infrastructure questions are just the beginning, however, of the issues that need to be addressed. The availability of the necessary infrastructure may bring with it further problems about protection, in the light of possibilities of data mining. Such scenarios support scepticism about the extent to which there is room for complacency about benefits. It is necessary, finally, to have a clear sense of what the 'object' is, which is a candidate for being regarded as a global public good: what is meant by 'information' and what is required not only to turn it into 'knowledge', concerning which there may be a right (not) to know?

References

Bhardwaj, M. (2004) 'Rich databases and poor people: opportunities for developing countries' *Trames* 8 (1/2): 90-105.

Bosnack, F. (1961) *Information, thermodianique, vie et pensée*, Paris: Gauthiers-Villars.

Castells, M. (2001) 'Informationalism and the network society' in Himanem, P. (2001) pp. 155-78.

Chadwick, R., Levitt, M., and Shickle, D. (eds) (1997) *The Right to Know and the Right not to Know* Aldershot: Avebury.

Chadwick, R. and Wilson, S. (2004) 'Genomic databases as global public goods?' *Res Publica* 10: 123-34.

Department of Health (2003) *Our Inheritance, Our Future: Realising the potential of genetics in the NHS* London: Department of Health.

Gezelter, D. (1999) 'Catalyzing Open Source development in science: the Open Science project http://www.openscience.org/talks/bnl/img0/htm (accessed 03/04/2003).

Holdsworth, D. (1999) 'The ethics of the 21st century bioinformatics: ethical implications of the vanishing distinction between biological information and other information' in Thompson, A.K. and Chadwick, R. (eds) *Genetic Information: Acquisition, Access and Control* New York: Kluwer/Plenum pp. 85-98.

Human Genome Organisation (HUGO) Ethics Committee, (2000) *Statement on Benefit-Sharing* London: Human Genome Organisation.

Human Genome Organisation (HUGO) Ethics Committee, (2002) *Statement on Human Genomic Databases* London: Human Genome Organisation.

Kaku, M. (1998) *Visions: How Science Will Revolutionize the Twentieth Century* Oxford: University Press.

Lewontin, R. (1999) *the Doctrine of DNA: Biology as Ideology* London: Penguin.

Lewontin, R. (2000) *It Ain't Necessarily So* London: Granta.

Marturano, A. (2003). 'Molecular Biologists as Hackers of Human Data: Rethinking IPR for Bioinformatics Research' *Journal of Information, Communication, and Ethics in Society*, 1: 207-215.

Schrödinger, E. (1944) *What is Life. The Physical Aspect of the Living Cell* Cambridge: University Press.

Sulston, J (2002) 'Heritage of humanity,' *le Monde Diplomatique* http://mondediplo.com/2002/12/15genome (accessed 15/02/2004).

Thorsteindottir, H. et al. (2003) 'Genomics – a global public good?' *The Lancet* 316: 891-2.

Trivedi, B. (2000) 'Sequencing the genome' *Genome News Network*, June 2, 2000; http://gnn.tigr.org/articles/06_00/sequence_primer.shtml (accessed 04/04/2003).

Vedder, A.H. (2000), "Medical data, new information technologies, and the need for normative principles other than privacy rules". In: M. Freeman, A. Lewis (ed.), Law and medicine. Current legal issues 2000. Volume 3., Oxford, Oxford University Press, p. 441-459.

Watson, J.D. and Crick, F.H.C. (1953) 'Molecular structure of nucleic acids,' *Nature* 171: 737-38.

5. Applying Genomic Technologies in Environmental Health Research: Challenges and Opportunities

David C. Christiani, Richard R. Sharp, Gwen W. Collman, and William A. Suk

For decades, it has been known that individuals differ greatly in their responses to chemicals, medicines, smoking, alcohol, ionizing radiation, and other environmental exposures. These varied responses are the result of complex interactions between several determinants, including genetic variability, age, sex, nutritional status, and comorbid conditions. In recent years, environmental health research has revealed that genetic variability in human populations can affect the entry, absorption, activation, and detoxification of environmental toxins. For example, genetic variation in the modulation of chemical exposures by polymorphic detoxification enzymes, such as P450s, glutathione transferases, sulfotransferases, and N-acetyltransferases, may be an important determinant of cancer risk in human populations.[1] Hence, these genetic variants may define subpopulations with increased sensitivities to exposure.

Our ability to examine how genetic characteristics affect response to environmental exposures offers exciting possibilities for the prevention and control of environmentally induced diseases. As new high-throughput technologies are developed for simultaneous analysis of multiple genes, many additional disease-related polymorphisms will be discovered. These new technologies, such as DNA microarrays and automated workstations capable of extracting, amplifying, hybridizing, and detecting DNA sequences, will allow large-scale, low-cost genotyping of both individuals and populations. Because nearly all human disease results from a combination of environmental exposures and genetic variation, these technologic advances will have a profound effect on disease prevention, allowing clinicians to identify individuals and groups who are most at risk and direct early intervention efforts to them.

Despite these potential benefits, the collection and analysis of genetic information in environmental health research presents many of the same ethical, legal, and social challenges found in other types of genetic research. For example, the

---

This essay originally appeared in the *Journal of Occupational and Environmental Medicine*, 2001; **43**:526–533. Copyright © 2001 by American College of Occupational and Environmental Health. Reprinted by permission.

detection of genetic sensitivities to occupational agents may affect employment opportunities and worker insurability.[2] Moreover, the collection of genetic information in environmental health research presents some new twists on familiar ethical and regulatory issues.[3] For example, in addition to the possible denial of employment or insurance on the basis of genetic susceptibility to an occupational agent, worker selection and job placement may supplant protective environmental controls and stricter standards for permissible exposure limits.

In response to these concerns, legal experts have proposed legislation to protect individual privacy with regard to genetic information.[4] Unfortunately, many of these proposed regulations could inadvertently curtail important epidemiologic studies employing genetic markers to examine gene-environment interactions.[5] Thus, two forces combine to support the need for addressing the ethical, legal, and social implications of research in environmental genomics: (1) growing public concern that genetic research presents substantial risks, and (2) increased concern among researchers that inappropriate regulatory standards are being brought to bear on public health research.

In this article, we describe several applications of genomic technologies in environmental health research and examine a number of ethical, legal, and social challenges in environmental genomics. We also suggest a way of advancing research in this area while simultaneously protecting the rights and welfare of individuals who choose to participate in such research.

Relevance of the Human Genome Project to Environmental Health Research

A program to sequence the entire human genome—the Human Genome Project—was initiated in 1988 when Congress appropriated funds to the Department of Energy and the National Institutes of Health. Project organizers set a 15-year time frame to sequence the approximately 80,000 genes scattered throughout the genome. Recent improvements in technology, success in early mapping, and growing demand for the human DNA reference sequence have accelerated the mapping project.[6] With the completion of this effort, researchers will have identified all, or nearly all, human genes.

The next phase of genomic research, which has already begun, is to elucidate the various functions of these genes. This includes identifying sequence variation within genes and determining the effect of such variation on the functioning of gene products. The current director of the Human Genome Project, Dr Francis S. Collins, has described the basic steps involved in understanding genetic contributions to disease and the project's potential for clinical medicine (Fig. 1).[7] Although establishing a complete human DNA reference sequence will not improve patient care directly, it can accelerate research efforts in pharmacogenetics, gene therapy, genetic diagnostics, and preventive medicine (Fig. 1).

Of special relevance to environmental health research is how the availability of genomic information might be used in early detection, intervention, and prevention programs. To promote and facilitate this area of research, the National Institute of Environmental Health Sciences (NIEHS) launched a multiyear project examin-

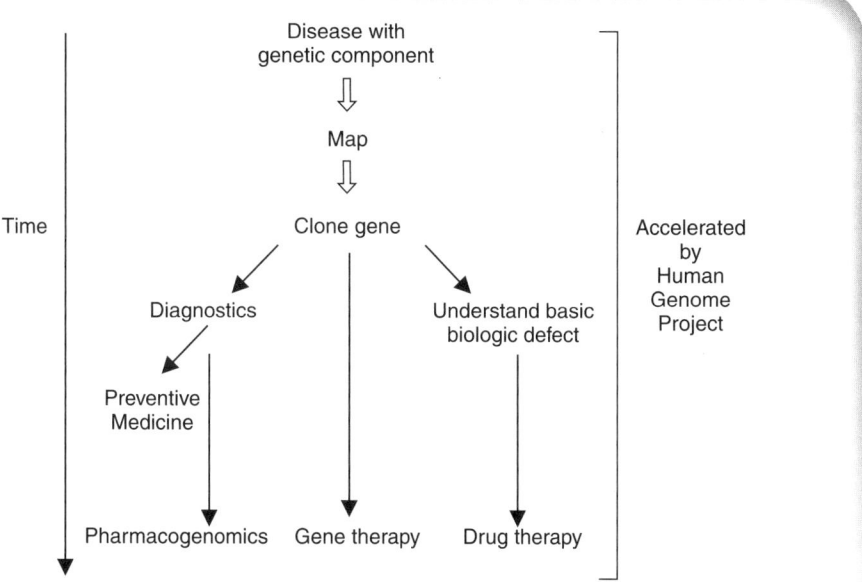

Figure 1 Steps involved in the genetic revolution in medicine.

ing these issues,[8,9] entitled *NIEHS Environmental Genome Project* (EGP), which focuses on common sequence variations in environmental-response genes. These genetic polymorphisms are not necessarily associated with increased disease risk. Some may have no function; some may be beneficial. However, we know that certain polymorphisms affect the metabolism of drugs and xenobiotics and thus seem to have important implications for human health. The EGP aims to identify such polymorphisms and clarify how polymorphic variation affects individual response to environmental exposures.

The EGP is examining polymorphisms that confer low-to-moderate relative risk of disease, and often only when accompanied by an adverse environmental exposure. An individual who possesses such a polymorphism may not develop the associated illness but may be at increased risk of developing the disease if particular environmental exposures are present. Thus, the EGP is interested in identifying and characterizing "susceptibility genes" (with low-to-moderate penetrance) and relevant "disease genes" (with high penetrance).

The potential benefits of this research are many. Determining how human genetic variation affects response to medications may eventually allow clinicians to subclassify diseases and tailor therapies to individual patients.[7] The effectiveness of therapy can be enhanced, and toxic side effects minimized, if we understand the genetic factors affecting these responses. This observation has prompted much interest in the field of "pharmacogenomics," a term referring to the potential of genetic assays to predict responsiveness to drugs.[10]

Although this vision of individualized medical care is exciting as long as proper protections of genetic information are in place and enforced, this scenario represents just one of the many opportunities presented by projects such as the EGP. Because many environmental-response genes exert an adverse health effect only in the presence of a specific environmental exposure, the opportunity for effective preventive interventions is great—identify and then minimize or eliminate the most dangerous environmental exposures for an at-risk population, and disease expression is prevented. The implications for public health are even more compelling—reduce toxic exposures to levels that are safe for the most vulnerable subgroups in a population, and disease expression in the entire population is controlled.

This suggests that the "genetic revolution" in medicine extends beyond the clinical care of individual patients. Because the majority of diseases are the consequence of both environmental exposures and genetic factors, understanding specific relationships between genetic variation and response to environmental exposures is important for understanding disease causation and for the development of effective disease-prevention strategies. Projects such as the EGP illustrate how the complete human DNA reference sequence can help guide research in public health and preventive medicine. Such applications hold forth the promise of preventing disease, or delaying disease onset, in large segments of the population without pharmacologic interventions (Fig. 2).

Gene–Environment Interactions: The Example of Lung Cancer

Studies of lung cancer illustrate how genetic analysis can be applied successfully in environmental health research. Lung cancer has a very high attributable risk to a single environmental exposure, namely, tobacco smoke. Nonetheless, individual variability in carcinogen metabolism seems to explain differential susceptibility to lung cancer.[11] In recent years, molecular techniques have been applied in the epidemiologic study of lung cancer, primarily in three areas: (1) determination of internal and biologically effective dose; (2) detection of early biological effects, particularly mutations and cytogenic changes; and (3) assessment of variation in individual susceptibility to carcinogens through the study of metabolic polymorphisms.

Determination of Internal and Biologically Effective Dose

Among the 3800 chemicals identified in tobacco smoke, a large number are biologically active compounds. The most important families of carcinogens found in tobacco smoke are polycyclic aromatic hydrocarbons (PAHs), aromatic amines, nitroso compounds, volatile organic compounds (eg, benzene, formaldehyde), and radioactive elements such as polonium-210. These carcinogens react with DNA to form adducts, covalently bound structures thought to be important in cancer development. Among the class of pulmonary carcinogens known as PAHs, benzo[α]pyrene forms DNA adducts in the lung. Recent research indicates that peripheral blood mononuclear cell adducts are quantitatively correlated with lung DNA adducts in cancer patients.[12] Such validation of the utility of surrogate tissue (blood) mark-

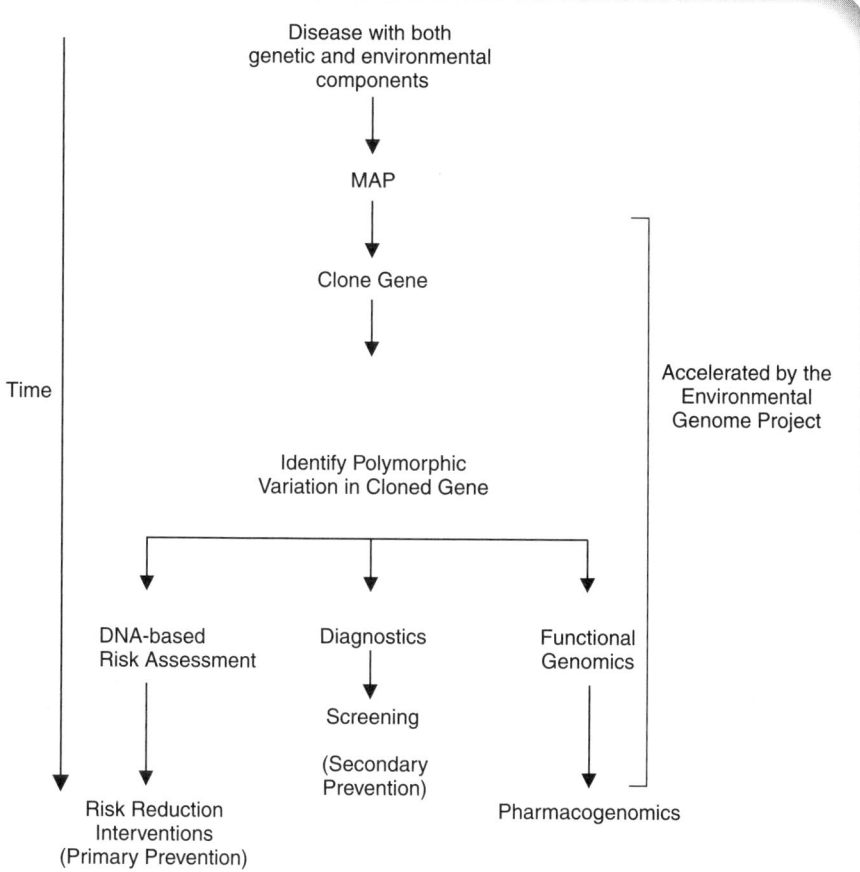

Figure 2 Steps involved in the genetic revolution in preventive medicine.

ers of target tissue (lung) damage will permit the assessment of interventions such as smoking cessation, chemoprevention, and elimination of passive-smoking exposures in reducing PAH DNA adducts in peripheral-blood mononuclear cells. Use of blood adducts will also permit epidemiologic studies of large populations exposed to PAHs in air pollution. For example, a recent application in a molecular epidemiologic study revealed an association between the initiation of smoking and evidence for PAH-DNA damage (adducts) in long-lived mononuclear cells.[13]

One aim of biological assessments of exposure is to improve estimates of environmental exposures in epidemiologic studies. DNA-based techniques, like adduct measurements, are one way of accomplishing this. These genetic tools thus complement existing methodologic approaches in environmental epidemiology.

Detection of Early Biological Effects

In recent years, genetic mutations have been used as "fingerprints" of specific exposures, as surrogate end points of cancer, and for subtyping cancers to clarify causal associations.[14] Epidemiologic evidence suggests an association among chemical exposures, the tumor-suppressor gene (p53), the oncogene (k-ras) mutation, and cancer occurrence.[15-19] Investigations concerning the ras oncogene family or the p53 tumor-suppressor gene have considered the association between lung cancer and tobacco smoking. Results of these studies suggest that the mutational spectrum of lung cancer may be associated with smoking and asbestos-exposure status.[20]

Although the presence of PAH-DNA adducts has not been definitively associated with lung cancer, these adducts seem to be important in the pathway leading to cancer,[21] and interventions aimed at reducing their presence in high-risk individuals are on the horizon. Identifying these "upstream" markers of dose and early effect is particularly important for diseases such as lung cancer, for which current treatments are an ineffective means of disease control.

Assessment of Variation in Individual Susceptibility to Carcinogens

Genetic susceptibility to lung cancer has been examined in a number of epidemiologic studies that incorporate molecular markers of susceptibility to carcinogens. It has been known for over 30 years that smoking and family history are independent risk factors for lung-cancer risk.[22] Additional evidence of genetic influences on cancer risk has been provided by the study of metabolic polymorphisms—genetic differences that affect an individual's ability to metabolize chemical carcinogens.[1]

Traditional monogenic susceptibility to cancer is related to rare diseases (xeroderma pigmentosa, Li-Fraumeni syndrome) or subgroups of common diseases (BRCA1 in familial breast cancer). Monogenic susceptibility is characterized by rare mutations that can be identified through linkage analysis. Polygenic susceptibility, by contrast, involves frequently occurring genetic polymorphisms and confers a low-to-medium elevation of risk for common diseases. This more frequent type of genetic susceptibility is identified through epidemiologic studies, often by using case controls. Examples of this include the CYP2D6 polymorphism phenotype and lung-cancer risk in Europeans[23] and the CYP1A1 and GSTM1 polymorphisms in lung and bladder cancer, respectively, among Japanese.[24]

This last example is noteworthy because recent work in the United States suggests that persons with the null GSTM1 genotype with a low dietary intake of antioxidants are at high risk of developing squamous-cell carcinoma of the lung, in contrast to persons with the null genotype and high antioxidant intake,[25] after adjusting for smoking intensity, duration, age, and time since quitting smoking. This kind of investigation may lead to antioxidant chemoprevention in ex-smokers, and it illustrates a dual clinical approach to disease control: exposure cessation combined with cumulative risk modification on the secondary level. If genotype-lung cancer risk is modified by a factor amenable to intervention, such as dietary intake of antioxidants,[26] then dietary changes and/or supplementation may be a useful intervention strategy for individuals who are attempting to quit or have already quit smoking.

The importance of this and related work in the molecular epidemiology of lung cancer promises to lead to the development of exciting new opportunities in environmental health research. This research could add greatly to our understanding of the health consequences not only of tobacco smoke but also of other harmful exposures.[27]

Ethical, Legal, and Social Challenges

Environmental genomic research presents many ethical and social challenges.[28-31] Like other types of genetic research, research in environmental genomics raises special concerns about possible discrimination and stigmatization. Similarly, common misperceptions of genetic information create challenges in obtaining informed consent and returning research findings to participants. In addition, although the scientific value of such studies is compelling, early research results often have unclear clinical value and may offer little direct benefit to individual study participants. As a result, these and other ethical considerations should prompt us to consider the adequacy of existing human-subjects protections and their applicability to environmental genomic research.

Genetic Information and Privacy

Genetic information is often treated as a separate category of health-related information.[32] This special treatment is most appropriate when genetic information is strongly predictive of an individual's disease risks.[33] A person who carries the gene for Huntington's disease, for example, will almost certainly develop that disease over the course of his or her lifetime. Because no treatment is currently available for Huntington's disease, individuals who carry this gene are concerned about their ability to secure health and life insurance. Genetic information may also demand special protection when DNA-based tests reveal genetic susceptibilities to occupational agents, because employers might choose to terminate or reassign workers on the basis of this information.

To date, however, virtually all of the scholarship on genetic privacy has focused on rare genetic conditions associated with highly penetrant genes.[3] This is unfortunate, because the protection of genetic privacy is very different in the context of low-penetrance genes—common genes that account for genetic variation in response to environmental exposures. Because this type of genetic information is much less predictive of future disease, and thus is more similar to other types of medical information, it is not clear whether individuals should have as much control over third-party access to this genetic information. Similarly, because these gene-disease associations are weaker and more difficult to validate,[34,35] it remains controversial as to when, how, and even whether research participants should be informed of the results of genotyping assays in environmental epidemiologic studies.[36] Therefore, in the context of epidemiologic research on low-penetrance genes, it is important to make the distinction between "susceptibility genes" and "disease genes" (for both subjects and Institutional Review Boards).

Additional research must be conducted to determine values and expectations with regard to learning and using information about one's genotype. Specific efforts that target education of study subjects and their families are necessary to deliver appropriate messages. In addition, guidelines must be developed with Institutional Review Boards that provide for the proper levels of protection when the clinical implications of such information are not clear.[5]

The protection of individual privacy is a major issue in medical law, and genetics is a major force in contemporary medical science, so it is logical that in the eyes of the public there may be potential conflicts between individual rights to privacy and biomedical progress. Although many states have attempted to address forms of "genetic discrimination" in health insurance and employment, federal legislation is necessary to provide comprehensive protection.[37] Without such protection in place, concern about the confidentiality of genetic information may make eligible research subjects reluctant to volunteer for epidemiologic studies of gene-environment associations because of the fear of losing their jobs and their health insurance and/or life insurance coverage.

The Pathologicization of Difference

In addition to the above-mentioned concerns about privacy and discrimination, another potential source of discrimination surrounds the incorporation of genomic technologies into environmental health research. As more is learned about how genetic factors influence our responses to environmental exposures, there may be a tendency to pathologize certain genotypic differences on the basis of perceptions of disease risk. Polymorphisms that place an individual at higher risk of developing an environmental disease may be viewed as "defective genes," even though key environmental triggers are required for the development of the associated disease. Each of us carries dozens of such risk-conferring polymorphisms. This pathologicization of genetic differences could result in discrimination against individuals with known genetic predispositions to environmental diseases. In addition, individuals who carry such polymorphisms may come to see *themselves* as deficient—even though they may be asymptomatic throughout their lives.

These risks present themselves not only to individuals with known risk-conferring polymorphisms, but also to broader ethnic groups. Because individual variation often is the result of heritable genetic traits, the ethnic groups of which individuals are members often possess these same polymorphisms, especially in historically isolated groups. Hence, the identification of disease-related polymorphisms can present risks to all members of a socially identifiable population.[38] All members of an ethnic group may suffer discrimination, stigmatization, or self-stigmatization because of an association between that group and a genetic predisposition to a disease.

Although nineteenth-century classifications of the major "races" are not biologically accurate, there is some basis for epidemiologic stratification by ethnicity because we live in social communities with ethnic identifiers. However, given the legacy of oppression and the ongoing social injustices encountered by some members of historically disadvantaged groups, there remains the potential for hardening of social categories and discrimination on the basis of socially constructed racial groupings.[39] These and other potential risks to individuals who opt not to participate as research subjects because of such fears are not currently factored into the risk-benefit assessments done by research sponsors and Institutional Review Boards.[40] Although a recent report by the National Bioethics Advisory Commission recommends that such social risks be addressed as a standard part of research reviews,[41] these potential long-term consequences of research in environmental genomics should be discussed in other public forums as well, with researchers consulting various stakeholders about the potential applications of the knowledge gained.

Regulatory Implications

Minimizing exposure to environmental and occupational toxins remains a central focus of regulatory policies relating to public health. Accurate and fair regulation regarding proper exposure limits requires appropriate scientific evidence. Some kinds of exposures should be limited for the entire US population (ubiquitous community exposures); some for particular populations (occupational groups); others for vulnerable populations (children). Some exposures may need to be eliminated altogether.

Clarifying gene-environment interactions will raise a host of difficult regulatory questions[42]: Should regulations be set for all on the basis of protecting the most susceptible among the population? Will toxic tort litigation increase as we learn more about genetic factors involved in environmentally induced diseases? With the availability of more information on particular genetic sensitivities to environmental agents, what standards should be used to define what constitutes an acceptable risk? These and many, many other questions have not been adequately discussed by legislators and policy-makers because information about genetic influences on environmental response has not been available until recently. As additional information becomes available, it will be necessary to examine how existing regulations should be modified to reflect this new knowledge.[43]

One issue of special note is how research in environmental genomics could inadvertently shift responsibility away from employers or governments onto individuals. If known genetic sensitivities to environmental agents are viewed as the salient factor in the development of disease—that is, as more important than the associated environmental triggers—then regulatory standards may be lightened. In other words, there could be a shift in the focus of regulatory policies away from eliminating harmful exposures and instead toward preventing vulnerable individuals from coming into contact with the harmful exposure. This is an issue both for individuals with known genetic vulnerabilities to particular environmental exposures and for the ethnic groups of which they are members.

Lastly, though it appears scientifically implausible at present, future concerns about a fine line between disease prevention and personal enhancement raises the specter of eugenics. For example, if the Environmental Genome Project led to the discovery that possessing a certain genotype predisposes some individuals to diseases caused by a particular exposure, the options available for disease prevention and control are to (1) lower exposure levels, or (2) eliminate the "deficient" genotype. Although there may be some legitimate applications of the second strategy, there is a danger that eugenic thinking could lead to more aggressive forms of government-supported approaches to the improvement of public health.

Recommendations

Suk and Anderson[44] describe a holistic approach to environmental health research that provides a framework for translational research in environmental health relevant to initiatives such as the Environmental Genome Project (EGP). Their approach stresses effective communication of research findings to policymakers and the public, with the building of a research infrastructure that includes input from stakeholders and active public participation in setting the research agenda. This approach helps to build social investment in research results and the public health interventions needed to protect the public's environmental health.

To further strengthen public confidence in the research enterprise, the ethical, legal, and social implications (ELSI) of research in environmental genomics should receive additional consideration. Just as these social issues were addressed in conjunction with the Human Genome Project, the EGP also should include a component that examines possible ethical and social implications.

These legal and ethical studies are needed in part because of growing public concern about the way in which science is done and whether research participants, and others, are adequately protected. In addition, there may be ethical and legal issues unique to environmental genomic research. that would not be addressed through other efforts (the ELSI Program of the Human Genome Project).[45] Thus, a research structure should be in place to address ethical and legal concerns as they are emerging, not simply in reaction to public concern or potential misuses of the information collected. Lastly, having a concurrent research program examining ethical and legal aspects of the EGP will help promote the development of environmental genomic research. The precedent of the ELSI Program of the Human Genome Project illustrates how the science can be strengthened by moving forward in an ethically reflective way, with potential research benefits and harms being weighed against each other as knowledge and experience are gained.

Clearly, the issues raised in this article are not easily resolved. Nevertheless, in an effort to move environmental health research forward, particularly in the area of environmental genomics, we make the following recommendations:

1. Environmental health scientists should be actively involved in policy debates regarding the ELSI implications of research in environmental genomics by increasing participation in national policy forums, developing consensus documents through professional societies, and informing policymakers through educational efforts.
2. The EGP should continue to serve as a focal point for discussions of the ELSI implications of the incorporation of genomic technologies into environmental health research. The EGP should follow the precedent set in connection with the Human Genome Project and support policy-related research in tandem with the basic science and epidemiologic studies supported by the NIEHS.
3. One focus of EGP work relating to ethical and social implications should be on educating potential research subjects and Institutional Review Boards in participating institutions. Although many of the ELSI challenges surrounding research in environmental genomics are not new,

familiar moral and legal perspectives on genetic research may not apply in this new context. Priority should be given to obtaining informed consent in environmental-genomic research, conveying research results to study participants, and assessing the potential impact of environmental-genomic research on socially identifiable groups.

This is an exciting era for researchers in the environmental health sciences. Genetic technologies promise to transform the way in which environmental health research is conducted. Ultimately, these advances could lead to exciting new developments in preventive medicine, successful treatment interventions, and more effective regulatory policies. Although adopting the recommendations we suggest here will not guarantee a positive outcome, we believe that taking these steps will help minimize the potential for research-related injuries and maximize the benefits offered by the genetic revolution in environmental health research.

Acknowledgments

Research was supported in part by NIH grants ES05947 and ES00002 (D.C.), the Division of Intramural Research at the NIEHS (R.S.), and the Division of Extramural Research at the NIEHS (G.C., W.S.).

References
1. Taningher M, Malacarne D, Izzotti A, Ugolini D, Parodi S. Drug metabolism polymorphisms as modulators of cancer susceptibility. *Mutat Res.* 1999;436:227–261.
2. Rothstein, MA. *Genetic Secrets: Protecting Privacy and Confidentiality in the Genetic Era.* New Haven: Yale University Press; 1997.
3. Sharp RR, Barrett JC. The Environmental Genome Project and bioethics. *Kennedy Inst Ethics J.* 1999;9:175–188.
4. Roche P, Glantz LH, Annas GJ. The Genetic Privacy Act: a proposal for national legislation. *Jurimetrics*. 1996;37:1–11.
5. Wilcox AJ, Taylor JA, Sharp RR, London SJ. Genetic determinism and the overprotection of human subjects [letter]. *Nat Genet.* 1999;21:362.
6. Bentley DR. The Human Genome Project—an overview. *Med Res Rev.* 2000;20:189–196.
7. Collins FS. Shattuck Lecture—medical and societal consequences of the Human Genome Project. *N Engl J Med.* 1999;341:28–37.
8. Albers JW. Understanding gene-environment interactions. *Environ Health Perspect.* 1997;105:578–580.
9. Kaiser J. Environmental institute lays plan for gene hunt. *Science.* 1997;278:569–570.
10. Evans WE, Reiling MV. Pharmacogenomics: translating functional genomics into rational therapeutics. *Science.* 1999;286:487–491.
11. Bartsch H, Nair U, Risch A, Rojas M, Wikman H, Alexandrov K. Genetic polymorphism of CYP genes, alone or in combination, as a risk modifier of tobacco-related cancers. *Cancer Epidemiol Biomarkers Prev.* 2000;9:3–28.
12. Wieneke J, Varkonyi A, Semey K, et al. Validation of blood mononuclear cell DNA-adducts as a marker of DNA damage in human lung. *Cancer Res.* 1995;55:4910–4915.

13. Wieneke J, Thurston SW, Kelsey KT, et al. Early age at smoking initiations and tobacco carcinogen DNA damage in the lung. *J Natl Cancer Inst.* 1999;91:614–619.
14. McMichael AJ. Molecular epidemiology: new pathway or new traveling companion? *Am J Epidemiol.* 1994;40:1–11.
15. Hollstein M, Sidransky D, Vogelstein B, Harris CC. p53 mutations in human cancers. *Science.* 1991;253:49–53.
16. Slebos RJC, Hruban RH, Dalesio O, Mooi WJ, Offerhaus JA, Rodenhuis S. Relationship between K-*ras* oncogene activation and smoking in adenocarcinoma of the human lung. *J Natl Cancer Inst.*1991;83:1024–1027.
17. Husgafvel-Pursiainen K, Hackman P, Ridanpaa M, et al. K-*ras* mutations in human adenocarcinoma of the lung: association with smoking and occupational exposure to asbestos. *Int J Cancer.* 1993;53:250–256.
18. Suzuki H, Takahashi T, Kuroishi T, et al. Mutations in non-small cell lung cancer in Japan: association between mutations and smoking. Cancer Res. 1992;52:734–736.
19. Vahakangas KH, Samet JM, Metcalf RA, et al. Mutations of p53 and *ras* genes in radon-associated lung cancer from uranium miners. *Lancet.* 1992;339:576–580.
20. Wang X, Christiani DC, Wiencke JK, et al. Mutations in the p53 gene in lung cancer are associated with cigarette smoking and asbestos exposure. *Cancer Epidemiol Biomarkers Prev.* 1995;4:543–548.
21. Wogan GN, Gorelick NS. An overview of chemical and biochemical dosimetry of exposure to genotoxic chemicals. *Environ Health Perspect.* 1985;62:5–18.
22. Sellers TA, Baily-Wilson SE, Elston RC, et al. Evidence for Mendelian inheritance in the pathogenesis of lung cancer. *J Natl Cancer Inst.* 1990;82:1272–1279.
23. Amos CI, Caporaso NE, Weston A. Host factors in lung cancer risk: a review of interdisciplinary studies. *Cancer Epidemiol Biomarkers Prev.* 1992;1:505–513.
24. Nakachi K, Imai K, Hayashi S, Kawajiri K. Polymorphisms of the CYP1A1 and glutathione S-transferase genes associated with susceptibility to lung cancer in relation to cigarette dose in a Japanese population. Cancer Res. 1993;53:2994–2999.
25. Garcia-Closas M, Kelsey KT, Wiencke J, Christiani DC. Nutrient intake as a modifier of the association between lung cancer and glutathione S-transferase μ deletion. In: Proceedings of the Eighty-Sixth Annual Meeting of the American Association for Cancer Research, Toronto, Canada; 1995:281.
26. Ames BN, Shinenaga MK, Hagen TM. Oxidants, antioxidants and the degenerative diseases of aging. *Proc Natl Acad Sci USA.* 1993;90:7915–7922.
27. Christiani DC. Utilization of biomarker data for clinical and environmental intervention. *Environ Health Perspect.* 1996;104(suppl 5):921–925.
28. Sharp RR, Barrett JC. The Environmental Genome Project: ethical, legal and social implications. *Environ Health Perspect.* 2000;108:279–281.
29. Samet JM, Bailey LA. Environmental population screening. In: Rothstein MA, ed. *Genetic Secrets: Protecting Privacy and Confidentiality in the Genetic Era.* New Haven: Yale University Press; 1997:197–211.

30. Soskolne CL. Ethical, social, and legal issues surrounding studies of susceptible populations and individuals. *Environ Health Perspect.* 1997;105:837–841.
31. Baird AP. Identifying people's genes. Ethical aspects of DNA sampling in populations. *Perspect Biol Med.* 1995;38:159–166.
32. Annas, GJ. Privacy rules for DNA databanks: protecting coded 'future diaries.' *JAMA.* 1993;270:2346–2350.
33. Murray TH. Genetic exceptionalism and "future diaries": is genetic information different from other medical information? In: Rothstein MA, ed. *Genetic Secrets: Protecting Privacy and Confidentiality in the Genetic Era.* New Haven: Yale University Press; 1997:61–73.
34. Caporaso N, Goldstein A. Cancer genes: single and susceptibility: exposing the difference. *Pharmacogenetics.* 1995;5:59–63.
35. Schulte PA. Validation. In: Schulte PA, Perera FP, eds. *Molecular Epidemiology: Principles and Practices.* New York: Academic Press; 1993.
36. Schulte PA, Singal M. Ethical issues in the interaction with subjects and disclosure of results. In: Coughlin SS, Beauchamp TL, eds. *Ethics and Epidemiology.* New York: Oxford University Press; 1996:178–196.
37. Reilly PR. Efforts to regulate the collection and use of genetic information. *Arch Pathol Lab Med.* 1999;23:1066–1070.
38. Foster MW, Sharp RR. Genetic research and culturally specific risks: one size does not fit all. *Trends Genet.* 2000;16:93–95.
39. Juengst EJ. Groups as gatekeepers to genomic research: conceptually confusing, morally hazardous and practically useless. *Kennedy Inst Ethics J.* 1998;8:183–200.
40. Sharp RR, Foster MW. Involving study populations in the review of genetic research. *J Law Med Ethics.* 2000;28:41–51.
41. National Bioethics Advisory Commission. *Research Involving Human Biological Materials: Ethical Issues and Policy Guidance.* Rockville, MD: US Government Printing Office; 1999.
42. Tarlock AD. Genetic susceptibility and environmental risk assessment: an emerging link. *Environ Law Reporter.* 2000;30: 10277–10282.
43. Yap W, Rejeski D. Environmental policy in the age of genetics. *Issues Sci Technol.* 1998;8:33–36.
44. Suk WA, Anderson BE. A holistic approach to environmental-health research. *Environ Health Perspect.* 1999;107:A338–A340.
45. Meslin EM, Thomson EJ, Boyer JT. The ethical, legal, and social implications research program at the National Human Genome Research Institute. *Kennedy Inst Ethics J.* 1997;7:291–298.

III: PERSONAL PRIVACY AND INFORMED CONSENT

The five chapters that comprise Section III examine a range of issues involving privacy, confidentiality, and informed consent in the context of computational genomics research. Perhaps no ethical issue associated with the field of computing/information technology or with genetics/genomics research has been more controversial than privacy. Thus it is not surprising that when the ELSI (Ethical, Legal, and Social Implications) Research Program was established to guide researchers working on the Human Genome Project, the first ELSI "program area" identified was entitled "Privacy and Fairness in the Use and Interpretation of Genetic Information."[1] Concerns in this particular ELSI program area include questions such as "who should have access to personal genetic information?" and "how will that information be used?"[2] Some "focus areas" pertaining to this particular ELSI program area are described in Chapter 1.

Because of concerns about the ways in which certain uses of personal genetic data can result in various kinds of psychosocial harm to persons, some genetic-specific laws and policies have been proposed. In the U.S., a patchwork of laws, enacted mainly at the state level, aim at protecting individuals against having their genetic information used in ways that can negatively affect their opportunities to gain employment and acquire health insurance. Thus far, 41 states have enacted legislation on genetic discrimination, and 31 states have enacted legislation affecting genetic discrimination in the workplace.[3]

While there are no federal laws in the U.S. that protect personal genetic data per se, the Health Insurance Portability and Accountability Act (HIPAA) of 1996, enacted into law on April 14, 2003, provides broad protection for personal medical information. That is, HIPAA includes standards for protecting the privacy of "individually identifiable health information" from "inappropriate use and disclosure." However, this act does not provide any special privacy protection for personal genetic information; HIPAA's "privacy rule" protects personal genetic information simply as one form of "health protected information." And the privacy rule does not preempt more stringent state laws that can trump HIPAA.

---

[1] See Chapter 1 for a description of this program. See also the web site for the ELSI Research Program at http://www.genome.gov/1000754.

[2] For example, should insurers, employers, courts, and schools have access to this information?

[3] See the "Policy & Ethics" Section of the NHGRI Web site (http://www.genome.gov/10002336).

On October 14, 2003, the U.S. Senate passed the Genetic Information Nondiscrimination Act. However, this bill has not yet been passed by the U.S. House of Representatives. So U.S. citizens living in states that have not enacted their own genetic privacy laws do not enjoy explicit legal protection for their genetic information.

For citizens residing in many nations in Western Europe, the European Union (EU) Directive 95/46/EC (sometimes referred to as the "privacy directive") protects "data subjects" in the processing (i.e., the collection and use) of personal data. Protections accorded to data subjects via this directive include the protection of personal genetic information as well.[4] However, as in the case of HIPAA, no special provisions in the EU Directive are included for protecting personal genetic data per se.

Does personal genetic data—i.e., data that identifies genetic information about specific individuals—deserve a special kind of normative protection? For that matter, does personal information of any kind warrant special laws or policies to protect it? What, exactly, is personal privacy, and why should it be protected? Answers to these questions presuppose an understanding of the concept of privacy and why it is important.

The Concept of Privacy

Because privacy is a concept that is not easily defined, we should not be surprised that many different theories or models of privacy have been suggested. Some privacy analysts have suggested that it is more useful to view privacy as a presumed *interest* that individuals have with respect to protecting personal information, personal property, or personal space than to think about privacy as a moral or legal right. In this sense, personal privacy can be thought of as an interest that is normatively stipulated. For example, personal privacy could be viewed in terms of an economic interest, in which information about individuals might be thought of in terms of personal property that could be bought and sold in the commercial sphere.[5] Many Western European nations view concerns involving individual privacy and confidentiality as issues of *data protection*—i.e., as an interest individuals have in ensuring the protection of their personal information—rather than in terms of a normative concept that needs philosophical analysis and justification. In the U.S., on the other hand, discussions involving the concept of privacy as a legal right are rooted in extensive legal and philosophical argumentation.

James Moor (1997) points out that in the United States, the concept of privacy has evolved significantly from the 18th century to the present. The evolution of this concept can be understood and analyzed in terms of three distinct kinds of privacy-related concerns: (1) protecting individuals from intrusion into their personal space; (2) guarding against interference with an individual's personal affairs; and (3) regulating

---

[4]For an excellent discussion of the EU Privacy Directive, see Elgesem (2004).

[5]For an analysis of this view, see Richard Spinello's discussion of privacy in terms of property in Chapter 12.

access to and use of personal information. The idea that privacy can be thought of primarily in terms of protecting against *intrusion* can be found in the Fourth Amendment to the U.S. Constitution, which protects against unreasonable government searches and seizures by the federal government. As a philosophical theory of privacy, however, this view was first articulated in a highly influential article by Warren and Brandeis (1890), which argued that privacy consists in "being let alone." By the 1960's, the concept of privacy had become more closely identified with concerns stemming from *interference* in personal affairs. This view of privacy is articulated in two U.S. Supreme Court decisions: one involving access to information about the use of contraceptives (*Griswold v. Connecticut*, 1965), and the other involving a woman's decision to have an abortion (*Roe v. Wade*, 1973). Because of concerns about the collection and exchange of personal information via electronic records stored in computerized databases, beginning in the 1970s much of the focus on privacy issues has centered on questions about how to protect one's personal *information*. These concerns influenced the passage of the Privacy Act of 1974.

Privacy analysts sometimes refer to "informational privacy" as a category of privacy with a set of issues that are distinguishable from privacy concerns related both to intrusion and interference, which have been described as "psychological privacy" (Regan, 1995) and "associative privacy" (DeCew, 1997). The discussions of privacy issues in the chapters in Section III are mainly concerned with informational privacy, because they tend to focus on the role that computers and information technology have played. But some privacy issues involving genetics and genomics also border on concerns affecting intrusion and interference. For example, opponents of genetic testing in the workplace argue that such a practice is intrusive in the lives of the persons affected, invading or violating their "personal space." Others argue that genetic testing can also interfere with a person's freedom to make certain decisions, thus denying those individuals autonomy. Nevertheless, the privacy issues examined in Section III tend to fall under the category of informational privacy because they address three kinds of computer-related privacy concern where one's personal genetic data can be: (1) stored indefinitely in computer databases; (2) exchanged between databases; and (3) mined for implicit patterns and then potentially used to make controversial decisions about individuals.

Two theories associated with informational privacy have received considerable attention in recent years: the "control theory" and the "restricted access theory." According to the control theory, one has privacy if and only if one has *control* over information about oneself (see, for example, Fried, 1984; and Rachels, 1995). According to the restricted access theory, one has privacy when access to information about oneself is limited or *restricted* in certain contexts (see, for example, Allen, 1988; and Gavison, 1984). One strength of the control theory is that it recognizes the aspect of choice that an individual who has privacy enjoys in being able to grant, as well as deny, individuals access to information about oneself. One strength of the restricted access theory, on the other hand, is that it recognizes the importance of setting up contexts or "zones" of privacy. However, critics have pointed out certain weaknesses in each theory.

The control theory has at least two major flaws: one which is practical in nature, and the other which is theoretical or conceptual. On a practical level, one is never able to have complete control over every piece of information about oneself. A theoretical or conceptual difficulty arises for control theorists who seem to suggest that one could conceivably reveal every bit of personal information about oneself and yet also be said to retain personal privacy. The prospect of someone disclosing all of his or her personal information and still somehow retaining personal privacy, merely because he or she retains control over whether to reveal that information, is indeed counter to the way we ordinarily view privacy. One problem with the restricted access theory, on the other hand, is that it tends to underestimate the role of control or choice that personal privacy also requires. So it would seem that neither of these two theories is fully adequate. However, each theory seems to offer an insight into what is essential for individuals to have privacy. Recognizing this, Moor (1997) integrates essential elements of both the control and restricted access theories into one comprehensive theory of privacy, which, appropriately perhaps, he calls the *control/restricted access theory*. His theory shows why it is important both to restrict access (i.e., set up "zones" of protection) and to grant limited controls to individuals.

In Chapter 6, James Moor applies his control/restricted access theory to controversies involving the collection and transfer of personal genetic data. He argues that, ethically, computers can help protect privacy by restricting access to genetic information in sophisticated ways. But Moor worries that computers will increasingly collect, analyze, and disseminate abundant amounts of genetic information made available through the "genetic revolution." He also worries that inexpensive computing devices will make genetic information gathering easier, which, Moor believes, underscores the need for strong and immediate privacy legislation in the area of genetics.

Chapter 7 continues with an analysis of privacy-and-technology issues in the context of genetic/genomic research. In this chapter, Judith DeCew begins with a discussion of the value of privacy and what we lose without it. She then examines some of the difficulties involved in preserving privacy for genetic information and other medical records in the face of advanced information technology. DeCew examines three distinct public policy approaches to the problem of protecting individual privacy and also preserving databases for genetic research: (1) governmental guidelines and centralized databases, (2) corporate self-regulation, and (3) her own "hybrid" approach (in which she introduces her notion of "dynamic negotiation"). For DeCew, none of these are unproblematic; she discusses strengths and drawbacks of each, emphasizing the importance of protecting the privacy of sensitive medical and genetic information. DeCew's approach defends privacy as a "priority through federal guidelines mandating *privacy* protection as the *default*." Her approach also would ensure that people are "educated, consulted, and allowed to give consent or refusal through the process of dynamic negotiation before their health information is gathered or disseminated." DeCew believes that protecting the privacy of patients need not undermine health care or medical and genetic research. She stresses the importance of protecting the privacy of genetic information for all individuals, while simultaneously "letting new information technology flourish to aid patient care, public health, and scientific research."

Data Mining and (Nonconsensual) Secondary Uses of Personal Information

Whereas the first two chapters in Section III examine conceptual and theoretical (as well as some practical) aspects of privacy, the remaining chapters focus mainly on identifying and analyzing specific privacy issues pertaining to medical data and genetic/genomic information. In particular, these chapters examine controversies surrounding some *secondary uses* of personal medical and genetic information. In Chapter 8, David Baumer, Julia Brande Earp, and Fay Cobb Payton point out that, increasingly, medical records are being stored in computer databases. On the one hand, the authors note that this practice allows for greater efficiencies in providing treatment and in the processing of clinical and financial services. They also note, however, that computerization of medical records has diminished patient privacy and has increased the potential for misuse. This is especially apparent in the case of *nonconsensual* secondary uses of personally identifiable records. For the most part, organizations that store and use medical records have had to establish security measures, prompted partially by what the authors describe as an "inconsistent patchwork of legal standards that vary from state to state." However, Baumer, Earp, and Payton also note that the Health Insurance and Portability Accountability Act (HIPAA) of 1996 (enacted into law in April 2003) was designed to provide a uniform standard of privacy protection for patient medical records. Although HIPAA provides more uniformity, the authors believe that it is not clear whether HIPAA also succeeds in solving all of the concerns affecting nonconsensual secondary uses of personal medical information.

Much of the recent controversy surrounding secondary uses of personal medical and genetic information has been associated with practices involving a computerized technique that is commonly referred to as *data mining*. What, exactly, is this technology, and why is it so controversial? Data-mining tools consist of algorithms that are used to "discover" information via analyses of certain kinds of patterns that are implicit in data. This newly discovered information is controversial because it can suggest "new facts" about persons based on associations and correlations that are revealed by data-mining algorithms. For example, individuals can become identified with new groups or new categories (of persons) generated through data mining. Furthermore, decisions can be made about individuals on the basis of their association with one or more of the "new" groups, despite the fact that these individuals may not possess the characteristics attributed to that group. For example, consider that an individual might be denied employment or health insurance based solely on his or her identification with a "risky" consumer group because of certain behavioral patterns revealed in one or more data-mining applications, even though the credit history of the individual in question is impeccable.

In Chapter 9, Bart Custers examines some privacy challenges that the use of data mining poses for research subjects in the context of epidemiologic research. Custers begins by noting that in order to get a better grip on the large amounts of data being generated and stored in databases, efforts have been made to find useful patterns and relations in the data. In the case of medical data, patterns may be used in epidemiology in order to solve etiological, diagnostic, prognostic, or therapeutic problems. However, the same information may also be used for determining selection criteria for insurance, jobs, loans, etc. Custers points out that patterns revealed from data-mining technology may result in the construction of both *individual profiles* and *group profiles*. He also points out that whereas data in individual profiles is protected by privacy law, data in group profiles lacks this protection. Furthermore, Custers worries that the group profiles generated by data mining may not be very reliable, especially when they are used to make decisions about persons assigned to groups on the basis of those profiles.

Chapter 10, the closing chapter in Section III, continues with an analysis of data-mining technology by focusing on its implications for informed-consent policies in the context of environmental genomics research.[6] In this chapter, Herman Tavani begins by examining some positive and negative aspects of data mining. He notes that on the one hand, data-mining tools have greatly assisted researchers in identifying certain "disease genes" common in specific populations, which, in turn, has accelerated the process of finding cures for diseases that affect members of those populations. On the other hand, the same technology now significantly threatens the principle of informed consent, at least in the sense in which that notion has traditionally been used in epidemiologic studies. For example, data-mining technology facilitates the construction of controversial new groups, which can be based on non-obvious patterns and statistical correlations that link together both sensitive and non-essential (or "trivial") information about persons. Because of this, Tavani argues that environmental genomics research may, unwittingly, contribute to the construction of these controversial new groups. Once research subjects become identified with one or more of these new groups, they may be vulnerable to increased risk for discrimination and stigmatization.

Tavani also argues that because of the use of data-mining technology in genomics studies, the consent process for research subjects has become increasingly "opaque" or non-transparent. This, in turn, poses a significant challenge for the possibility of "valid informed" and "fully informed" consent procedures needed for research subjects. He concludes by suggesting that unless genetic/genomic researchers, Institutional Review Board (IRB) representatives, and policy/law makers work together to frame adequate informed-consent policies that respond to challenges posed by data-mining technology, the future of environmental genomics research, as well as population genomics studies in general, may be in jeopardy.

---

[6]This reading also responds to some challenges and issues raised in Chapter 5, by David Christiani, et al.

Review Questions for Chapters in Section III
1. What is "personal privacy," and why is privacy such a difficult concept to define?
2. Briefly identify and describe three "traditional" theories of privacy.
3. Describe the main features of James Moor's "control/restricted access" theory of privacy. How is this theory an improvement over traditional theories of privacy?
4. Why does Moor believe that computers can be used to protect personal privacy, as well as pose a significant threat to it?
5. Identify and briefly describe the three distinct kinds of public policy approaches that Judith DeCew considers. Which one does DeCew believe is the most promising for resolving the "problem of protecting individual privacy and also preserving databases for genetic research"? Do you agree with DeCew? Explain.
6. What does DeCew mean by "dynamic negotiation" in the context of the privacy debate? How can DeCew's principle be applied in particular cases involving genetic/genomic research?
7. What is HIPAA? Does this act offer any significant improvements over previous privacy-protection schemes for personal medical information?
8. Does the enactment of HIPAA ensure the protection of patient medical records in general? Can it also ensure the confidentiality of personal genetic information? Explain.
9. Why are Baumer, Earp, and Payton concerned about the nonconsensual secondary uses of personal medical information? Are their concerns justified?
10. What, exactly, is *data mining*? Why is the use of data-mining technology in the context of genomic research so controversial?
11. Why does Bart Custers believe that data mining poses significant problems for epidemiologic research?
12. In which ways does Custers distinguish between *individual* and *group* profiles?
13. Describe some of the challenges that data mining poses for informed-consent policies and practices in the context of genomic research.
14. What kinds of risks do "new facts" that can be generated from data-mining tools pose for research subjects in environmental genomics studies?
15. How can human research subjects contribute, perhaps unwittingly, to the potential construction of controversial new groups and subgroups?
16. Are the ideals of *fully informed* and *valid informed* consent possible for research subjects who participate in genetics/genomics studies that use data mining? Explain.

References

Allen, Anita (1988). *Uneasy Access: Privacy for Women in a Free Society*. Totowa, NJ: Rowman and Littlefield.

DeCew, Judith Wagner (1997). *In Pursuit of Privacy: Law, Ethics, and the Rise of Technology*. Ithaca, NY: Cornell University Press.

Elgesem, Dag (2004). "The Structure of Rights in Directive 95/46/EC on the Protection of Individuals With Regard to the Processing of Personal Data and the Free Movement of Such Data." In R.A. Spinello and H.T. Tavani, eds. *Readings in CyberEthics*. 2nd ed. Sudbury, MA: Jones and Bartlett, pp. 418–435.

Fried, Charles (1984). "Privacy." In F. D. Schoeman, ed. *Philosophical Dimensions of Privacy*. New York: Cambridge University Press.

Gavison, Ruth (1980). "Privacy and the Limits of the Law," *Yale Law Journal*, Vol. 89, 1980.

Moor, James H. (1997). "Towards a Theory of Privacy for the Information Age," *Computers and Society*, Vol. 27, No. 3, pp. 27–32. Reprinted in R. A. Spinello and H. T. Tavani, eds. (2004). *Readings in CyberEthics* 2nd ed. Sudbury, MA: Jones and Bartlett, pp. 407–417.

Rachels, James (1995). "Why Privacy Is Important." In D.G. Johnson and H. Nissenbaum, eds. *Computing, Ethics and Social Values*. Upper Saddle River, NJ: Prentice Hall, 351–357.

Regan, Priscilla M. (1995). *Legislating Privacy: Technology, Social Values, and Public Policy*. Chapel Hill, NC: University of North Carolina Press.

Spinello, Richard A., and Herman T. Tavani (2004). "Privacy in Cyberspace." In R. A. Spinello and H. T. Tavani, eds. *Readings in CyberEthics*. 2nd ed. Sudbury, MA: Jones and Bartlett, pp. 397–406.

Tavani, Herman T. (1999). "KDD, Data Mining, and the Challenge to Normative Privacy," *Ethics and Information Technology*, Vol. 1, No. 4, pp. 265–273.

Warren, Sammuel, and Louis Brandeis (1890). "The Right to Privacy," *Harvard Law Review*, Vol. 14, No. 5.

Westin, Anthony F. (1967) *Privacy and Freedom*. New York: Atheneum Press.

Suggested Further Readings

Alpert, Sheri A. (1998). "Health Care Information: Access, Confidentiality, and Good Practice." In K. W. Goodman, ed. *Ethics, Computing, and Medicine: Informatics and the Transformation of Healthcare*. New York: Cambridge University Press, pp. 75–101.

Annas, George J. (1993). "Privacy Rules for DNA Databanks: Protecting Coded Future Diaries," *Journal of the American Medical Association*, Vol. 270, pp. 2346–2350.

Bennett, Colin J., and Rebecca Grant, eds. (1999). *Visions of Privacy: Policy Choices for the Digital Age*. Toronto: University of Toronto Press Incorporated.

Custers, Bart (2004). *The Power of Knowledge: Ethical Legal, and Technological Aspects of Data Mining and Group Profiling in Epidemiology*. Nimijen, The Netherlands: Wolf Legal Publishers.

DeCew, Judith W. (1999). "Alternatives for Protecting Privacy While Respecting Patient Care and Public Health Needs," *Ethics and Information Technology*, Vol. 1, No. 4, pp. 249–255.

Elgesem, Dag (1996). "Privacy, Respect for Persons, and Risk." In C. Ess, ed. *Philosophical Perspectives on Computer-Mediated Communication*. New York: State University of New York Press.

Etzioni, Amatai (1999). *The Limits of Privacy*. New York: Basic Books.

Garfinkel, Simson (1999). *Database Nation. The Death of Privacy in the 21st Century*. Cambridge, MA: O'Reilly and Associates, Inc.

Gostin, Lawrence, et al. (1996). "The Public Health Information Infrastructure: A National Review of the Law on Health Information Privacy," *Journal of the American Medical Association*, Vol. 275, pp. 1921–1927.

Khoury, Muin J., and Janice S. Dorman (1998). "The Human Genome Epidemiology Network (HuHENet^{TM})," *American Journal of Epidemiology*, Vol. 148, pp. 1–3.

Levine, Robert J. (1999). "Informed Consent: Some Challenges to the Universal Validity of the Western Model." In T. Beauchamp and L. Walters, eds. *Contemporary Issues in Bioethics*. 5th ed. Belmont, CA: Wadsworth Publishers.

Markman, Maurie (2004). "What Must Research Subjects Be Told Regarding the Result of Completed Randomized Trials?" *IRB: Ethics and Human Research*, May–June.

Murray, Thomas H. (1997). "Genetic Exceptionalism and Future Diaries: Is Genetic Information Different From Other Kinds of Medical Information?" In M. Rothstein, ed. *Genetic Secrets: Protecting Privacy and Confidentiality*. New Haven, CT: Yale University Press, pp. 61–73.

Nissenbaum, Helen (1998). "Protecting Privacy in an Information Age," *Law and Philosophy*, Vol. 17, pp. 559–596.

Office of Technology Assessment (OTA). (1993). *Protecting Privacy in Computerized Medical Information*. Washington, DC: U.S. Government Printing Office.

O'Neill, Onora (2002). *Autonomy and Trust in Bioethics*. Cambridge: Cambridge University Press.

Parent, W. (1983). "Privacy, Morality, and the Law," *Philosophy and Public Affairs*, Vol. 12, No. 5, pp. 269–288.

Powers, Madison (2002). "Privacy and Control of Genetic Information." In R. Sherlock and J. D. Murray, eds. *Ethical Issues in Biotechnology*. New York: Rowman and Littlefield, pp. 439–460.

Reiman, Jeffrey (1984). "Privacy, Intimacy, and Personhood." In F. D. Schoeman, ed. *Philosophical Dimensions of Privacy*. New York: Cambridge University Press.

Rothstein, Mark ed. (1997). *Genetic Secrets: Protecting Privacy and Confidentiality*. New Haven, CT: Yale University Press.

Samet, J. M. and L. A. Bailey (1997). "Environmental Population Screening." In M. Rothstein, ed. *Genetic Secrets: Protecting Privacy and Confidentiality*. New Haven, CT: Yale University Press, pp. 197–211.

Tavani, Herman T. (2004). "Genomic Research and Data-Mining Technology: Implications for Personal Privacy and Informed Consent," *Ethics and Information Technology*, Vol. 6, No. 1, 2004, pp. 15–28.

Vedder, Anton H. (2004). "KDD, Privacy, Individuality, and Fairness." In R.A. Spinello and H.T. Tavani, eds. *Readings in CyberEthics*. 2nd ed. Sudbury, MA: Jones and Bartlett, pp. 462–470.

6. Using Genetic Information While Protecting the Privacy of the Soul

James H. Moor

Introduction

In the last half of the twentieth century two technologies that potentially threaten privacy have developed dramatically—genetics and computing. These informational technologies are closely connected. It is, for example, only through the "eyes" of computers that we can hope to map and sequence the human genome in a practical period of time. Our understanding of the details of genetics is becoming increasingly important when providing the best medical recommendations and treatment. And yet, this genetic information, which is so helpful to the medical practitioner and to the epidemiological researcher, is a significant threat to the privacy and security of individuals. Genetic information can too easily be used to discriminate against individuals to deny them health benefits, educational programs, and employment opportunities. The problem is exacerbated by recent advancements in computer technology and genetics that makes genetic testing easy to do in non-medical contexts.

Not all personal information is equally sensitive and some personal information should be treated with more care and protection than the rest. Genetic information concerning particular individuals demands special protection. In a large information society in which we tend to have a lot of information about individuals, without actually knowing them personally, discrimination can come all too easily. Given recent atrocities in the world including genocide, it is not difficult to imagine that in the coming century genetic information could be used to promote a genetic caste system conducive to unfair discrimination or even the extermination of a group of people.

This essay originally appeard in *Ethics and Information Technology* **1**: 257–263, 1999. © 2000 by *Kluwer Academic Publishers.* Reprinted by permission.

All of us, especially those involved in medical information gathering, have an obligation to act quickly to place enforceable safeguards around genetic information so that, as the store of genetic information and understanding of it grows, individuals are not placed in harm's way. Electronic medical record keeping can be designed and adjusted to enhance individual privacy. But, the problems of invasions of privacy and discrimination regarding genetic information are not limited to the medical community and cannot be solved simply with technological fixes. No nation has sufficiently developed its privacy laws to adequately protect its citizens against abuses of genetic information. We need to reassess the importance of privacy, particularly genetic privacy, in the information age. Our goal should be to find policies that permit the proper and effective use of genetic information while safeguarding the privacy of our souls.

Gene Machines

James Watson and Francis Crick were to genetics what Alan Turing was to computing. They described a simple arrangement of simple components following simple rules that captured a mechanism, a biological computer, that can generate incredible diversity and complexity. When in 1953 Watson and Crick, benefiting from the insights of many others, postulated a double helix structure for Deoxyribose Nucleic Acid (DNA), the stage for the explosive growth in molecular genetics was set. Because of their efforts it is now well-known that DNA molecules are made of two backbones containing sugars and phosphates that spiral around each other. Genetic information is stored in the molecule by the order of the four kinds of bases that project inwards from the backbones. These bases (or nucleotides) are adenine (A), cystosine (C), guanine (G) and thymine (T). The coding of the information is structured so that whenever cystosine occurs on one of the backbones of the double helix it will be paired with guanine on the other backbone and vice versa. Similarly, adenine is always paired with thymine. Thus, when the strands of the helix unwind, each serves as a template for a new strand which will be its complement. Although copying errors sometimes occur, typically new cells of the organism will contain an exact match of the original DNA code after the cell has duplicated by "mitosis."

The sequence of nucleotides can be regarded as a kind of computer program for constructing an organism. Genes are activated when special enzymes attach to the DNA molecule. When genes are switched on or expressed, information in the DNA is transcribed into a complementary strand of ribonucleic acid (RNA), so called "messenger RNA", which is translated into proteins. The order of the bases in the DNA fixes the order of the bases in the RNA which in turn fixes the order of the amino acids composing the proteins. The bases are read systematically in threes with each triplet (codon) coding for a particular amino acid or a "stop" instruction. The story of the genetic code is exquisite and, of course, can be told in far more detail (e.g., Lewin, 1997), but my point here is to emphasize its informational content and computational nature. Indeed, researchers sometimes describe the operation of genes as straightforward computational programs (Yuh et al., 1998).

The computational nature of genetics can explain why it works so well when it does and why things can go so tragically wrong when it doesn't. For example, a common cause of Tay-Sachs disease, a disease characterized by neural degeneration, occurs when a mutation inserts four base pairs in a DNA strand. When this happens, the genetic computer continues to read the bases three at a time but no longer is reading the usual sequence of triplets and does not construct the proper protein. If both of the baby's genes for the formation of this protein are mutated, the child will not produce the protein and will die from the disease within a few years.

A common cause of mental retardation is Fragile X syndrome. Those who suffer from this condition have an abnormally long trinucleotide repeat, a gene on the X chromosome has too many recurring copies of the Triplet CGG. Normal people have 6–54 copies but those afflicted may have well over 300 copies. These unnecessary copies of the triplet make the X chromosome prone to breaking during mitosis, leaving the descendent cells short on genetic material. Girls with the excessive repeats are sometimes mentally handicapped and boys almost always are. IQs are low and these patients frequently lack appropriate responses to emotions expressed by others.

Cystic fibrosis is another genetic disease, which occurs when an individual inherits two particular recessive genes. The patient's lungs becoming clogged with mucus typically characterizes the disease. The defect may be caused by the deletion of merely one triplet of bases—the codon that codes for the 508th amino acid in a standard protein is missing! When the genetic machine, properly programmed, runs well, the results are inspiring. When the genetic machinery breaks down or there is a bug, even a very small one, in its program, disastrous results may occur. Substantial medical payoffs in the understanding of human development and the treatment of diseases will follow when we know more about the nature and operation of the entire human genome. This desire to know the full information contained in the human genome has become a Holy Grail in molecular medicine.

Significant progress has been made in mapping the human genome. A genome map indicates the locations of genes on various chromosomes. By the 1970s only several hundred human genes had been mapped. During the eighties a number of genes causing diseases were mapped including the locus of Huntington's disease, adult polycystic renal disease, and muscular dystrophy. By the late eighties plans were forming for an international Human Genome Project whose goal would be to map and sequence (the precise order of the bases in) the human genome over a fifteen-year time frame as well as map and sequence selected other genomes such as those of a bacterium, fruit fly, worm, and mouse. Because the human genome contains approximately 3 billion bases and at that time a good scientist took a day to sequence several dozen bases, the proposal to sequence the entire human genome seemed, to say the least, extraordinarily ambitious in terms of time and money. But new techniques in biology and the use of robotic analyzers greatly increase the pace and reduced the cost per base. By 1995, the bacterium *Haemophilus influenzae* had its 1,830,137 base pairs sequenced and genes mapped by a team headed by Craig Venter.

In May, 1998, a private venture associated with Craig Venter declared that it would sequence nearly the entire human genome in 3 years for as little as $300 million.

This proposal would beat the U.S. government's original plan by four years and at one tenth of the cost. What makes such a claim plausible is the planned use of small, automated DNA sequencers. This project will replace manually controlled DNA sequencers with machines that perform these tasks continuously inside hair-thin capillaries several inches long. This approach requires considerable computing power to compare resulting chains of bases and clearly would not be possible practically without the existence of contemporary computers.

Benefits of Genetics

That genetic knowledge is and will continue to be extremely useful to medicine is abundantly clear. For example, diabetics require injections of insulin to live. For some patients human insulin is more beneficial than that harvested from cows and pigs. If the right piece of human DNA along with the regulatory sequence for the manufacture of human insulin is inserted in the common bacterium E. coli, human insulin is produced. Similar techniques can produce other essential hormones and vaccines to fight human diseases. Or consider the well-known test on babies for phenylketonuria (PKU). PKU is a dreadful disease, which leads to mental retardation if untreated. If a child has two recessive alleles (genes in the same location) for this disease, the child will not produce an enzyme that metabolizes phenylalanine into another amino acid, tyrosine. This imbalance leads to serious disruption of proper mental development. If the child is given a special diet (high in tyrosine and low in phenylalanine) through adolescence, the child will develop nearly normally. It is crucial to have an accurate test for PKU, for, if a normal child without PKU is given the diet that remedies PKU, the normal child will suffer serious retardation. Prior to very accurate genetic testing for PKU, the testing for PKU (high levels of phenylalanine in the blood) produced false positives with tragic results when the normal children were given the diet for PKU. The existence of genetic testing greatly increases the chances that the right treatment will be given to—and only to—the children who have PKU. Moreover, the promise of gene therapy, still in its infancy, is that more direct remedies for patients with genetic diseases such as diabetes and PKU may be possible if new genes can be implanted to produce the missing chemical substances that cause or permit the disease.

Even without gene therapy or complete cures the simple knowledge of the presence of a genetic risk factor can be helpful in planning for one's health. For example, in 1994 researchers discovered the gene BRCA 1 (Breast Cancer 1) whose function apparently is the suppression of tumors of the breast. Researchers have also located a similar gene, BRCA 2, on another chromosome. When there is a mutation or problem with one of these genes, there is cancer of the breast. Although genetic factors are only known to account for 5% of cancer, those who are at higher risk (e.g., Eastern European Jews have a higher rate of deletion of base pairs in the breast cancer suppression gene) may wish to have more frequent testing for the possible onset of the disease.

Building a case for the medical benefits of genetics is easy and it explains why privacy considerations have often been given rather minimal consideration. The medical benefits are potentially so great who would want to stand in the way of such obvious progress? Some money was set aside for ethical considerations when the Human Genome Project was established, but the bulk of it was for science, and eyes were focused there. In reference to the early stages of the Human Genome Project Dr. George Cahill, a leader in medical genetics, made it clear how the tantalizing promises of medical marvels of genetics research distracted people from its ethical ramifications.

> The potential benefits for cancer, genetic diagnosis in pregnancy, and the genetic prediction for diseases in midlife and later, and even the proper prevention of genetic problems by gene manipulation were all so attractive that considerations of the privacy of the data, their possible prejudicial use in insurance and employment, and many other problematic applications were essentially obscured by the enthusiasm of the protagonists. (Cahill, 1996, p. 10)

The amount of information resulting from the human genome project is already enormous and certainly will continue to climb. Intellectually this is a very exciting time in biology. Scientists justifiably sing its praises, for the human DNA sequence will be an essential reference for all biologists and will be the central organizing principle for human genetics in the next century as Waterston and Sulston (1998) suggest. The commercialization of the knowledge of DNA, not to mention possible patent rights, will be worth billions of dollars. And, some of the future Nobel laureates surely will be selected from those who are now pioneers in exploring the details of the double helix for the discovery of which Watson and Crick have already won Nobel Prizes. Knowledge, money, and prestige are powerful forces. When one adds to them the kind of medical benefits already discussed, genetic research is a scientific juggernaut, which can not and should not be stopped. But it can and should be controlled.

The Nature of Privacy

Is there a right to genetic privacy? How could such a right be justified? What is a right to privacy anyway? The right to privacy is a normative claim about protection from the intrusion of others. Privacy is, thus, a feature of social situations. Imagine someone alone on a desert island. She is in a naturally private situation but does not require normative privacy. Nobody else is there to intrude. Privacy is a reality but not a normative issue for her. As people come to the island to form a society, privacy still may not be a normative concern, at least at first. If the society is small enough and people trust one another, the society may function well without any privacy concerns or normative restrictions. It depends a lot on the personalities and preferences of the people. Imagine a couple or small, gregarious group on the island who share experiences and keep no secrets from each other. From their point of view the lack of privacy is a manifestation of their trust and respect for each other. But, as a society grows it is unlikely that everyone will know everyone else equally well or would want to. Privacy concerns will arise. People will pair off and subgroups will form. People will pursue their own projects and resent intrusion by others. Barrington Moore makes an interesting anthropological observation.

> If social concerns take precedence, one might ask if any human societies without privacy exist. At first glance the Siriono Indians in Boliva, among whom all physiological activities can and do occur in the presence of other people, suggest a positive answer to this question. But a closer examination qualifies the answer by giving evidence of at least a desire for privacy. There are frequent grumblings about the noise and disturbances that are due to shared living quarters. Lovers seek assignations in secluded areas away from the camp. (Moore, 275)

Privacy, then, is a concern that arises from social interactions. When we seek a right to privacy we seek protection from the intrusion of others. A small, intimate society may have minimal or no privacy needs or rights. As societies grow, the need for privacy and the right to privacy will grow as well. Privacy is a kind of shield that protects individuals from the harmful demands and idiosyncrasies of other members of society and in some cases protects other members of society from individuals. Eskimo families that must live together in igloos over the winter have developed normative rules for privacy though they live in very crowded quarters. For example, the Utku Eskimos "have a heavy emphasis on privacy that prevents unseemly curiosity about the feelings that lie beneath the polite and friendly exterior" (Moore, p. 9). Such privacy rights protect individuals and foster social cohesion. As societies grow still larger, the need for privacy rights is likely to expand. In large, informationally rich societies privacy rights are crucial for the protection of individuals. Hence, privacy is a concept whose content may evolve over time and will be determined in part by the details of the structure of that society including political and technological features of the society's environment.

On this account of privacy what counts as an intrusion will more than likely vary from one culture to the next and will evolve over time. Consider the path of development in the United States (Moor, 1990). Privacy is not mentioned explicitly in foundational documents such as the Declaration of Independence or the United States Constitution. But it is present in them indirectly in that the founders of the United States sought protections against unwarranted intrusion by the government. In a seminal *Harvard Law Review* article in 1890 Samuel Warren and Louis Brandeis argued to broaden the conception of privacy protection in the U.S. to include non-governmental intrusions into private lives such as intrusion by the press. Court decisions have expanded the concept of the right to privacy still further in the 1960's and 70's. Courts maintained that governmental interference in individual reproductive choices is intrusion. This included the right to information about and the use of contraceptives (*Griswold v. Connecticut*) and the right of a woman to have an abortion (*Roe v. Wade*). Although these court decisions are arguably as much about the right to know and free choice as they are about privacy, they are about privacy to the extent that the courts were marking off a broader terrain, human reproductive activity, that should be beyond governmental intrusion. Finally, the right to privacy has been extended, particularly with the advent of computer technology, by statute and case law to include protection of personal information. In some situations the gathering of personal, computerized information is regarded as an intrusion into the lives of individuals. Indeed, concerns about informational privacy have become increasingly important in the U.S. and elsewhere to the extent that protecting personal information is now a paradigm of the right to privacy. The concept of privacy has become informationally enriched by the explosive growth and use of computer technology (Moor, 1998).

To understand the right to privacy as a right against intrusion by others while allowing the exact specification of intrusion to vary among societies and to develop as societies develop gives the notion of the right to privacy stability without rigidity. Because the notion of privacy, at least the notion of what counts as an intrusion, is an evolving concept in societies we should not be surprised if the concept continues to evolve during the oncoming genetic revolution. If one thinks of the right to privacy as a fixed notion or as something that is binary (we have it or we don't), then one would not be alert to the need to reassess the situation as technology shifts. Genetic technology is shifting rapidly and some substantial policy vacuums have emerged concerning the proper use of genetic technology.

The theory of privacy I wish to advocate is the control/restricted access theory (Moor, 1997). According to this theory we should ask for a given situation "Who should have access to what or whom in what ways under what circumstances?" For example, in a hospital setting a physician may have legitimate access to his own patients' records but not to others. Some nurses may have access to some of these records and some of the records of patients of other physicians. Hospital financial officers might have access to patient business records but not records of their medical histories. And so forth. The right to privacy can be a very fine-grained notion. When asking "Who should have access to what or whom in what ways under what circumstances?" one is asking for the details of how a privacy right is elaborated for a given situation. But how does one decide who should have access to what or whom in what ways under what circumstances? If we grant the importance of human autonomy, then the answer might be that the individual should have as much control as possible and should make the decision in his or her own case. In the control/restricted access theory of privacy, all things being equal, the default position is to give control of access to the individual potentially being accessed either directly or indirectly. However, various factors can conspire to make things unequal. In reality there is only so much control one can have and still devote time to other matters. In a computerized world in which information about people flows around continually, it is not realistic to assume people can exercise full control over all of their personal information all of the time; and therefore it is important to consider the conditions under which access is permitted or not permitted when personal control is not realistically possible.

To determine who should have access to what or whom in what ways under what circumstances it is useful to ask the question who has a *legitimate need* to access what or whom in what ways under what circumstances. Again, this approach protects individual autonomy by shifting the burden of proof to the party trying to acquire access. Physicians certainly need access to their own patients' records for patient care but not to the records of other patients. And in cases in which there is some disagreement about the legitimacy of access, e.g., medical researchers want access to patient records, some "dynamic negotiations," as Judith DeCew (1997) puts it, among the parties (patients, physicians, and researchers) might yield a level of access and identification that was mutually acceptable.

Before giving the case for the right to genetic privacy let me point out that information technology can be very useful in the support of privacy, as well as a threat to it. Again using the hospital example, many people want to have information about patients, including physicians, nurses, residents in training, outpatient service providers, financial staff, researchers, risk managers, personnel officers, insurance companies, employers, family members, and friends. Electronic medical patient records can offer increased and selective protections over their paper counterparts. For example, electronic records can identify patients using numbers to avoid confusion and to protect anonymity. Computer programs that access these records can set access levels for different people (who, what, and when) and they can keep audit trails of users. Computers offer the possibility of biometric sensors (face, fingerprint, iris, etc.) to identify personnel and proximity sensors which automatically close access on terminals when users with the appropriate badges walk away. By one insider's estimate, 80% of the invasions of privacy within a hospital come from one employee improperly reading another employee's medical records. Computers can provide employees with instant information on who has accessed their own medical files. Computers can help medical researchers who use bots (agent programs) that can search out and compare data from medical records without revealing identities of patients to anyone (cf. Moor, 1989).

Obviously, computers are not the complete answer to protecting patient privacy, but privacy, as characterized by the control/restricted access theory of privacy, is well served by the fine-grained possibilities of applying computer technology. Ultimately, humans in hospitals and other situations have to have respect for others and their rights in order for any right of privacy to be well guarded. In the case of electronic medical records this requires well-trained, sensitive health care providers as well as public statements of current hospital policies about patient records so that patients and health care providers can make informed, ethical choices.

Genetic Privacy and the Need for Strong Protection

Genetic information provides insights into who we are, where we came from, and where we are going. Obviously, genetics does not provide the whole story but it does lay out some significant features of the story of each of our lives. What we see in the mirror and that we see in the mirror is grounded in genetics. And much of what we don't see in the mirror about ourselves is grounded in genetics. Thus, genetic information is particularly sensitive information and potentially damaging.

Genetic information as part of the electronic record of a patient should be given at least the level of protection of routine medical records, for, if such information about a patient were known by others, it might affect the patient's insurance, employment, and financial future. Restricting access to genetic information about people is extremely important to avoid discrimination against them. Using computers to protect privacy would also give us the power to elevate the security of genetic records if it were so desired. Some genetic testing is diagnostic. A patient shows symptoms of some disease and genetic testing is done on the patient to confirm or disconfirm the genetic basis for the disease in that patient. But, some genetic testing is predictive. That is, a patient who may be asymptomatic is tested for a genetic condition that may appear at some future time in the patient's life. An example of such a predictive test is the genetic test for Huntington's disease. Huntington's disease appears later in life, usually in middle age, and produces devastating neural degeneration. The illness is inexorably progressive. Patients lose the physical and mental abilities to care for themselves and eventually severe dementia occurs. There is no known treatment and the suffering of the patient and the patient's family and friends can be traumatic. Until the last couple of decades no reliable test for the disease was known but now genetic testing does provide a reliable test. A patient who seeks predictive testing for Huntington's disease may not want the information to be accessible in her own medical records. Perhaps, that the test was given should be indicated, but the results might be kept under very restricted access. Even the patient's regular physician might not have access to the test results if there were nothing to be done about it even if the test results were positive. The physician would not have a legitimate need to know. The patient might decide that only she would have access to the test results given that no treatment is possible. Such genetic information, if known by others, could devastate the patient's life although no symptoms of the disease would manifest themselves for decades!

My claim is that we need very strong privacy laws to protect genetic information in the information society. First, as already discussed, the benefits of genetic research are and will be considerable. There will be irresistible scientific projects proposed to gather more and more information about human genetics. Strong genetic privacy laws are needed to counterbalance this, not to stop the research, but to keep it within ethical boundaries. Second, many organizations can use genetic information to cherry pick clients. It is not simply a problem limited to the U.S. system of private medical care underwritten by insurance companies. Educational or financial institutions under any medical system can use such information adversely against people. Hence, strong, general laws protecting genetic privacy are needed. Third, genetic information can now be gathered outside of the traditional medical contexts and, thanks to computers, the gathering of such information is becoming more common. Traditional safeguards, even sophisticated computerized safeguards, of the electronic patient records will not be effective in controlling the dissemination of genetic information. Strong, general protection of genetic privacy is needed. Fourth, discrimination on the basis of genetic information will become increasingly common. A person's genetic code may become a scarlet sequence of letters that brands him in a way that makes him undesirable for somebody's purposes. It would be all too easy to use genetics to reinforce existing prejudices or to create new ones. Imagine that a "gay gene" were actually found. Would parents be justified in selecting it out or using gene therapy to "fix" it?

Assuming gays are less likely to have children than non-gays, the gay population could dramatically be reduced. In general, caste systems could develop in which people were stratified by their genetic makeups. Improved anti-discrimination laws are essential, but strong, general genetic privacy protection laws also would be essential in preventing discrimination from occurring. Fifth, the problem of latent information exists. We do not really know what we are about to learn about humans based on a growing understanding of genetics. This is not the fear of genetic determinism, but the acknowledgement that genetic factors do play a role not only in what we look like, but how we function, and how we behave. To what extent, nobody yet knows. But someday we will. Stored in our blood samples in the hospital from our PKU tests at birth or in our hair that falls on the floor when our hair is cut is our genetic code. When the latent information contained in these genetic samples is understood as usable knowledge, what will it tell us about ourselves? Who will be able to know it? Until we do know what knowledge will manifest itself, is it not prudent to be cautious, to enact strong, general genetic privacy laws to protect ourselves?

If we have the right theory of privacy (I offer the control/restricted access theory as a candidate), we can better understand what privacy is, and how it can be tailored and justified. We have in the genetic revolution the potential for great benefits. Only a fool would not want the fruits of that. But, we also want policies whose consequences are just (Moor, 1999). Discrimination based on genetic information or misinformation is surely unjust. The current danger is that privacy concerns will be overwhelmed or not heard at all. Thus, I have argued for the need to enact strong genetic privacy laws. Perhaps, mentioning the "soul" is a bit old fashioned, but to the extent that genetics may tell us about who we are as individuals and what lives we are about to lead, by protecting genetic privacy we may indeed be protecting the privacy of the soul.

References

G. F. Cahill. A Brief History of the Human Genome Project. In B. Gert, et al., eds. *Morality and the New Genetics*. Jones and Bartlett Publishers, Sudbury, Massachusetts, 1996.

C. Culver, J. Moor, W. Duerfeldt, M. Kapp, and M. Sullivan. Privacy. *Professional Ethics*, 3 (3 & 4): 3–25, 1994.

J. Wagner DeCew. *In Pursuit of Privacy*. Cornell University Press, Ithaca, 1997.

C. Fried. Privacy. In F. D. Schoeman, editor, *Philosophical Dimensions of Privacy*. Cambridge University Press, New York, 1984.

B. Gert, E. M. Berger, Jr. G. F. Cahill, K. Danner Clouser, C. M. Culver, J. B. Moeschler, and G. H. S. Singer. *Morality and the New Genetics*. Jones and Bartlett Publishers, Sudbury, Massachusetts, 1996.

Griswold v. Connecticut, 381 U.S. 479, 1965.

P. Kitcher. *The Lives to Come*. Simon & Schuster, New York, 1996.

B. Lewin. *Genes VI*. Oxford University Press, Oxford, 1997.

J. H. Moor. What is Computer Ethics? *Metaphilosophy*, 16 (4): 26–75, 1985.

J. H. Moor. How to Invade and Protect Privacy with Computers. In C. C. Gould, editor, *The Information Web*. Westview Press, Boulder, 1989.

J. H. Moor. Ethics of Privacy Protection. *Library Trends*, 39 (1 & 2): 69–82, 1990.

J. H. Moor. Towards a Theory of Privacy in the Information Age. *Computers and Society*, 27 (3): 27–32, 1997.

J. H. Moor. Reason, Relativity, and Responsibility in Computer Ethics. *Computers and Society*, 28: 14–21, 1998.

J. H. Moor. "Just Consequences and Computing," Ethics and Information Technology, 1 (1): 65–69, 1999.

B. Moore. *Privacy: Studies in Social and Cultural History*. M. E. Sharpe, Inc, Armonk, New York, 1984.

J. Rachael. Why is Privacy Important? *Philosophy and Public Affairs*, 4 (Summer): 323–333, 1975.

J. H. Reiman. Privacy, Intimacy, and Personhood. In F. D. Schoeman, editor, *Philosophical Dimensions of Privacy*. Cambridge University Press, Cambridge, 1984.

Roe v. Wade, 410 U.S. 113, 1973.

F. D. Schoeman. *Philosophical Dimensions of Privacy*. Cambridge University Press, Cambridge, 1984.

Roland Somogyi, and Carol Ann Sniegoski. Modeling the Complexity of Genetic Networks: Understanding Multigenic and Pleiotropic Regulation. *Complexity*, 1 (6): 45–63, 1996.

Samuel D. Warren and Louis D. Brandeis. The Right to Privacy. *Harvard Law Review*, IV (5), 1890.

R. Waterson, and J. E. Sulston. The Human Genome Project: Reaching the Finish Line. *Science*, 282 (2): 53–54, 1998.

Chiou-Hwa, Yuh, Hamid Bolouri, and Eric Davidson. Genomic Cis-Regulatory Logic: Experimental and Computational Analysis of a Sea Urchin Gene. *Science*, 279 (20): 1896–1902, 1998.

7. Privacy and Policy for Genetic Research

Judith Wagner DeCew

Introduction: The Value of Privacy

I shall focus my essay on privacy and its implications for medical records and genetic research data. Individuals care about and guard their privacy intensely in many areas. With respect to patient medical data, people are exceedingly concerned about privacy protection, recognizing that health care, and especially pharmaceutical and genetic data, generates the most sensitive sorts of personal information. In an age of advancing technology, privacy is threatened, and privacy protection for medical and genetic information poses a dramatic challenge.

As a moral philosopher, I have defended the view that privacy acts as a shield to protect us in various ways, and that its value lies in the freedom and independence it provides for us, nurturing our creativity and allowing us to become better people.[1] Privacy shields us not only from interference and pressures that preclude self-expression and the development of relationships, but also from intrusions and pressures arising from others' access to our persons and details about us. Threats of information leaks, as well as threats of control over our bodies, our activities, and our power to make our own choices, give rise to fears that we are being scrutinized, judged, ridiculed, pressured, coerced, or otherwise taken advantage of by others. Protection of privacy enhances and ensures the freedom from such scrutiny, pressure to conform, and exploitation that we require so that as self-conscious beings we can maintain our self-respect, develop our self-esteem, and increase our ability to form a coherent identity and set of values, as well as our ability to form varied and complex relationships with others. Thus privacy is a shield protecting us from scrutiny, prejudice, coercion, pressure to conform, and the judgment of others.

Loss of privacy leaves us vulnerable and threatened. We are likely to be more conformist, less individualistic, and less creative. Loss of a private sphere can be

---

This essay originally appeared in *Ethics and Information Technology*, **6**: 5-14. Copyright © 2004 by Kluwer Academic Publishers. Reprinted by permission.

[1]For a full defense of this view, see J.W. DeCew, *In Pursuit of Privacy: Law, Ethics, and the Rise of Technology* (Ithaca: Cornell University Press, 1997). This paper is adapted from J.W. DeCew, "Alternatives for Protecting Privacy While Respecting Patient Care and Public Health Needs," *Ethics and Information Technology* 1, 4 (1999), 249–255, and J.W. DeCew, "The Priority of Privacy for Medical Information," *Social Philosophy and Policy*, 17, 2, (June 2000), 213–234.

stifling. In the context of medical information, the possibilities of exploiting, aggregating, or misusing genetic testing results, prescription records, drug test data, mental health records, information on pregnancy, and results from tests for sexually transmitted diseases and HIV status, to name just a few, make it obvious how important it is to preserve the protection that privacy affords individuals. In such cases, the potential harms from disclosure range from embarrassment, loss of self-esteem, social stigma, isolation, and psychological distress to economic loss and discrimination in such areas as employment, child custody, insurance, housing, and immigration status. Thus it is necessary to determine how much to defend privacy for medical data.

How can we protect privacy for medical information and also preserve databases for genetic research? I shall examine three alternative public policy approaches to the protection of medical records and genetic data: reliance on governmental guidelines, the use of corporate self-regulation, and my own hybrid view on how to maintain a presumption in favor of privacy with respect to medical information. None of the three models I examine are unproblematic, yet it is crucial to weigh the strengths and weaknesses of these alternative approaches to reconcile the polarization between those who approve of unlimited data flow and those who defend the privacy of genetic and other medical information as absolute.[2]

Note first that everyone has a certain expectation of privacy, and at least one recent poll shows that Americans do not believe that they have adequate privacy protection over how medical information about them is used and circulated. Eighty-five percent of those questioned in this study put a far higher priority on confidentiality of medical data than on providing universal coverage, reducing paperwork, and gaining better data for medical research on disease and treatments.[3]

Second, it is worth noting that different perspectives on medical privacy may depend on existing health care systems. Perhaps there is more concern about privacy among individuals in the United States—where people still fear possible loss of medical insurance coverage despite recent legislation intended to prevent that loss—than in countries with guaranteed national health care.

Medical Records and Access: The Scope of the Problem

While paper records and copying machines have never been particularly secure, computerized records introduce new risks and new opportunities for abuse. At every stage of the process of collection and storage, dangers can arise, including entry errors, improper access, exploitation, and unauthorized disclosure. Secondary use and aggregation of data are all easier, faster, and less expensive, and thus pose additional threats to an individual's control over the disposition of medical and genet-

---

[2] M. Rigby, I. Hamilton, and R. Draper, "Finding Ethical Principles and Practical Guidelines for the Controlled Flow of Patient Data," presented at an international conference, "Electronic Patient Records in Medical Practice," Rotterdam, The Netherlands, October 7, 1998.

[3] A. Westin et al., "Health Care Information Privacy: A Survey of the Public and Leaders" 23 (1993), survey conducted for Equifax, Inc., cited in L.O. Gostin, "Health Information Privacy," 80 *Cornell Law Review* 101 (1995), 104.

ic information. In addition to records on individuals kept by primary health care providers in computer databases and pharmacy records on individuals' prescriptions, there are also state-wide, regional, and population-wide health databases which include information such as genetic blueprints from blood samples, data on communicable diseases such as AIDS and tuberculosis, various test results, health policy research, specific diseases such as cancer, and more.[4]

Health database organizations (HDOs), operating under the authority of the government, private for-profit, and non-profit organizations, have access to massive databases of health information on a defined population or a specific geographical area. Their central mission is the public release of data and of analyses performed on the data. HDOs acquire data from individual health records as well as secondary sources including financial transactions from private insurance companies and government programs; public health surveillance and tracking systems; epidemiologic, clinical, behavioral, and health services research; and surveys conducted by government, academics, and private foundations.[5]

Medical data is often stored in a form termed "personally identifiable." This refers to information that includes any uniquely identifiable characteristic such as a name, Social Security number, fingerprint, or genetic link. Some data may contain no such unique identifier, but may include sufficient details on age, sex, race, and other personal information to make connection with a specific individual possible; such data is thus viewed as identifiable data as well. All other data stored is viewed as "anonymous." However there is a major difficulty in determining how to classify data that is not directly identifiable, but which is linked to a named person with a confidential code. If this code were somehow accessed, the data, which was previously anonymous, would of course become personally identifiable.

Because electronic records can be accessed in combination with other databases, and in diverse geographic locations, linking capacity makes it possible for data to be compared, matched, and aggregated so that even data with neither personal identifiers nor a confidential code can be linked with other data to get a profile of an identifiable person or population. The rapid and sophisticated ways that data can be updated, changed, and configured with few restrictions on dissemination and use, combined with the difficulties of getting rid of data that is obsolete or inaccurate, make privacy concerns for medical information appear virtually intractable.[6] It is no surprise that people are becoming ever more distrustful of computerized medical databases.

There are additional concerns compounding the privacy problems associated with medical data. First, technological advances allowing easy access to such data make it difficult to determine who is the "owner" of the computer record. Many find it obvious that the patient owns his or her genetic record and should contin-

---

[4]L.O. Gostin, "Health Information Privacy," 80 *Cornell Law Review* 101 (1995), 116-117, footnotes omitted.

[5]Gostin, "Health Information Privacy," 115.

[6]See J. Moor, "Towards a Theory of Privacy in the Information Age," *Computers and Society* (September 1997), 27–32, and C. Culver, J. Moor, W. Duerfeldt, M. Kapp, and M. Sullivan, "Privacy," *Professional Ethics* 3, 3&4 (1994), 4–25, for descriptions of the problems and for some general guidelines for establishing privacy protection guidelines.

ue to be named as owner. Others urge, however, that in an electronic world, especially when some data is compiled anonymously, privacy protection is needed from wherever the data may flow, namely the source, without designation of an owner with privacy rights.

Second, it is widely recognized that the risks of fraud and abuse of individual medical information comes not from outside hackers, but mainly from those described as "authorized" users. Consequently, the perhaps obvious tactic of restricting access to medical and research data only to those with authorization is by itself unlikely to provide adequate safeguards.

Third, in America most legal protection of informational privacy in medicine relies heavily on a relationship view of doctor-patient confidentiality that is outdated in an era of HMOs and group practices.

Fourth, federal rules protecting the privacy of medical records were proposed by President Clinton in December 2000, and called the Health Insurance Portability and Accountability Act (HIPAA). These guidelines were initially endorsed by President Bush, but were altered by the Bush administration before being set as law in August 2002. Prior to these rules, most legislation proposed in an attempt to provide some privacy protection for medical information formed a patchwork of guidelines with no consistency and many gaps. It often included vague language allowing broad authority for disclosure of information without a patient's consent for treatment, reimbursement, public health, emergencies, medical research, and law enforcement.

Governmental Guidelines and Centralized Databases

The European Union (EU) has endorsed detailed privacy guidelines to restrict dissemination and use of many kinds of personal data.[7] These privacy guidelines, currently being implemented by members of the European Union, require companies to register all databases containing personal information, require that subjects be told and give consent before their personal data can be collected or used, and require that any information gained for one purpose not be used for any other purpose unless the subject consents to the sharing of the information. The strict guidelines also prevent the transfer of information from one country to another unless the latter country also has adequate protection of records, and they do not allow collection of data on race, ethnic origin, political or religious affiliation, health status, or sexual orientation.[8]

---

[7]See J.W. DeCew, *In Pursuit of Privacy: Law, Ethics, and the Rise of Technology*, 151–152.

[8]J. Markoff, "Europe's Plans to Protect Privacy Worry Business," *New York Times*, April 11, 1991, A1+; L. Tye, "EC May Force New Look at Privacy," *Boston Globe*, September 7, 1993, 10. An excellent summary of the European approach is supplied in P.M. Schwartz, "European Data Protection Law and Restrictions on International Data Flows," 80 *Iowa Law Review* 471 (1995). On the domestic approaches in Germany and Sweden, see C. Bennett, *Regulating Privacy: Data Protection and Public Policy in Europe and the United States* (Ithaca: Cornell University Press, 1992). See also C. Franklin, ed., *Business Guide to Privacy and Data Protection Legislation* (Dordrecht: Kluwer Law International, 1996), where the text of the Council of Europe OFCD and national laws in Europe are summarized and explained, with relevant portions translated.

The theme underlying the European Union approach echoes a dominant thesis I endorse: the initial presumption must be that privacy protection is important and guidelines are essential. Moreover, the European Union plan helps individuals retain control over information about themselves by providing knowledge about the data banks and access to the information, and by requiring permission and consent for collection and transfer of data.

Consider applying these guidelines to medical information. There would be centralized medical data files, with the potential benefits of easier access to medical histories and genetic information. This type of program was proposed by the Clinton administration and former Secretary of Health and Human Services (HHS) Donna Shalala, requiring a universal health care card. What was actually enacted in 1996 was the Kennedy–Kassebaum health insurance reform bill. What is called the "Administrative Simplification" clause of this bill requires the creation of a national electronic data collection and data transfer system for personal health care information, along with a computer code as a "unique health identifier" to trace each citizen's medical history.

There are certainly benefits to such a program, including quality assurance in health care, monitoring of fraud and abuses, the ability to use tracking to evaluate access to health services, and lower health care costs. For patients who move or for other reasons change health care providers, centralized databases of medical records can also lead to better individualized medical care (for example, by decreasing the likelihood of problematic drug interactions and allergic responses, or by helping to track diseases more effectively to enhance prompt diagnosis and treatment). Indeed, without appropriate sharing of information, patients are at risk for uninformed and suboptimal care. Data can also aid consumers in making more informed choices regarding health care plans and providers. In addition, medical records—including genetic information—in centralized databases aid society as a whole. They are particularly useful for medical labs and research aimed at disease prevention and treatments aimed at reducing unnecessary mortality. Public health is enhanced when information allows health problems to be identified and publicized, and allows funding decisions to be well informed. A centralized data system is potentially even more of a boon to insurance companies, pharmaceutical companies, biotechnology companies, and employers.

Yet the "Administrative Simplification" clause of the Kennedy–Kassebaum bill was widely attacked,[9] after hearings by a Health and Human Services panel on the "unique health identifier" began in July 1998. The reason for the uproar was that the U.S. program did not initially include any of the privacy safeguards endorsed by the European Union, and Congress was never able to agree on such standards. In the 1996 law, Congress had instructed the Health and Human Services Secretary to issue rules on medical privacy if Congress itself did not take action, and this led to the HIPAA comprehensive standards on medical privacy issued by Clinton in late

---

[9]B. Woodward, "Intrusion in the Name of 'Simplification'," *The Washington Post*, August 15, 1996; "Medical identifier hearings to begin," The Boston Globe, July 20, 1998, A5.

2000. These rules apply to virtually every patient, doctor, hospital, pharmacy, and health plan in the U.S. As initially proposed by the Clinton administration, they preserve portions of the European Union guidelines and broadly safeguard the confidentiality of medical records, by requiring, for example, that patients know how their records will be used, and have a federal right to inspect and copy their records and the ability propose corrections.[10] Nevertheless, the rules do not cover most purchases, searches, or other actions on health-related Web sites,[11] and key provisions were contested as "unworkable" by the Bush administration in March 2002.[12] The rules were modified by the Bush administration in August 2002, relaxing health privacy protection. This newest version carries the force of law without congressional approval, and took effect in April 2003.[13]

We can consider a centralized database system, with the addition of new privacy rules, and see there remain serious disadvantages to storing medical information in that form. One worry is that whether the individual health identifier is a Social Security number or a less traceable number, it will make it even easier for medical and genetic information to be linked through technological networks to other databases, like employment data; tax and credit records; and insurance, welfare, and custody files.[14]

A related worry with centralization of such sensitive information is that it places too much power in a single public agency. Although a statutory right to see one's files can place a check on the government, having access to information does not guarantee control over the information. It is still necessary to have procedures for those who find erroneous information or want data eliminated from their record. Furthermore, even with the addition of guidelines for protecting privacy, other questions remain about who or what group should oversee enforcement of the guidelines and especially how effective such enforcement could be.

The National Committee on Health Information Privacy, funded by the Center for Disease Control and the Carter Center, and chaired by Lawrence Gostin of Georgetown University Law Center, drafted a model statute recommending the adoption of some level of privacy protection.[15] This model statute (other statutes have been proposed) is a welcome attempt to find a compromise between the free flow of medical information advocated by many and the absolute privacy protection defended by those who believe virtually all data collection and disclosure must be severely limited. Nevertheless, most of the statute regulations are intended to apply only to personally identifiable data. The Committee has acknowledged that

---

[10] J. Eilperin, "New Rules Planned on Health Privacy," *The Boston Globe*, December 20, 2000, A3; R. Pear, "Bush Acts to Drop Core Privacy Rule on Medical Data," The New York Times, March 22, 2002, A1 and A22.

[11] L. Richwine, "Privacy Laws for Patients Don't Extend to Internet," *The Boston Globe*, November 11, 2001.

[12] R. Pear, "Bush Acts to Drop Core Privacy Rule on Medical Data," *The New York Times*, March 22, 2002, A1 and A22.

[13] B. Ladine, "Medical Privacy Rules are Relaxed," *The Boston Globe*, August 10, 2002, A1 and A5.

[14] J. Foreman, "Your Health History–Up for Grabs?," *The Boston Globe*, July 20, 1998, C1.

[15] L.O. Gostin, "Making Tradeoffs Between the Collective Good of Human Health and the Individual Good of Personal Privacy," presented at an international conference, "Electronic Patient Records in Medical Practice," Rotterdam, The Netherlands, October 6, 1998.

access to most data would be granted to many, and thus, through database links and aggregation, privacy would be compromised. Gostin and others have suggested that a unique health identifier, distinct from one's Social Security number and used only for the health care system, can adequately protect patient privacy. They urge further that ethical arguments for the importance of privacy protection are balanced by equally compelling arguments supporting a more efficient health information system. Moreover, Gostin himself has allowed that "one of the burdens of achieving cost effective and accessible [health] care is a loss of privacy."[16] The Committee's suggestions are welcome for advancing the debate, but may well allow too much privacy to be sacrificed for administrative efficiency.

Corporate Self-Regulation

A second alternative focuses on corporate self-regulation, and this appears to be the model that was defended by the Clinton administration for computer and Internet technology, if not for medical records. At a conference in May 1998,[17] Ira Magaziner, then Special Assistant to President Clinton, wryly observed that "government understands enough to know it doesn't understand developing technology." Current policy in the communications field, he argued, views privacy protection as a way of empowering people to protect themselves, for example, through the use of filtering software on computers that gives users choices, rather than mandating federal guidelines that deprive individuals of choice. Privacy guidelines, on this view, (1) should be led by industry and the private sector, (2) should be market-driven and not regulated, and (3) should maximize consumer choice as well as governmental restraint. According to Magaziner, the government is generally too slow to make or change policies quickly enough to keep pace with the rapid changes in technologies. Thus, he argued, when government does act, its role should be minimal and transparent. Magaziner envisioned the use of corporate and consumer pressure to develop privacy "Codes of Conduct," which could then be backed by some sort of enforcement agencies (though it was unclear whether these agencies would be led by corporations or consumers). Nevertheless, the result is private management of the Internet and other communications technologies.

Consider how this second model applies in the field of medical data. Presumably it reflects what had been largely happening in the U.S. prior to the Health Insurance Portability and Accountability Act (HIPAA): hospitals, HMOs, and insurers had basically been left to regulate themselves in their handling of patient medical records and genetic data. It is arguable that the fast pace of genetic research developments supports the claim that government is too slow and unwieldy to be involved. Moreover, such private self-regulation has the potential benefit of enhancing patient choice over what data is collected and how it is stored, used, and accessed.

For years, the self-regulation model was the U. S. norm. There were virtually no patient controls over how personal health information could be used, disseminated,

---

[16]Gostin, "Health Information Privacy," 171.

[17]ACM Policy '98, Renaissance Hotel, Washington D.C., May 10-12, 1998.

and sold. Personal medical files could be accessed not only by doctors and nurses, but also by insurers, self-insured employers, and public health and law enforcement officials, to name a few. In reality, individual control was minimal and there was no consistency, uniformity, or certainty of protection. Indeed, medical data was largely unprotected in the U. S., and what few protective measures there were formed a complex patchwork of different regulations at the federal level and in different states and different organizations. Even the National Committee on Health Information Privacy continued this patchwork approach by endorsing regional enforcement of its proposed privacy guidelines. There was no coherent policy on privacy and medical data, and "U. S. citizens often [had] no legal recourse if they [were] harmed by inappropriate disclosure of their medical records."[18] Health organizations, researchers, and insurers may welcome the freedom from external regulations, but it seems clear that from the patient's point of view, the disadvantages of a corporate self-regulation system for medical data far outweigh the advantages.

Reaction to the "unique health identifier" demonstrates that many Americans are distrustful of giving government more power, and more access to their data, through centralized databases and the federal regulation model, but studies show the public is also becoming more and more skeptical about the self-regulation model.[19]

A Hybrid Approach: Dynamic Negotiation

A third model makes it possible to protect privacy stringently in the face of a variety of advancing information technologies, requiring some federal regulations to mandate the presumptive importance of privacy protection, yet allowing people dynamically to negotiate the degree of privacy they wish to sacrifice or maintain.[20]

This approach to medical and genetic information is one I first defended in print prior to the release of the Clinton HIPAA guidelines, which incorporate some of the same aspects.[21] This model would require federal guidelines mandating the priority of privacy, so that the collection, storage, and use of medical records would require maximal privacy protection as the *default*. Thus the presumption of privacy would be *mandated*. The theoretical principles underlying governmental privacy guidelines would include commitments to preserving anonymity of data when at all possible, establishing fair procedures for obtaining data, requiring that proposed collections of data have both relevance and purpose, and specifying the legitimate conditions of authorized access. These principles would also demand commitments to developing systematic methods for maintaining data quality,

---

[18] D.E. Detmer and E.B. Steen, "Shoring Up Protection of Personal Health Data," *Issues in Science and Technology* (1996), 76.

[19] Electronic Privacy Information Center, e-mail correspondence, 1998.

[20] The term "dynamic negotiation" was introduced by R.E. Mitchell and first appeared in R.E. Mitchell and J.W. DeCew, "Dynamic Negotiation in the Privacy Wars," *Technology Review* 97, 8 (1994), 70–71.

[21] J.W. DeCew, "Alternatives for Protecting Privacy while Respecting Patient Care and Public Health Needs," *Ethics and Information Technology* 1, 4 (1998), 249–255, and J.W. DeCew, "The Priority of Privacy for Medical Information," *Social Philosophy and Policy*, 17, 2, (June 2000), 213–234.

requirements that data collected for one purpose not be used for another purpose or shared with others without the consent of the subject, and limiting the retention time of data to what is necessary for the original purpose of the data collection.[22] Clear and public articulation of these principles would be essential, to emphasize that privacy is a fundamental part of a positive strategy for addressing health care information technology rather than merely a defensive and *ad hoc* reaction to threats. This will also help restore public confidence in the use of electronic medical and research records.

Some might prefer to endorse proposals from the National Committee on Health Information Privacy and require privacy guidelines solely for identifiable data. Certainly identifiable patient genetic records pose the greatest risk, and may need the greatest guarantee of privacy protection. Nevertheless, there are dangers in focusing just on privacy for identifiable health records. There is a false security in the apparent anonymous nature of information that lacks a name or other identifier.[23] The linkage potential that allows nonidentifiable information to be aggregated and linked in ways such that it becomes identifiable, eliminates any clear line between identifiable and nonidentifiable data, making it more imperative that privacy guidelines apply to databases of anonymous information as well. Furthermore, protecting all databases, rather than just those containing identifiable information, will enhance public confidence in the technological safeguards.

Others have described the details of a computerized system that can incorporate the theoretical principles I have articulated. Such a system would require user authentication, different levels of access control, a method of coding, a process by which to audit the trails of users, some type of cryptography, a system of data ownership that would allow authors of documents to set access controls and to shadow records, and so on.[24] *Security* would be provided through the technology, and the security measures would then be used to protect confidentiality and *privacy*.

The challenge is no longer technological; the issues are ethical and political, namely, how to incorporate and implement the available security safeguards in the computer program designs and how to determine which features to include or exclude, prioritize, use as a default, and so on. The options for protecting privacy need no longer be total denial or total access to a medical record or database. Well-designed computer systems can incorporate and differentiate customized access controls such as layering or partial restriction of very sensitive data items such as genetic test results, enhancing security for medical information.[25]

---

[22]These principles echo the European directives. See M. Rigby, I. Hamilton, and R. Draper, "Towards an Ethical Protocol in Mental Health Informatics," *MedInfo '98: Proceedings of the 9th World Congress on Medical Informatics*, ed. B. Cesnick, A.T. McCray, J.-R. Scherrer (Amsterdam: IOS Press, 1998), 1223–1227.

[23]M. Rigby has suggested to me in correspondence that although downloading of rich data sets of de-identified data should not be allowed, perhaps if the simplicity of the data set or the large size of the study, for example, would preclude indirect identification, or even suspicion of linkage and identification, then use of the database for public health research may be morally justifiable.

[24]R.C. Barrows, Jr. and P.D. Clayton, "Privacy, Confidentiality, and Electronic Medical Records," *Journal of the American Medical Informatics Association* 3, 2 (March/April 1996).

[25]Barrows et al., "Privacy, Confidentiality, and Electronic Records," 146.

The first feature of the principles underlying this approach, which I have sometimes termed "dynamic negotiation," is the requirement that overarching guidelines mandating privacy protection be the default. On this view, privacy protection is expected, individual access to one's own data is assured, and access and use of medical data by others must be justified and approved. The model takes patient privacy—rather than clinician needs, research needs, or administrative efficiency—as fundamental. (Contrast dominant systems in the U.S.—e.g. Charles Safran's explanation a few years ago of Boston's Beth Israel Deaconess Medical Center's way of dealing with the conflict between confidentiality and availability in the use of electronic medical records. They adopted *access* as the *default*, augmented with accountability *after the fact* through audit trails and monitoring.)[26]

Second, a major theme of a dynamic negotiation system includes emphasis on individual choice through a negotiation, more accurately referred to as "counseling," between patients and those wanting or needing access to their medical histories and genetic data. The priority of privacy would extend both to primary uses of medical records for health care delivery as well as to secondary uses by insurers, employers, public health agencies, medical researchers, educational institutions, etc. Medical data would *not* be *unavailable* to third parties, however they would be expected to gain consent for the use of the data from the subject or to demonstrate their need to know to a data supervisory committee or other comparable gatekeeper. Parties with different stakes in the information would be required to ask for access and demonstrate a need to know.[27] This might at first sound like the process known as "informed consent to disclosure of information," which includes telling patients what information is to be disclosed, ensuring they understand what is being disclosed, and requiring that the patient is competent and willingly consents. Dynamic negotiation goes further, not merely presenting patients with a consent form to sign, but requiring health providers and secondary users to have a meaningful dialogue with subjects about the implications of allowing access, the importance of the research study for which data is needed, etc. The goal is to educate patients so that they understand the benefits of the release of their data—not only for themselves, but for medical research or public health as well—so they may choose voluntarily. This goal emphasizes the importance of autonomy, allowing patients to retain control over their health data.[28]

Third, a dynamic negotiation system would insist on protection for physicians as well, allowing them to explain that they are unable to provide services for

---

[26]C. Safran, "The Introduction of EPRs in the Beth Israel Deaconess Medical Center, Boston," presented at an international conference, "Electronic Patient Records in Medical Practice," Rotterdam, The Netherlands, October 6, 1998. D. Friedman suggested to me that federally mandated systems have failed in the past to allow for experimentation with different solutions to problems. He urged that it would be preferable to have variety in information control, in order to see what system is best. To the contrary, I believe we have already had years of experimenting with a patchwork of programs, and the Beth Israel system as well as others indicate that no self-regulating system has emerged that adequately protects patient privacy.

[27]For a defense of a "restricted access" view of privacy see R. Gavison, "Privacy and the Limits of Law," 89 *Yale Law Journal* 421 (1980); see also J. Moor, "The Ethics of Privacy Protection," *Library Trends* 39, 1&2 (1990), 69–82.

[28]See T. Beauchamp and J. Childress, *Principles of Biomedical Ethics*, 3rd ed., (New York: Oxford University Press, 1989) for their stress on four ethical principles, namely respecting autonomy, beneficence, non-maleficence, and justice (emphasis mine).

patients who refuse to divulge essential medical information. It is likely most or all patients would gladly release the information to gain care.

Fourth, control over the information would need to include the ability of patients to access their records or database information. Patients would be provided ways of correcting misinformation or information that is incomplete and therefore misleading.

The advantages of such a system are that the presumption in favor of privacy is protected, and when patients choose to release their medical data they *know* exactly what they are doing and *why*. They will not be losing their privacy by default. The burden is placed on those who want the genetic data, for example, to provide assurance that release and use of the information is for a genuinely useful public health study or pharmaceutical development, or for a third party much in need of the genetic information to make appropriate health care decisions, not merely for commercialization or profit. The government need not micro-manage the details, but neither can it be passive and allow total self-regulation as long as the health information marketplace is apparently unwilling or unable to protect individual privacy claims without intervention.

The disadvantages of this approach are not insignificant, however. First, there will be difficulties similar to those arising for informed consent, concerning the ability of patients meaningfully to understand, discuss, or consent if they are infirm, confused, or unaware of the content of their medical records, or if they have died. There need to be guidelines concerning who or what supervisory group has custody or oversight of a medical or genetic record when a patient is unable to be consulted.

Second, some doctors and secondary users of health information are better communicators than others. Even with forms to aid them and database prompts, those with less-than-gracious bedside manners may find initiating the necessary counseling and conversations extremely difficult, and they may do a poor job. The end result may be neither a "negotiation" nor "dynamic."

Third, some system of enforcement must be mandated.

Fourth, the time, effort, potential paperwork, and computer entries needed for the system of dynamic negotiation are likely to be eschewed by physicians and researchers used to a system allowing free flow of so much data. They may also lack the time to educate patients—especially given the economic stresses of managed health care—even when the possibility of quick confirmations for consent are built into the computer programs. Some have argued that this will likely lead to (1) the authorizations approved in forms to be substantially broadened to avoid tracking down patients later, and (2) doctors simply avoiding inconveniences by not attempting to get consent for the release of information for secondary purposes. Both outcomes would work to the detriment of medical and genetic research and public health.[29]

Fifth, some might argue that many individuals, even with the best information, will be comfortable sharing sensitive medical data with only some of the many

---

[29] D.E. Detmer and E.B. Steen, "Shoring Up Protection of Personal Health Data," 77.

involved in their direct patient care who can help with their own personal health needs. They may be unwilling to part with their data, even if encrypted, for medical research and public health studies, often just as crucial for the medical profession as individual patient care. If such unwillingness is common, research samples may be skewed and less useful than data collected without the restraint of patient consent. This may well lead to inadequate or statistically invalid research databases and declines in other public health activities.

To the contrary, however, I believe that many or even most people, when educated, will be willing to share more data than is anticipated. Many researchers and funding agencies already require consent forms. More information sharing has been documented in West Virginia and elsewhere, where "some patients are allowing their physicians to put unencrypted data about their health on the Internet's World Wide Web," indicating they believe the benefits outweigh "the potential risks of privacy invasion."[30] These individuals have made precisely the choices doctors and researchers would have made on their behalf. Moreover, they might even actively consent to do so for all their medical information, reversing the default so that their data is provided routinely. Another example familiar to many occurred in Iceland, where people voluntarily gave their information for genetic research. Some might agree that many people would so choose to release their data, and then paternalistically conclude that such broad information sharing should be allowed automatically and as a matter of course. I believe it preferable to endorse a public policy allowing individuals to decide for themselves whether it is worthwhile to relinquish some of their privacy for higher-quality medical care, for the potential for future cures, for public health, and for genetic research.

The Clinton-era federal HIPAA guidelines for protecting the privacy of medical records mirror some of the features of this model, and the Bush administration initially took political credit for those rules and planned to allow those rules to take effect. However, President Bush and current Secretary of Health and Human Services Tommy Thompson, apparently with support from much of the medical and health care industry including pharmacists and hospitals, later objected to the guidelines, and proposed in March 2002 to repeal the core provision on consent. This provision required doctors, hospitals, and other health care providers to obtain written consent before using or disclosing medical information for treatment, payment of claims, setting insurance premiums, and other health care operations. The Bush position was that the consent requirement would impair and delay quality health care.[31]

The version of the HIPAA guidelines settled on by Bush in August 2002 eliminates the "Clinton-era proposal that patients must give written consent before their medical records are shared with doctors, health-care facilities, pharmacies, or insurance companies. The new rule[s] ... instead will require that health providers

---

[30]Detmer and Steen, "Shoring Up Protection of Personal Health Data," 78.

[31]R. Pear, "Bush Acts to Drop Core Privacy Rule on Medical Data," *The New York Times*, March 22, 2002, A1 and A22.

notify patients of their privacy rights and make a 'good-faith effort' to obtain written acknowledgment of the notice."[32] The Bush administration, health care providers, and insurance companies insist this change will enhance care for patients, providing patient privacy as well as high quality health care. Many privacy advocates reply that this is a major setback for medical privacy and patient privacy rights. HMOs and insurance companies will have broad access to sensitive medical information. Moreover, pharmacies may sell data on patient prescriptions to drug manufacturers, allowing commercialization of sensitive medical information.

Most health-care professionals agree on the need for patient privacy protection, and believe it will encourage patients to share more honestly with their health-care providers, enhancing patient care. Yet there is anxiety and confusion over what counts as compliance with, or violation of the regulations.[33] Meanwhile, the information technology industry is continuing to work on systems changes required by HIPAA that are massively more complex and costly (in the billions of dollars) than those for Y2K. We do not yet know how implementation and enforcement of the new guidelines will work out. But the main arguments on each side remain the same. Some privacy protections do remain in the now settled guidelines, and thus overall they provide an important improvement over the self-regulation of the past. Nevertheless, the elimination of the patient consent provision, which I believe is crucial, undermines patient control over medical and genetic information. Providing notice of disclosure policies, all that Bush has required, does not ensure either that individuals understand or that they gain privacy protection.

Conclusion: Guiding Criteria

In sum, it is essential to weigh the benefits and drawbacks of these different public policy approaches. Several criteria guide the approach I have defended: (i) the need to protect the privacy of sensitive health and genetic information for all individuals, (ii) the value of emphasizing individual autonomy and education to encourage consent to release data given convincing justifications, (iii) the importance of letting new information technology flourish to retain the benefits electronic databases bring to patient care, to public health, and scientific research, and (iv) the need to replace a conglomeration of conflicting state or local guidelines with national guidelines to provide consistency in handling medical information. My approach defends privacy as a priority through federal guidelines mandating privacy protection as the *default* and then ensuring that people are educated, consulted, and allowed to give consent or refusal through a process of dynamic negotiation before

---

[32] B. Ladine, "Medical Privacy Rules are Relaxed," *The Boston Globe*, August 10, 2002, A1.
[33] A. Barnard, "Doctors Brace for Changes in Privacy," *The Boston Globe*, January 11, 2003, A1, A10 & A11.

their health and genetic information is gathered or disseminated. Privacy protection for patients need not undermine health care or medical and genetic research.

References

ACM Policy '98, Renaissance Hotel, Washington D.C., May 10–12, 1998.

A. Barnard. Doctors Brace for Changes in Privacy. *The Boston Globe*, January 11: A1, A10 & A11, 2003.

R.C. Barrows, Jr. and P.D. Clayton. Privacy, Confidentiality, and Electronic Medical Records. *Journal of the American Medical Informatics Association* 3(2): 1996.

T. Beauchamp and J. Childress. *Principles of Biomedical Ethics*, 3rd ed. Oxford University Press, New York, 1989.

C. Bennett. *Regulating Privacy: Data Protection and Public Policy in Europe and the United States.* Cornell University Press, Ithaca, 1992.

C. Culver, J. Moor, W. Duerfeldt, M. Kapp, and M. Sullivan. Privacy. *Professional Ethics*, 3 (3&4): 4–25, 1994.

J.W. DeCew. *In Pursuit of Privacy: Law, Ethics, and the Rise of Technology.* Cornell University Press, Ithaca, 1997.

J.W. DeCew. Alternatives for Protecting Privacy While Respecting Patient Care and Public Health Needs. *Ethics and Information Technology*, 1(4): 249–255, 1999.

J.W. DeCew. The Priority of Privacy for Medical Information. *Social Philosophy and Policy*, 17(2): 213–234, 2000.

D.E. Detmer and E.B. Steen, Shoring Up Protection of Personal Health Data. *Issues in Science and Technology*, 76, 1996.

J. Eilperin. New Rules Planned on Health Privacy. *The Boston Globe*, December 20: A3, 2000.

J. Foreman. Your Health History—Up for Grabs?. *The Boston Globe*, July 20: C1, 1998.

C. Franklin, editor. *Business Guide to Privacy and Data Protection Legislation.* Kluwer Law International, Dordrecht, 1996.

R. Gavison. Privacy and the Limits of Law. 89 *Yale Law Journal*, 421: 1980.

L.O. Gostin. Health Information Privacy. 80 *Cornell Law Review* 101: 104, 1995.

L.O. Gostin. Making Tradeoffs Between the Collective Good of Human Health and the Individual Good of Personal Privacy, presented at an international conference, "Electronic Patient Records in Medical Practice," Rotterdam, The Netherlands, October 6, 1998.

B. Ladine. Medical Privacy Rules are Relaxed. *The Boston Globe.* August 10: A1 and A5, 2002.

J. Markoff. Europe's Plans to Protect Privacy Worry Business, *New York Times*, April 11: A1, 1991.

R.E. Mitchell and J.W. DeCew. Dynamic Negotiation in the Privacy Wars. *Technology Review* 97(8): 70–71, 1994.

J. Moor. The Ethics of Privacy Protection. *Library Trends* 39 (1&2): 69–82, 1990.

J. Moor. Towards a Theory of Privacy in the Information Age. *Computers and Society*, September: 27–32, 1997.

R. Pear. Bush Acts to Drop Core Privacy Rule on Medical Data. *The New York Times*, March 22: A1 and A22, 2002.

L. Richwine. Privacy Laws for Patients Don't Extend to Internet. *The Boston Globe*, November 11: 2001.

M. Rigby, I. Hamilton, and R. Draper. Finding Ethical Principles and Practical Guidelines for the Controlled Flow of Patient Data, Presented at an International Conference, *Electronic Patient Records in Medical Practice.* Rotterdam, The Netherlands, October 7, 1998a.

M. Rigby, I. Hamilton and R. Draper. Towards an Ethical Protocol in Mental Health Informatics. In B. Cesnick, A.T. McCray, J.-R. Scherrer, editors, *MedInfo '98: Proceedings of the 9th World Congress on Medical Informatics*, pp. 1223–1227. Amsterdam, IOS Press, 1998b.

P.M. Schwartz. European Data Protection Law and Restrictions on International Data Flows, 80 *Iowa Law Review* 471: 1995.

L. Tye. EC May Force New Look at Privacy, *Boston Globe*, September 7: 10, 1993.

A. Westin et al. Health Care Information Privacy: A Survey of the Public and Leaders. 23 (1993).

B. Woodward. Intrusion in the Name of 'Simplification.' *The Washington Post*, August 15, 1996; Medical Identifier Hearings to Begin, *The Boston Globe*, July 20: A5, 1998.

8. Privacy of Medical Records: IT Implications of HIPAA

David Baumer, Julia Brande Earp, and Fay Cobb Payton

I. Introduction

Passage of the Health Insurance and Portability Accountability Act (HIPAA) in 1996 required the Clinton Administration, via the Office of Health and Human Services (HHS), to compose uniform standards for electronic exchanges of health information if Congress failed to enact a comprehensive privacy act by August 21,1999.[1] When Congress failed to act by the August 1999 deadline, HHS responded with proposed regulations on November 3, 1999.[2] We examine the regulations offered by the Office of Health and Human Services (HHS) (Standards for Privacy of Individually Identifiable Health Information; Proposed Rule 45 CFR Parts 160 through 164), which became law, February 21, 2000.[3] These regulations, however, do not preempt current state laws that provide greater protection for the privacy of medical records.[4]

As might be expected, there are numerous controversies on both sides of the privacy debate. Some think that the protections contained in the proposed records are inadequate, while others contend that privacy safeguards will erode the efficiencies that computerized medical records create, impede medical research and, in some cases, interfere with law enforcement.[5] We (Drs. Earp and Payton) conducted a survey of healthcare workers at a healthcare provider whose employees have access to personally identifying medical records.[6] The healthcare provider, whose employees are the respondents for the survey, will be regulated under the HHS Rule discussed in this paper. We compare the Administration's proposed regulations (the HHS Rule) with the survey responses of those who have access to medical records. There is a strong correlation between what healthcare workers regard as objectionable and what is outlawed by the proposed HHS Rule.

This essay originally appeared in *Computers and Society*, **30**: 40-47. Copyright © 2000 by David Baumer, Julia Earp, Fay Payton. Reprinted by permission.

II. Background of the Privacy Debate Regarding the Privacy of Individually Identifiable Medical Records

A. Current Federal Protection for the Privacy of Medical Records

Legal protection for the privacy of medical records is a crazy patchwork that is in desperate need of reform.[7] To date there is no federal statute that protects the bulk of personally identifiable medical records. The Americans with Disabilities Act (ADA) and the Privacy Act of 1974 provide some protection under some circumstances.[8] Also for government employees, the Fourth Amendment requires a showing of probable cause when their employer (the government) searches their medical records. For the vast bulk of medical records, however, there is currently little or no effective legal protection for the privacy of individually identifiable medical information at the federal level.[9] When medical records are not protected, a number of abuses can and do take place, including:

- unauthorized secondary use of medical records,
- inaccuracies that are not corrected,
- discovery and disclosure of medical records by hackers and commercial vendors,
- use of medical records by employers for employment decisions, and
- revelation of medical records by employees of insurance companies, who may be among our neighbors and who do not have medical training.

The Privacy Act of 1974 provides some protection for medical records that are held by *federal agencies*, but does not cover medical records held by *private* groups where most of the medical records are created and stored.[10] Likewise, the Privacy Act has numerous exceptions so that its protection is leaky at best. In its time, the Privacy Act was heralded as a huge step forward, but currently it has been labeled as the "most outdated" privacy act in the world.[11]

Some protection for medical records is also contained in the Americans with Disabilities Act (ADA) of 1990, which outlaws discrimination based on disabilities. As long as a disabled person can perform the essential functions of the job, employers cannot refuse to hire an applicant based on disability and must make reasonable accommodations for them to perform the job functions. The protection for medical records of employees under ADA is also filled with loopholes. First, the ADA does not protect the medical records of those who are not defined under the Act as "disabled." If the employee is considered "disabled," according to one of the three criteria detailed in the ADA, disclosure of medical records is permitted when:

(1) the supervisor needs to be informed regarding the necessary restrictions on the duties of the employee,
(2) the employer's medical staff needs to be informed for purposes of emergency treatment, and
(3) government officials are seeking to determine compliance with the ADA.[12]

A recent case illustrates the vulnerability of employees who are considered disabled under the ADA. In *Doe v. SEPTA* a self-insured, public employer hired Pierce to audit medications taken by employees to determine if waste and fraud were taking place.[13] Doe was infected with the AIDS virus and was reluctant to apply for benefits under the company plan because he feared detection. Doe specifically asked company officials whether Pierce would have access to his records and was told by the company that there was no need for Pierce to match Doe's name with the treatment he was receiving. Nevertheless, Pierce did match Doe's name up with his prescriptions and soon Doe's supervisor and co-workers learned of his condition. Doe sued under 42 U.S.C. 1983 for deprivation of his constitutional right to privacy, but after winning at the District Court level, Doe lost a Court of Appeals decision, which held that the greater good outweighed Doe's privacy concerns. Pierce had saved SEPTA more than $42,000,000 in medical and dental bills, and the court indicated that the benefits of Pierce's actions, "outweighs the minimal intrusion into Doe's privacy."[14]

B. Privacy Protection at the State Level

Much of the legal protection for the privacy of medical records occurs at the state level. Unfortunately, there is no uniformity across state jurisdictions so that compliance with state statutes by those creating, storing and using medical records is difficult given the wide disparities in protection. To add to the confusion, plaintiffs in medical disclosure cases must rely on state common law claims that are riddled with anomalies and exceptions. In 30 of the 50 states invasion of privacy by the unreasonable publicity given to the private facts of a person's life is actionable.[15] Attempts to fashion uniform state laws regarding medical records have been a spectacular failure. To date, only two states have adopted the Uniform Health Care Information Act (UHCIA).[16] The UHCIA itself only applies to healthcare providers and does not apply to third parties such as insurers and claims processors. The National Association of Insurance Commissioners (NAIC) has endorsed the Health Information Privacy Model Act, but this bill does not apply to healthcare providers.[17]

C. Impact of Technology

The debate regarding the privacy of medical records has been sharpened by several long-term trends. First, there is intense pressure to contain costs of medical treatment, not only among Medicare patients, but also by private insurers and employer health plans. There is increased scrutiny by third-party payers of medical treatments, tests and all kinds of psychiatric care. HMOs are caught in middle between the demands of patients for appropriate medical treatment and the costs they can recover from private and public insurers. Computerization of medical records can yield significant savings, but with it comes increased opportunities for disturbing disclosures. Second, much of the current privacy protection is based on paper records, which are being replaced by computerized files. The laws that were adequate for paper records are often inadequate to provide protection for computerized records. Computerized medical records are much more amenable to abuse on a much larger scale. Additionally, the exposure to detection of snoopy employees and others with access to medical files is much less when operating through a computer, than when they have to access paper records stored in visible file cabinets. The impacts of security breaches of company protocols that inadequately protect stored records are much more significant than with paper records. Third, as technology progresses, the potential for more intrusions into personal medical records will grow, particularly in the area of DNA testing. The potential use of DNA test results by insurers and employers to exclude "undesirables" from risk pools is becoming more and more evident.[18]

D. Need for Reform

There is general recognition that legal protection for the privacy of medical records is inadequate and disorganized. The debate about the privacy of medical records pits privacy advocates and patients against insurers, healthcare providers, public health organizations, researchers, law enforcement, government agencies, educational institutions and many others. Not surprisingly, each special interest has a rationale as to why it deserves access to personally identifiable medical records without obtaining consent of patients.

In some cases, personally identifiable medical records are sold to commercial interests (often at large profits) for direct marketing campaigns. Given the large amounts of revenue at stake, health organizations that are often labeled, "pro-patient" are enticed to traffic in sales of such information.[19]

E. Fair Information Principles

The basic building blocks of federal confidentiality law are not in dispute. In the *GeoCities* case, the Federal Trade Commission prosecuted a business that maintained a website that collected information from website visitors.[20] Following the *GeoCities* decision, the FTC issued a report that stated its views on fair information principles.[21] According to the FTC privacy report, there are generally accepted fair information principles that all websites (and repositories for medical information) should subscribe to:

(1) Notice/Awareness—consumers (patients) should be notified as to who is gathering the data and the uses that will be made of that data.
(2) Choice/Consent—consumers (patients) should consent to any secondary use for the data. There should be opt-in and opt-out provisions.
(3) Access/Participation—consumers (patients) should have the right to contest the accuracy of the data collected.
(4) Integrity/Security—there should be managerial mechanisms in place to guard against loss, unauthorized access, or disclosures of the data.
(5) Enforcement/Redress—there should be remedies available to victims of information misuse. The FTC envisions self-regulation by industry groups, private rights of action based on invasion of privacy, and government enforcement as in the *GeoCities* case.

There is, however, considerable debate as to how to translate these principles into practice. Healthcare advocates of privacy question the need for circulation of individually identifiable data beyond that necessary for healthcare providers and treatment. Others involved in the healthcare sector favor exceptions for various reasons. Among the exceptions are: facilitation of approvals and third-party payment systems, medical research, and desires of public healthcare associations. Hospitals, insurers, claims agents, managed care organizations, health researchers and law enforcement view the *sharing* of healthcare information as essential to the efficient functioning of the healthcare system.[22] These groups desire uniform federal laws that are understandable and which eliminate obstacles, including consent of the patients, to the sharing of healthcare information.

III. Department of Health and Human Services: Standards for Privacy of Individually Identifiable Health Information; Proposed Rule

In an effort to deal comprehensively with the issue of privacy of medical records, HHS developed a proposed Rule that occupied 145 pages of the Federal Register on November 3, 1999.

A. Making the Case for Additional Privacy Standards

According to the HHS proposed regulations, under the title *Need for Privacy Standards*,

> *The maintenance and exchange of individually identifiable health information is an integral component of the delivery of quality healthcare. In order to receive accurate and reliable diagnosis and treatment patients must provide healthcare professionals with accurate, detailed information about their personal health, behavior, and other aspects of their lives. Healthcare providers, health plans and healthcare clearinghouses also rely on the provision of such information to accurately and promptly process claims for payment and for other administrative functions that directly affect a patient's ability to received needed care, the quality of that care, and the efficiency with which it is delivered.*[23]

The foregoing statement by HHS recognizes (1) that effective medical care requires transfer of medical (and other) information from patient to healthcare professionals and (2) that other health organizations have legitimate needs to access healthcare information to facilitate payments and enhance treatment. According to the HHS, "Efforts to provide legal protection against the inappropriate use of individually identifiable health information has been, to date, undertaken primarily by the States."[24] Further, HHS states that "The number of entities maintaining and transmitting individually identifiable health information has increased significantly over the last 10 years... [t]he expanded use of electronic information has had clear benefits for patients and the healthcare system as a whole."[25] Finally, according to HHS, "The absence of national standards for the confidentiality of health information has, however, made the healthcare industry and the population in general uncomfortable about this primarily financially driven expansion in the use of electronic data."[26]

B. Statutory Background

According to HHS regulations, "section 262 [of HIPAA] directs HHS to issue standards to facilitate the electronic exchange of information with respect to financial and administrative transactions carried out by health plans, healthcare clearinghouses, and healthcare providers who transmit electronically in connection with such transactions."[27] Also, the regulations recite Section 264(c)(1) of the HIPAA which provides that:

> *If legislation governing standards with respect to the privacy of individually identifiable health information transmitted in connection with the transactions described in section 1171(a) of the Social Security Act (as added by section 262) is not enacted by (August 22, 1999), the Secretary of Health and Human Services shall promulgate final regulations containing such standards not later than (February 21, 2000).*[28]

Since Congress did not enact such legislation, HHS is required to act. Among the goals of the proposed HHS standards are:

- Allow for the smooth flow of identifiable health information for treatment, payment and related operations, and for purposes related to healthcare that are in the public interest.
- Prohibit the flow of identifiable information for any additional purposes, unless specifically and voluntarily authorized by the subject of the information.
- Put into place a set of fair information practices that allow individuals to know who is using their health information and how it is being used.
- Establish fair information practices that allow individuals to obtain access to their records and request amendment of inaccurate information.
- Require persons who hold identifiable health information to safeguard that information from inappropriate use or disclosure.
- Hold those who use individually identifiable health information accountable for their handling of this information, and provide legal recourse to persons harmed by misuse.[29]

Clearly, these goals have much in common with the report by the Federal Trade Commission that discusses Fair Information Principles following the *GeoCities* action.

C. Limitations and Protections Contained in HHS Proposed Regulations

Although the report of HHS recommends that "'everyone in this chain of information handling be covered by the same rules.' HIPAA limits HHS authority to *covered entities*, which include health plans, healthcare clearinghouses and any healthcare provider who transmits health information in electronic form..."[30] "In particular, the proposed regulation does not directly cover many of the persons who obtain identifiable health information from the covered entities."[31] HHS, however, attempts to expand their authority by requiring those receiving health records to apply the same fair information principles as the covered entities. Those receiving protected healthcare information from a covered entity may be classified as "business partners."

The basic thrust of the HHS Rule is to prohibit covered entities from using or disclosing protected health information except as provided in the proposed rule. Covered entities could use or disclose protected health information without authorization (from the patient) *for treatment, payment and healthcare operations*.[32] Covered entities would be allowed to disclose healthcare information without authorization only in very narrowly specified public health and public policy-related purposes, "including public health, research, health oversight, law enforcement..."

According to HHS, "A central aspect of this proposal is the principle of 'minimum necessary' disclosure."[33]

D. Permissible Uses and Disclosures for Purposes Other Than Treatment, Payment and Heathcare Operations

Although the proposed HHS Rule adheres to a "minimum necessary" disclosure principle, it does elucidate a long list of exceptions to the requirement to obtain individual authorization before disclosure of medical records. Most of these exceptions are based on public policy considerations, but the terms are potentially very elastic. Much of the controversy involves the details of the exceptions to the general principles of privacy that require individual authorization for secondary use of medical information. HHS considered permitting only those disclosures required by law, but decided that these exceptions are so important to public health and other purposes, that some unauthorized secondary disclosures should be permitted.[34] Although the sale of medical records for commercial purposes is not included among the exceptions, secondary use of the information might, and probably will, take place. The following sections describe some of the exceptions for disclosure without individual authorization.

1. Public Health Activities

The HHS Rule proposes to allow disclosure of medical records without individual authorization by covered entities for purposes of "carrying out public health activities authorized by law, to non-governmental entities authorized by law to carry out public health activities, and to persons who may be at risk of contracting or spreading a disease (when other law authorizes notification)."[35] The (HSS) Rule goes on to note that the covered entity could also be a public health agency and the same rules would apply. The Rule indicates that traditional public health activity to combat the spread of communicable diseases is so important that individual interests in privacy are overridden by community interests. If the HHS Rule interfered with the transmission of such information, it would require major changes in the ordinary practices of public health officials and would potentially expose people to communicable diseases.

2. Disclosure for Health Oversight Purposes

Chief among health oversight purposes are "combating fraud in the healthcare industry, ensuring nondiscrimination, and improving quality of care..." according to the HHS Rule.[36] According to HHS, "[O]versight activities are a national priority in part because of the losses in the healthcare system due to error and abuse..." which the HHS Office of Inspector General estimates at about 7 percent.[37] The Rule indicates that much of the work can be done with statistical tests that do not require disclosure of individual records without authorization, but not all of the oversight activities can be carried in that manner.

3. Use for Judicial and Administrative Proceedings and Use by Coroners and Medical Examiners

The HHS Rule envisions disclosures in personal injury and medical malpractice cases, "in which the medical condition of a person is at issue..." and a judicial or administrative proceeding is taking place to determine the cause of the injury or medical condition.[38] The HHS Rule indicates these disclosures "are clearly necessary to allow the smooth functioning of the legal system."[39] In addition, the HHS Rule would allow disclosure of medical records to coroners and medical examiners.[40] Again, public interest in determining causes of death is cited as well as the disclosure requirement by state law.

4. Disclosures to Law Enforcement

The HHS Rule would permit unauthorized secondary disclosure of individually identifiable medical records if the request by law enforcement is pursuant to judicial or administrative process. In addition, disclosure of medical information to law enforcement could occur without a search warrant if the disclosures by covered entities were given in good faith for fraud detection or to reveal a criminal action. The disclosure of healthcare information under this exception would generally be pursuant to a search warrant, absent exigent circumstances.[41] HHS admits that it has not figured out all of the boundaries of appropriate releases to law enforcement.

5. Government Health Data Systems and Health Directories

HHS proposes to allow secondary use of protected health information without individual authorization if the disclosure is "authorized by State or other law to support policy, planning, regulatory, or management functions."[42] According to HHS, governmental (including federal) examination of individually identifiable health data plays an essential role in examining the effectiveness of various policies. Moreover, if such transfers of information were prohibited, then HIPAA would negatively impact ongoing procedures that are routinely used by governments to evaluate policies that require detailed information that can only be obtained by examining protected health information. Unless the patient objects, the HHS Rule proposes to allow a health facility to include patients in a directory. For those patients who are not disabled at the time, the HHS Rule would require the facility to obtain information from the patient. The patient could specify to the health facility who would be entitled to receive the protected health information and who would not be so entitled.

6. Disclosure for Banking and Payment Purposes

According to the HHS Rule, checks and credit card payments of necessity will disclose protected health information.[43] The HHS contends that "Failure to allow this kind of disclosure of protected health information would impede the efficient operations of the healthcare system."[44] Recognizing what is likely to become commonplace in the future, the Rule states that "We understand that financial institutions may also provide covered entities that accept payment via credit card with software that, in addition to fields for information to process the transaction, includes blank fields in which health plans or healthcare providers may enter any type of information regarding their patients, such as diagnostic and treatment information, or other information that the covered entity wished to track and analyze."[45] Going further, the HHS Rule suggests that banks could become "business partners of covered entities in accordance with and subject to the conditions of 164.506(e)."[46] By becoming *business partners* of a covered entity, banks would become subject to the same prohibitions regarding protected health information as the covered entities.

7. Medical Research

HHS permits covered entities to disclose (to entities pursuing medical research) protected medical information without authorization as long as "the covered entity receives documentation that the research protocol has been reviewed by an Institutional Review Board or equivalent body—a privacy board—and that the board found that the research protocol meets specified criteria (regarding protected health information) designed to protect the subject."[47] In the absence of such documentation, medical researchers would have to obtain permission from the individuals who supplied the information, i.e., the patient. HHS stresses the importance of medical research that has led to many breakthroughs that have had dramatic impacts on the nation's health.

E. Concluding Thoughts about the Proposed HHS Rule

The absolutist wing of privacy lobby would limit secondary use of individually identifiable medical information to situations in which individual authorization has been obtained. On the other hand, at present, there is no overall protection for the privacy of privately held medical information that is individually identifiable. The Administration's approach, through the HHS Rule, is to split the difference between nearly a complete absence of protection and prohibitions on distribution of medical information for purposes other than treatment (the absolutist approach). Reasonable people can disagree about whether the proposed HHS Rule achieves its objective of "the minimum disclosure required."

The Administration's regulations would allow for unauthorized disclosure of medical information, not only for treatment, but also for payment and healthcare operations. The second goal of the Administration's Rule is to require individual authorization for secondary uses of medical information that is individually identifiable. The HHS Rule, however, allows for a lengthy list of exceptions to the requirement for individual authorization for secondary use of the protected medical information. Most of these exceptions reflect commercial practices within the healthcare industry currently extant. Most of the exceptions are justified on public policy grounds from prevention of the spread of communicable diseases, to detection of fraud, facilitating the justice system, and medical research where obtaining permission adds significantly to costs.

IV. Views of Healthcare Workers Who Have Access to Medical Records

A recent survey by Earp and Payton reveals the concerns of healthcare workers regarding the privacy of medical records.[48] In order to be eligible to take the survey, the respondent had to be a healthcare worker who had access to patient records. The survey was administered to a diverse group of 163 respondents (133 females, 30 males, 114 whites, 40 African-Americans, 9 Hispanics) who have extensive experience in healthcare (average number of years in healthcare 10.6, average age 44.6, and average number of years with current employer 5.2).

A. Frequency Data

The survey data consist of the responses of healthcare workers to statements made about healthcare issues involving privacy of, access to, uses made of, and the accuracy of medical records. The survey instrument and response data are available in the Appendix as well as some descriptive statistics. The responses varied from 1 to 7, with 1 indicating strong disagreement with the statement and 7 indicating strong agreement.

B. Gathering Patient Information

In Table 1 of the Appendix, survey data are arranged according to weighted-average responses to the statements in the questionnaire, from lowest to highest. Statements A, B, and F all relate to requests by their employer to gather information from patients. There are three variations of the same statement, which is, "I am troubled by requests from my employer to gather information from patients." The response data clearly indicate that healthcare workers are *not* troubled by the *gathering* of information from patients. Part of this lack of concern could, of course, be motivated by self-interest since knowledge by healthcare workers of any communicable diseases will enable them to take preventive measures. In addition, more knowledge of medical and other conditions of the patents enables healthcare workers presumably to better evaluate and treat patients.

C. Accuracy of Records

Responses to statements G, C, I, and M all concern the accuracy of the files containing medical records. Some variations of the statement suggest that their employer should devote more time and resources to making sure that the records are accurate. The response data indicate that healthcare workers recognize that the accuracy of medical files is a problem and that more resources should be devoted to enhancing the accuracy of these records.

D. Access to Records

Statements E, O, and J refer to access to patient records by others, not including the patient. Survey responses indicate that healthcare workers are very cognizant about possible abuses in the form of unauthorized or inappropriate access to medical records that contain personal information. Healthcare workers appear very concerned about the inappropriate and unauthorized access to medical records that is made possible by computerized medical records maintained by their employer. Note that it is much less obvious for a healthcare worker or others to invade the privacy of computerized medical records than for the same person to rummage through paper records in file cabinets. On a scale that varies between 1 and 7, the average response across statements E, O and J was 3.72—thus, agreeing with the statement that unauthorized and inappropriate access to personal medical records is a serious problem.

E. Unauthorized Secondary Use

Statements H, D, L, and N all refer to unauthorized secondary access and use of personal medical records. Again, using the same scale, the average response was 5.57 condemning unauthorized secondary use of patient information. Statement H essentially says that employers should refrain from using information given by patients for any other reason than medical treatment. The average response of 5.07 indicates that healthcare workers were undecided about that statement. The word, "never" seems too strong in this context. Probably the greatest source of resentment among patients is that vendors in the medical sector will somehow access their medical records and that they will be targeted for marketing campaigns based on what is in their medical files. Note that the highest agreement among healthcare professionals with access to medical records is for the statement that their employer "should never sell the patient information in its computer databases to other companies."

F. Implications of Analysis of Frequency Data

Based on the frequency data, the following implications appear to be critical:

- Healthcare professionals do not regard the *gathering* of patient information as an abuse or source of concern.
- Healthcare workers recognize that information received from patients may sometimes be required for purposes other than treatment.
- There are significant concerns among healthcare workers about the *accuracy* of medical files and that employers should do more to improve accuracy of medical records.
- There are also very significant concerns among healthcare workers about *inappropriate and unauthorized* access to medical records.
- Healthcare workers take most seriously *unauthorized secondary use* of medical information. The most egregious abuse, in the view of healthcare workers, occurs when a company sells computerized medical files for money.

G. Factor Analysis

A factor analysis reveals much the same results. Based on the factor loadings the following four variables were created (in order of significance): **Errors** (concern about the accuracy of patient medical records), **Unauthorized Secondary Use**, **Improper Access**, and **Collection** (of data).

V. Conclusion

Most major advances in technology also entail unintended consequences. As computerized medical records have enabled healthcare providers to efficiently gather and evaluate medical information, via modern database and database-enabled technologies, the potential for misuse of this information has also increased. Among the major players in the health sector are insurers, employers, hospitals, pharmaceutical companies, healthcare researchers, public health administrators, vendors of medical equipment, various levels of government, medical researchers, law enforcement, and of course, healthcare providers. Each of these players has their own rationale as to why they are entitled to individually identifiable medical records. In most cases, the rationales put forward by these groups are well-thought out, but the combined impact is one that leaves patients feeling exposed, as too many people are seeing patients' medical records.

The principles of fair use of information have been agreed upon for at least 25 years. Implementation, however, has been contentious because various groups argue that there should be exceptions. The most fundamental principle of fair use of information is that no secondary use of medical information should take place unless authorized by the patient. The HHS Rule, promulgated by the Clinton Administration, allows numerous exceptions to this principle based on public health policy considerations, commercial practices, and state laws. Although the proposed regulations of HHS allow for exceptions to the principles of fair use of information, they still represent a major step forward in protecting medical records.

Healthcare workers are on the frontlines of the privacy battle. They know that the gathering of information is essential for effective treatment and for processing claims. Healthcare workers also know that medical records are not always accurate and that too many have access to these records. Finally, healthcare workers know there is something inherently wrong about unauthorized secondary use of medical information that does not involve treatment, payment, or operations and that the sale of medical information without authorization should be prohibited.

Endnotes

1. 42 U.S.C.A. 1320d to d-8 (West Supp. 1998).
2. Federal Register 59918 et seq., Dept. of Health and Human Services, Office of the Secretary, 45 CFR Parts 160 through 164, Standards for Privacy of Individually Identifiable Health Information (November 3, 1999), (hereinafter Fed. Reg.).

3. Fed. Reg. 59921.
4. Fed. Reg. 59926. The essence of the proposed federal Rule is that it would provide a floor for the privacy of medical records. The Rule is not intended to replace state common law or statutory laws affecting privacy.
5. Bartley L. Barefoot, Enacting a Heath Information Confidentiality Law: Can Congress Beat the Deadline?, 77 *N.C.L. Rev.* 283 (1998).
6. Earp and Payton, Information Privacy Concerns Facing Healthcare Organizations in the New Millennium, under review at the *Journal of American Medical Information Association*.
7. Barefoot, 1998.
8. 42 U.S.C. 12112 (1994); 5 U.S.C. 552a (1994).
9. HHS notes that, "Efforts to provide legal protection against the inappropriate use of individually identifiable health information have been, to date, undertaken primarily by the States." Fed. Reg. 59919-20.
10. 5 U.S.C. 552a (1994).
11. Barefoot, 294.
12. 42 U.S.C. 12112(c)(3)(B).
13. 73 F.3d 1133 (3rd Cir. 1995).
14. Id., at 1143.
15. Causes of action in the other 20 states for disclosure of medical records are unknown to the authors. Twomey, LABOR AND EMPLOYMENT LAW, 584. In 22 states patients do not have the right to view their own medical records, CNN Broadcast of Congressional Hearings regarding HHS Rule, February 17, 2000.
16. Barefoot 304.
17. *Id.*, 305.
18. Mark A. Rothstein, Betsy D. Gelb, and Steven G. Craig, Protecting Genetic Privacy by Permitting Employer Access Only to Job-Related Employee Medical Information: Analysis of a Unique Minnesota Law, 24 Am. J. L. and Med. 399 (1998).
19. Barefoot 288.
20. GeoCities; Analysis to Aid Public Comment,Thursday, August 20, 1998, 63 Federal Register 44624 AGENCY: Federal Trade Commission.
21. www.ftc.gov/reports/privacy3/index.htm.
22. Barefoot 309.
23. HSS Regs, I.A. The Need for Privacy Standards, Fed. Reg. 59919.
24. Fed. Reg. 59919-20.
25. Fed. Reg. 59920.
26. Id.
27. Id.
28. Fed. Reg. 59921.
29. Fed. Reg. 59923.

30. Fed. Reg. 59923-4.
31. Id.
32. Fed. Reg. 59924.
33. Fed. Reg. 59943.
34. Fed. Reg. 59955.
35. Fed. Reg. 59956.
36. Fed. Reg. 59957.
37. Id.
38. Fed. Reg. 59958.
39. Fed. Reg. 59959.
40. Fed. Reg. 59960.
41. Fed. Reg. 59961.
42. Fed. Reg. 59964.
43. Fed. Reg. 59966.
44. Fed. Reg. 59966.
45. Id.
46. Id.
47. Fed. Reg. 59967.
48. Julie Earp and Fay Cobb Payton, Information Privacy Concerns Facing Healthcare Organizations in the New Millennium, Working Paper, NCSU, College of Management.

Appendix

Table 1 Frequency data for survey

	SD 1	2	3	4	5	6	SA 7	WtdSum	WtdAvg
E. XXXXX should devote more time and effort to preventing unauthorized access to patients' personal and financial information.	48	25	16	39	20	7	1	451	2.89
A. It usually bothers me when XXXXX asks patients for their medical information.	35	39	12	45	16	3	6	469	3
O. XXXXX should take steps to make sure that unauthorized persons cannot access patient information in its computers.	38	22	14	64	10	4	4	482	3.09
G. XXXXX should take more steps to make sure that the patient information in its files is accurate.	19	17	22	30	15	17	36	668	4.28
H. When patients give personal information to XXXXX for some reason, XXXXX should never use the information for any reason (other than its original intent).	2	5	9	52	25	22	41	791	5.07
J. Computer databases that contain patient information should be protected from unauthorized access – no matter how much it costs.	2	4	10	44	27	27	42	807	5.17
D. XXXXX should not use patient information for any purpose unless it has been authorized by the patient who provided the information.	3	4	8	42	23	26	50	824	5.28
B. It usually bothers me when XXXXX asks patients for their personal and financial information	5	7	8	22	31	41	42	826	5.29
L. XXXXX should never sell the patient information in its computer databases to other companies.	2	3	2	32	41	35	41	844	5.41
F. When XXXXX asks patients for personal and financial information, I sometimes think twice before recording it.	4	3	7	26	30	33	53	854	5.47

Table 1 (continued)

	SD 1	2	3	4	5	6	SA 7	WtdSum	WtdAvg
C. All the patient information in computer databases should be double-checked for accuracy – no matter how much this costs.	5	6	7	10	11	20	97	932	5.97
I. XXXX should have better procedures to correct errors in patient information.	2	3	2	11	22	37	79	943	6.04
M. XXXX should devote more time and effort to verifying the accuracy of the patient information in its databases.	3	1	1	5	8	23	115	1011	6.48
N. XXXX should never share patient information with other companies unless it has been authorized by the patients who provided the information.	2	1	1	5	9	21	117	1017	6.51

9. The Risks of Epidemiological Data Mining

Bart Custers

Introduction

Epidemiology is the study of the distribution and determinants of disease frequency in man.[1] It tries to give insight into the prevalence and incidence of diseases by quantifying diseases and their cure and/or mortality. The purpose of studying diseases is to (help to) solve etiological, diagnostic, prognostic or therapeutic problems.[2]

Epidemiology tries to describe diseases as a function of *determinants*. A determinant is a possible factor of influence on the frequency of a disease.[3] It should be noted that in epidemiology, determinants are not necessarily natural factors, such as age and gender; social factors, such as residence and income, may be possible determinants as well.

The basic method to find determinants is the same as in the natural sciences: by comparison. This may be *observational*, i.e. by comparing heart attack frequency between persons with and without hypertension. Or the comparison may be *experimental*, in which case one factor is changed (while the other factors remain constant) and the effects of this change are studied.[4] One of the drawbacks of the latter approach is that experimental manipulation of determinants is often limited. This may lead to discussions about the causality of discovered relations. I will go into this later on.

The study of determinants goes back to the time of the Greek physician Hippocrates (460-380 BC), who is considered to be the founder of medical science. Hippocrates already knew there were links between the incidence of diseases and

---

This essay originally appeared in the *Proceedings of the Fourth International Conference on Computer Ethics–Philosophical Enquiry (CEPE: 2001)*, Lancaster University, UK, pp. 61-70. Copyright © 2001 by Bart Custers. Reprinted by permission.

[1] MacMahon and Pugh (1970), p.1.

[2] Vandenbroucke and Hofman (1993).

[3] Bunge (1959).

[4] Vandenbroucke and Hofman (1993).

the conditions of water and air, and living conditions. However, he never took a quantified approach; this was first possible with the rise of the natural sciences, which took place in the 16th and 17th centuries. But the real rise of epidemiology came in the second half of the 20th century, when large-scale population research was performed, for instance, on the effects of smoking, or the screening programmes for cancer were carried out.

Epidemiology in the Information Age

The increasing possibilities of epidemiology are related to the rise of information and communication technology.[5] Because of these technologies, enormous amounts of data are being generated and stored in large databases by companies, governments, universities, hospitals, etc. Epidemiology is based on statistics, and it is well known that statistics requires certain minimum amounts of data.[6] In addition, a general rule of thumb is that statistics become more reliable when the (representative) samples from which they are derived are larger, i.e. contain more data.

Not only the generating of new data is important, also the increased accessibility of both the existing and the new data is an important factor in the development of epidemiology. With information and communication technologies, access to data has become much easier. For instance, databases of hospitals can be linked and useful information can be found on the Internet.

However, the statement that the chances of finding useful information increase when more information is being generated and more information becomes accessible, should be qualified. Nowadays, because of the enormous amounts of available data, people often sense an information overkill. When your Internet browser finds one million hits on your search key, you are no closer to your goal than if it had found zero hits. The problem is, therefore, to distil the useful pieces of information. Another problem is to know what information is useful and what is not.

To get a better grip on the large amounts of data, there are several methods. For instance, one could try to describe the data with the use of metadata.[7] When averages and deviations are investigated, a database may become more understandable. For instance, when 80 per cent of the records of a patients database are records on women, this does not correspond to the expected average of 50 per cent. Such metadata may reveal a relationship between gender and the particular disease.

A more interesting method is to find patterns in the data. Researchers traditionally try to find patterns and relationships by forming hypotheses and testing these against a background of data resulting from experiments or data that are already available. After testing the hypothesis, it is either rejected or accepted.[8] An accepted hypothesis can still be rejected afterwards when new data proves the

---

[5] On the rise of information and communication technology, see Castells (1996).

[6] Note that a statistical relation is not necessarily a causal relation. But statistical relations are often used to find causal relations.

[7] Metadata are data describing the data, i.e. data about data.

[8] Modification of a hypothesis is also possible, but this may be considered as a new hypothesis.

hypothesis wrong.⁹ When a hypothesis holds for some time, it will often be labelled a theory.

The capacities and capabilities of researchers have increased by information and communication technology, but the amounts of available data have increased much faster. In order to keep track, the process of finding patterns and relationships has been automated. One of the most important techniques for this is *data mining*. Data mining is an automatic analysis, using mathematical algorithms, to find patterns and relations in data.[10] The search for patterns can be done in various ways, such as regression, clustering or classification.[11] A description of these techniques is beyond the scope of this paper.[12]

Profiling Individuals and Groups

The patterns resulting from data mining may concern people, in which case I will speak of *profiles*. Profiles concerning an individual are called *personal profiles*, sometimes also referred to as individual profiles or customer profiles. A personal profile is a property or a collection of properties of a particular individual.

An example of a personal profile is the personal profile of Mr. John Doe (44), who is married, has two children, earns 25,000 Euro a year, has two credit cards and no criminal record. He was hospitalized twice in his life, once for appendicitis and last year for a heart attack. Expanding this example even more should not really be difficult: recent research showed that about 150 characteristics of each Dutch citizen are available on the Internet.[13]

Profiles concerning a group of persons are referred to as *group profiles*. Thus, a group profile is a property or a collection of properties of a particular group of people. Group profiles may contain information that is already known, for instance people who smoke live on average a few years less than people who do not smoke. But group profiles may also show new facts, such as, for instance, people living in ZIP code area 8391 run a (significantly) greater risk of developing asthma. But group profiles do not necessarily describe a causal relation. For instance, people driving a red car may run a (significantly) greater risk of getting colon cancer than people driving a blue car.

The creation and use of profiles may be experienced as an intrusion of privacy or a loss of autonomy. This may happen, for instance, when someone is confronted with the fact that there is a great deal of information available about him or her. For instance, people often wonder where this or that personally directed mail or e-mail with special offers comes from.[14] Possible effects of the creation and use

---

[9] This is the so-called falsification theory developed by the philosopher Karl R. Popper.

[10] Fayyad, Piatetsky-Shapiro and Smyth (1996).

[11] Adriaans and Zantinge (1996).

[12] For an overview of data mining techniques, see for instance Adriaans and Zantinge (1996), Fayyad, Piatetsky-Shapiro and Smyth (1996) and Frawley, Piatetsky-Shapiro and Matheus (1993).

[13] M. van Nieuwstadt, 'Digitale sporen', *NRC*, 9 Januari 1999.

[14] Custers (2001).

of profiles will be discussed more elaborately in the section "Effects of Using Group Profiles."

Legal Protection of Privacy

In Europe, the collection and use of personal data is protected by a European Directive (the so-called privacy directive), which has been implemented in national law in the member countries of the European Union.[15] Privacy principles that are safeguarded in the European privacy directive correspond to the principles in the Organization for Economic Cooperation and Development (OECD) guidelines,[16] which were also included in the Council of Europe Treaty of Strasbourg.[17]

These principles are:

- the *collection limitation principle*, stating that "[t]here should be limits to the collection of personal data and any such data should be obtained by lawful and fair means and, where appropriate, with the knowledge or consent of the data subject";
- the *data quality principle*, stating that "[p]ersonal data should be relevant to the purposes for which they are to be used, and, to the extent necessary for those purposes, should be accurate, complete and kept up to date";
- the *purpose specification principle*, stating that "[t]he purposes for which personal data are collected should be specified and that the data may only be used for these purposes";
- the *use limitation principle*, stating that "[p]ersonal data should not be disclosed, made available or otherwise used for purposes other than those specified, except a) with the consent of the data subject; or b) by the authority of law";
- the *security safeguards principle*, stating that reasonable precautions should be taken against risks of loss, unauthorized access, destruction, etc., of personal data;
- the *openness principle*, stating that the subject should be able to know about the existence and nature of personal data, its purpose and the identity of the data controller;
- the *individual participation principle*, stating, among others, that the data subject should have the right to have his personal data erased, rectified, completed or amended;
- the *accountability principle*, stating that the data controller should be accountable for complying with measures supporting the above principles.

These privacy principles for fair information practices are based on the concept of *personal data*, which is described in article 2 sub a of the European privacy directive as 'data concerning an identified or identifiable natural person', a definition that also stems from the OECD guidelines. Personal profiles contain personal data

---

[15] European Directive 95/46/EG of the European Parliament and the Council of 24th October 1995, [1995] OJ L281/31.

[16] See <http://s3-hq.oecd.org/scripts/pwv3/pwhome.htm>.

[17] See <http://www.coe.fr/dataprotection/Treaties/Convention%20108%20E.htm>.

and are therefore protected by the (national implementation of the) directive, but group profiles do not necessarily contain personal data and may therefore lack this protection. Since group profiles have significantly lower levels of protection, this paper will focus on group profiles rather than on personal profiles.

What Is New about This?

Group profiles have been made in the past without data mining. One of the easiest and most common ways of creating group profiles is by observation. Simply see if it is possible to distinguish a particular group and investigate its characteristics. Thus, it is often questioned what is new about group profiling through data mining as compared to traditional group profiling through observation. Are we not dealing with an old problem in a new context? Is it not true that we have always had the stigmatization of groups?

I will show that group profiling through data mining is very different from group profiling through observation. The result of this difference is that the known problems of group profiling through observation are more serious. At the same time there are also totally new types of problems that are only related to group profiling through data mining.[18] However, this does not mean that nothing can be learned from the ways in which previous efforts were made to tackle the problems concerning group profiling through observation.

The first and most obvious reason why group profiling through data mining may cause more serious problems than group profiling through observation is a scale argument. Data mining is an automated analysis and, thus, it can be very fast. It is even possible to run parallel sessions. Since it is very easy to apply, much more group profiling can be done compared with group profiling through observation. With growing amounts of data and increasing potential of data mining tools, there will be an overload of profiles.

A second difference is that data mining investigates every possible relation, whereas observation and research usually only consider causal relationships. The relations found by data mining are not necessarily causal. Or they may be causal without the researcher knowing why. In this way, many more profiles may be discovered—unexpected profiles in unexpected areas.

An example of this may be the following. Suppose a certain gene deficiency causes both large hands and thrombosis, but this gene and the implications of its deficiency are not known. Since there is no logical causal relation between large hands and thrombosis, no one will expect a relationship. Only a physician specialized in thrombosis may notice some day that many of his patients have particularly large hands.

Data mining is not dependent on such a coincidence. Data mining tools automatically generate hypotheses, independent of whether a relationship is (expected to be) causal or not, and each generated hypothesis is then checked against the data in the database. Thus, it systematically checks all relations and if the data are

---

[18] It is important to make a distinction between technology-specific and technology-enhanced concerns; see Tavani (1999).

in the database, the relation between large hands and the thrombosis will be found immediately.

Of course the fact that not only causal relationships are searched for results in many more patterns and relations. In this way it contributes to the first reason as well: the number of group profiles increases largely because non-causal relations can be found as well.

A third difference is the fact that trivial information may be linked to sensitive information. This may be illustrated by an example. Suppose it is found by mining a database that there is a relation between driving a red car and developing colon cancer. Thus, a trivial piece of information, the colour of one's car, becomes an indicator of one's health, which is sensitive information.

In this way it becomes very difficult for individuals to protect their personal data against group profiling. People who only provide trivial information may be unaware of the fact that they are also providing sensitive information about themselves because they belong to a group about which sensitive information is known.

Besides, the mentioned (implementations of) European privacy laws contain a special regime for sensitive data, offering additional protection. But when sensitive data is linked to trivial data, the additional protection can be evaded by using the trivial data, for which the higher standards are not required.

A fourth difference is the lack of forgetfulness of information and communication technology.[19] Once a piece of information has been disclosed, it is practically impossible to withdraw it again. Computer systems do not forget things, unless you explicitly delete information, and even then it can often be retrieved. But since disclosed information is usually not contained, it spreads through computer systems by copying and distributing.[20] Thus it becomes difficult to trace every copy and delete it.

Effects of Using Group Profiles

As was mentioned in the introduction, epidemiology studies the frequency of disease by describing it as a function of determinants. With the help of group profiles (in particular, group profiles through data mining), it is possible to (help to) solve etiological, diagnostic, prognostic or therapeutic problems.

When the causes or causing mechanisms of a disease are known, it becomes easier to establish a prognosis of the incidence of that particular disease, in which case preventive measures can be taken. It also becomes possible to establish a diagnosis sooner and start treatment or therapy. In general, a treatment has better results if the diagnosis is made in an early stage of the development of a disease.

Group profiles may not only enhance therapeutic results by earlier diagnoses or by prevention on the basis of prognoses, but also by providing information on the results of the therapy itself. Group profiles may be used to compare the results of a particular therapy or medicine with results of groups having other treatment

---

[19] Blanchette and Johnson (1998).

[20] Think of Napster as an example.

or medicines or groups without treatment or medicines at all. And with a better understanding of the etiology of diseases, new medicines will be found.

Note that not only natural determinants of a disease can be investigated; it is also possible to find social determinants. When a profile shows that in a certain district there is a higher incidence of asthma, it may be investigated whether the air in this district is less clean, for instance because it concerns a highly industrialized area. If this is the case, measures may be taken against the air polluting factories.

All these applications make group profiles very useful. But although the well-being of patients may be enhanced and lives may be saved, group profiles should be treated with caution because there are risks involved in the use of group profiles.

Harvey (1990) states that group claims are often treated as discovered instead of as created, i.e. as if it were 'knowledge'.[21] According to him, this may lead to three inference errors that should be avoided. First, there is the risk that subjects sharing some characteristics included in the group profile will be supposed to have many other attributes in common. Such an inference is not valid as long as the profile has not been checked on these additional characteristics. Especially in the case of group profiling through data mining, where causal relations may be absent, inferring additional characteristics will quickly lead to mistakes.

Second, there is the risk of inferring causal relations between the characteristics in the group profile. As in the example mentioned before about the gene deficiency causing both large hands and thrombosis, it is clear that there is not a *causal* relationship between the large hands and the thrombosis, although there certainly is a relationship.

Third, the risk of inferring group characteristics to individual group members should be avoided, unless adequate evidence is found which bears directly on the individual. This brings us to a difficult problem. Whether or not group properties are valid for individual members depends on whether one takes an internal or external perspective with regard to the group. Actually, it depends on the amount of available information, but the perspective determines whether or not additional information is available. Let me try to make this clear with an example. Take a randomly composed group with 50 per cent being men and 50 per cent being women. From an external perspective, i.e. from outside the group, each member will have the property of a 50 per cent chance of being a man. From an internal perspective, i.e. from within the group, being a group member, it is immediately clear whether or not the group member is a man or a woman. This may seem to depend on the perspective, but it actually depends on the additional information that a group member has, namely the information about his or her personal data.

Another risk of group profiles is that they may be used as selection criteria in an unjustified way. Selection is one of the main applications of group profiles. As mentioned before, selection to trace the incidence of diseases and provide earlier and better therapy to patients is very useful. But selection may also be used for less noble motives. When selection for jobs is performed on the basis of medical group profiles, this may quickly lead to discrimination. For instance, an employer who

---

[21] What Harvey (1990) defines as a group claim is similar to what I call a group profile.

does not want to spend too much money on personnel that is ill most of the time may want to know the probability that a new applicant will often be absent. But in the Netherlands a medical examination for employees is forbidden by the 'Wet op de medische keuringen' (Medical Examinations Act).[22] Unjustified selection may not only take place for jobs, but also for purchasing products, acquiring services, applying for loans, etc.

Some of the group profiles constructed by companies, government or researchers may also become 'public knowledge', which may then lead to stigmatization of that particular group.[23] Recently, in the Dutch city of Groningen an internal police report stated that criminal behaviour in asylum seekers was five times that in local people.[24] After the mayor had used this conclusion to start a public debate, the report had to be made public. Although the research showed many methodological errors, the conclusions are still in people's minds.

When group profiles are used for early diagnosis or the prevention of diseases, people in risk groups may be approached with warnings. Thus, people are confronted with their health prospects without having requested to get such information about themselves. Especially in the case of very negative group profiles this may have a great impact on people's lives. Healthy people may be confronted with the fact that they will only have a limited lifetime left, which may turn their life and that of others upside down. In some cases people may prefer not to know their prospects as long as they are healthy.

A Note on Reliability

The reliability of a group profile may influence the effects, both positive and negative, of the use of the profile. The reliability of a group profile may be divided into two different factors. The first is the reliability of the profile itself and the second is the reliability of its use.

The creation of group profiles consists of several steps, in which errors may occur.[25] First, the data on which a group profile is based may contain errors, or the data may not be representative for the group it tries to describe. Furthermore, to take samples, the group should be large enough to give reliable results.

In the second step the data is prepared for data mining. In this preparation phase, data may be aggregated, missing data may be searched for, superfluous data may be deleted, etc. All these actions may lead to errors. For instance, missing data is often made up, which is proved by the fact that a significantly large number of people in databases tend to have been born on the 1st of January (1-1 is the easiest to type).[26]

---

[22]There are some exceptions to this general rule, in cases where a medical examination is considered absolutely necessary.

[23]Vedder (2000).

[24]'Roep om extra studie naar criminele asielzoekers', *De Volkskrant*, 26 Januari 2001.

[25]Frawley, Piatetsky-Shapiro and Matheus (1993).

[26]Denning (1983).

The actual data mining consists of a mathematical algorithm. There are different algorithms, each having its strengths and weaknesses. Releasing different data mining programmes on the same database may lead to different resulting group profiles. The choice of algorithm is very important and the consequences of this choice for the reliability of the results should be realized. For instance, in case of a classification algorithm, the chosen classification criteria determine most of the resulting distribution of the subjects over the classes.

As far as the reliability of the use of group profiles is concerned, this depends on the interpretation of the group profile and the actions that are taken upon (the interpretation of) the group profile. Both the interpretation and the determined actions are dependent on whether an internal or external perspective is taken with regard to the group. Take as an example a particular group in which 90 percent of the group members wear glasses. From an external point of view, inferring that every individual group member has a 90 percent chance of wearing glasses may be useful information. For instance, an optician may be interested in sending this group promotional material or an epidemiologist may be interested in investigating this unusually high rate of persons wearing glasses. From an internal perspective, a group member can immediately tell if he or she is wearing glasses or not. The 90 percent chance is not applicable to this individual since he has additional information. Still, to exclude the people not wearing glasses from the group may lead to more work and costs than including them, because they have to be traced individually.[27] Thus, this lower reliability may be taken for granted.

It should be noted that a perfectly reliable use of a group profile, i.e. 100 percent of the group members sharing the characteristic, does not necessarily imply that the results of the use are fair or desirable. Especially in the case of negative characteristics this may occur, for instance when a group with 100 percent handicapped people are all being refused insurance. Although the use of the group profile is now perfectly reliable, it is not justified.

Note that the difference between an internal and an external perspective is not applicable to the case of future properties. For instance, an epidemiological group profile with the characteristic that 5 percent of a particular group will die from a heart attack does not provide any information on the question whether Mr. Smith, who is a member of this group, will die from a heart attack. And since Mr. Smith himself has no additional information on this, his internal perspective is no different from the external perspective.

If group profiles are valid for the group and for individuals as members of that group, though not for those individuals themselves, we speak of *non-distributive properties*.[28] On the other hand, properties that are valid for each individual member of a group are called *distributive properties*. The fact that in non-distributive profiles not every group member shares the group characteristic, has different consequences depending on whether the characteristic is generally regarded as negative or positive. This is illustrated in Figure 1.

---

[27] In groups of over a thousand persons, tracing 10 per cent person by person may be a tough job.
[28] Vedder (1997).

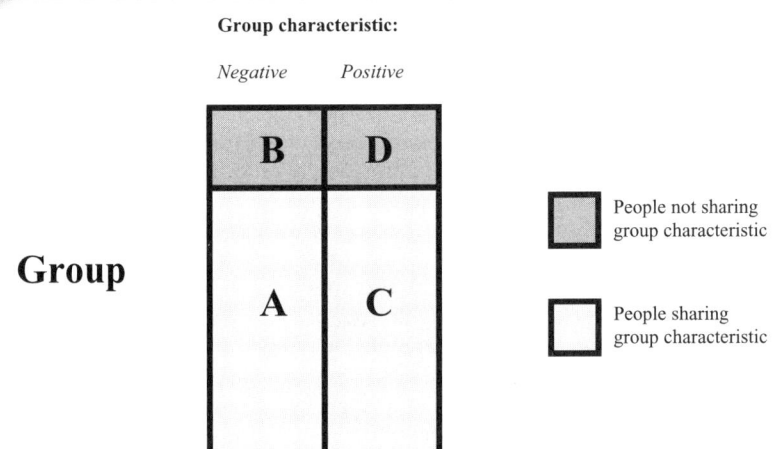

Figure 1: *Not every group member necessarily shares a group characteristic. This has different consequences depending on whether the characteristic is negative or positive.*

People in category A have the disadvantages of sharing the negative group characteristic and of being treated on the basis of this negative profile. This may result in an accumulation of negative things: first, there is the negative health prospect; next, stigmatization and selection for jobs, insurance, etc. may follow.

In category B people have the disadvantage of being treated as if they share the negative characteristic, although they do not actually do so. There may be an opportunity for these people to prove or show they do not share the characteristic, but they are 'guilty until proven innocent'. Sometimes, proving exceptions is useless anyway, for instance when a computer system does not allow exceptions or when handling exceptions is too costly or time consuming.

Sometimes people in category B may have an advantage. This is the case when measures are taken to improve the situation of the people with the negative characteristic. For instance, when the government decides to grant some extra money to a group with a very low income, some group members not sharing this characteristic may profit from this.

People in category C have the advantage of sharing as well as being treated on a positive characteristic. Similar to people in category A, this may be accumulative. The group may get the best offers on jobs, insurance, loans, etc.

Finally, in category D are the people who do not share the positive group characteristic. Their advantage may be that they are being treated on a positive characteristic, but the disadvantage is that they are not recognized as not sharing the positive characteristic or even having a negative characteristic. Lack of such recognition may become a problem when measures are taken to help the people with negative characteristics. For instance, people in category D may not be recognized as people running a great risk of getting colon cancer and are thus easily forgotten in government screening programmes.

From Figure 1 it becomes clear that there is a difference between correct treatment and fair treatment. People in categories A and C can be said to be treated correctly, since they are treated on a characteristic they in fact share. Whether this treatment is also fair remains to be seen. Accumulation of negative things for people in category A and of positive things for people in category C may lead to polarization.

People in categories B and D do not share the group profiles of the groups they belong to and are therefore being treated incorrectly. Incorrect treatment very probably also implies unfair treatment since it does not take into account the actual situation people are in.

Concluding Remarks

An inventory was made of possible risks of epidemiological data mining. Data mining methods are used to search for patterns and relations in data. Patterns concerning people are called profiles. Profiles may invade someone's privacy. Personal profiles contain personal data and are protected by privacy laws. Group profiles, however, do not necessarily contain personal data and may therefore lack this protection.

The question arises of what can be done about these possible risks of data mining in epidemiology. Possible solutions to avoid or confront these risks should be searched for in different directions.

First, informing and educating the public about these problems may create awareness. When people want to protect their privacy, they have to know what is done with their personal data before they are able to decide whether they want to give away their personal data or not. People often fail to recognize the link between direct mail and the fact that they filled out some forms asking for personal data.[29]

Information is a prerequisite for free market mechanisms. If people have a choice between two insurance companies, one offering no privacy protection and the other offering good privacy protection but at a slightly higher price, it could be argued that people will have a choice. However, I would argue that people only have a real choice when they are aware of the implications of poor privacy protection.

Second, technology may play an important role in decreasing the risks of epidemiological data mining. The most important technology that can be thought of is cryptography. Cryptography is the study of secret writing,[30] and can be used for hiding or encoding information. This is often done with the use of secret keys. Only the person possessing this secret key can decode the information, while for others it is practically impossible to do so.[31]

---

[29]People who are aware of this may on purpose include a spelling mistake in their name or address to trace how a company obtained the personal data.

[30]Koops (1999).

[31]Van der Lubbe (1997).

With the use of cryptography it is possible to control information. According to Denning, the control of information can be divided into three types of controls.[32] *Access controls* handle the authorization of direct access to objects. The holder of information can decide for whom the information is available. *Information flow controls* are concerned with the authorization to disseminate information. Even when access controls are present, information can be leaking. Flow controls specify valid channels along which information may flow. Finally, *inference controls* ensure that released statistics do not disclose confidential data. Whereas access and flow controls are mainly concerned with the raw data, the contents of databases, inference controls focus on the data mining process and its results.

Third, law and regulations may play an important role. As has been pointed out before, the current privacy laws protect personal data, but group profiles are not protected by these laws. It should be investigated whether the current privacy laws may be extended in a way that they also comprise group profiles. If the shortcomings in privacy laws cannot be changed or if such changes are not desirable, self-regulation, for instance by the use of codes of conduct, may be a solution. But since self-regulation is a free market mechanism, people should be informed and educated about the risks involved.

Acknowledgments

The author wishes to thank the cooperation centre of Tilburg and Eindhoven Universities (SOBU) and the Schoordijk Institute of the Faculty of Law of Tilburg University, which funded the research that was at the basis of this paper.

References

Adriaans, P., and D. Zantinge (1996) *Data Mining*, Harlow: Addison-Wesley-Longman.

Blanchette, J.F., and D.G. Johnson (1998) Data Retention and the Panopticon Society: The Social Benefits of Forgetfulness, in: L. Introna (ed.), *Computer Ethics: Philosophical Enquiry (CEPE 98). Proceedings of the Conference held at London School of Economics, 13-14 December 1998*, London: ACM SIG/London School of Economics, pp. 113-120.

Bunge, M. (1959) *Causality: The Place of the Causal Principle in Modern Science*. Cambridge, Mass.: Harvard University Press.

Castells, M. (1996) *The Information Age*, Cambridge, Mass.: Blackwell Publishers.

Custers, B.H.M. (2001) Data Mining and Group Profiling on the Internet, in: Anton Vedder (ed.), *Ethics and the Internet,* Antwerpen Groningen Oxford: Intersentia.

Denning, D.E. (1983) *Cryptography and Data Security,* Reading, Mass.: Addison-Wesley.

Fayyad, U.M., G. Piatetsky-Shapiro and P. Smyth (1996) From Data Mining to Knowledge Discovery: An Overview, in: U.M. Fayyad et al. (eds.), *Advances in Knowledge Discovery and Data Mining*, Menlo Park, Calif./Cambridge, Mass.: AAAI Press/MIT Press.

Frawley, W.J., G. Piatetsky-Shapiro and C.J. Matheus (1993) Knowledge Discovery in Databases: An Overview, in: G. Piatetsky-Shapiro and W.J. Frawley (eds.), *Knowledge Discovery in Databases*, Menlo Park, Calif./London: AAAI Press/MIT Press.

---

[32]Denning (1983).

Harvey, J. (1990) Stereotypes and Group Claims: Epistemological and Moral Issues, and Their Implications for Multi-Culturalism in Education, *Journal of Philosophy of Education*, 24, 1, 39-50.

Koops, B.J. (1999) *The Crypto Controversy: A Key Conflict in the Information Society*, The Hague: Kluwer Law International.

MacMahon, B. and T.F. Pugh (1970) *Epidemiology: Principles and Methods*, Boston: Little, Brown.

Tavani, H. (1999) Internet Privacy: Some Distinctions between Internet-Specific and Internet-Enhanced Privacy Concerns, in: A. D'Atri, A. Marturano, S. Rogerson, T. Ward Bynum (eds.), *Ethicomp 99, Proceedings of the 4th Ethicomp International Conference on the Social and Ethical Impacts of Information and Communication Technologies*, Roma: Centro di Ricerca sui Sistemi Informativi della Luiss Guido Carli (CeRCIL).

Vandenbroucke, J.P. and A. Hofman (1993) *Grondslagen der epidemiologie*, Utrecht: Wetenschappelijke Uitgeverij Bunge.

Van der Lubbe, J.C.A. (1997) *Basismethoden cryptografie*, Delft: Delftse Universitaire Pers.

Vedder, A.H. (1997) Privatization, Information Technology and Privacy: Reconsidering the Social Responsibilities of private organizations, in: G. Moore (ed.) *Business Ethics: Principles and Practice*, Sunderland: Business Education Publishers Ltd., pp.215-226.

Vedder, A.H. (2000) Medical Data, New Information Technologies and the Need for Normative Principles Other than Privacy Rules, in: M. Freeman (ed.), *Law and Medicine*, Oxford: Oxford University Press.

10. Environmental Genomics, Data Mining, and Informed Consent

Herman T. Tavani

What is environmental genomics? What is data-mining technology? Why does the use of this technology in environmental genomics studies put research subjects at increased risk for discrimination and stigmatization? Are the traditional informed-consent policies that are used in the collection of medical data also adequate in the case of genomics studies that employ computational techniques such as data mining? These and similar questions are examined in this chapter.

In Chapter 1 we examined some controversies surrounding the case of deCODE Genetics, a privately owned genomics company in Iceland. We saw that deCODE's practice of cross-referencing information in its genetic database with information it had acquired from Iceland's healthcare and genealogy databases raised ethical concerns, some of which challenge key objectives articulated in the ELSI (Ethical, Legal, and Social Implications) Research Program.[1] For example, many individuals who agreed to donate samples of their DNA to deCODE were not aware of how their personal genetic data would be cross-referenced with other kinds of data about them and then subsequently "mined." In this chapter, we question whether individuals can make "fully informed" decisions when consenting to have their personal genetic data used in research that employs computational tools such as data mining. In particular, we focus on challenges to the principle of informed consent that arise because of the use of data-mining technology in the context of environmental genomics research.

This essay, which draws from and expands upon material previously published in *Ethics and Information Technology* (**6**:15–28 2004), appears for the first time in *Ethics, Computing, and Genomics*. I am grateful to Kluwer Academic Publishers for permission to reproduce parts of the original text. Copyright © 2005 by Herman T. Tavani. Printed by permission.

[1] We examined the ELSI Research Program in Chapter 1 and we considered some ELSI objectives in conjunction with our analysis of the deCODE Genetics case. For example, we saw that the deCODE case raised ethical concerns affecting personal privacy, informed consent, and intellectual property.

1. The Environmental Genome Project: Challenges for Environmental Genomics Research

In Chapter 1 we noted that the Human Genome Project's initial goal of mapping the entire human genome by 2003 was successfully accomplished and that, as a result, the basic *structure* of the human genome has been revealed. We also noted that HGP has recently initiated a series of specialized projects, some of which aim at understanding the *functions* of genes. One of these initiatives is the Environmental Genome Project (EGP), which was recently launched by the National Institute for Environmental Health Science (NIEHS). Operating on the premise that most human diseases result from a combination of environmental exposures and genetic variation, EGP research focuses on understanding common sequence variations in "environmental-response genes" (Christiani, Sharp, Collman, and Suk, 2001). EGP researchers, in their attempt to understand how genetic variation affects response to environmental exposures, identify polymorphic[2] variation in genes that appear to play an important role in environmentally associated diseases (Sharp and Barrett, 2000).

Scientists working on EGP initiatives, many of whom refer to their specialized research field as *environmental genomics*,[3] hope that the information learned from this field will be instrumental in improving public health. For example, Sharp and Barrett point out that a better understanding of genetic influences on environmental response could lead to "more accurate estimates of disease risks," and they note that this information could also provide a "basis for disease prevention and early intervention programs directed at individuals and populations at increased risk." But the authors also point out that ethical, legal, and social issues are raised by environmental genomics research, arguing that the "most immediate" of these concerns has to do with protecting individual research participants. Sharp and Barrett note that, as in the case of genetics studies in general, environmental genomics studies present research subjects with various kinds of "psychosocial" risks.[4] They also note that environmental genomics research poses additional risks to protecting research subjects because some ethical concerns that can arise in this research area are difficult to anticipate and thus can also be difficult to convey to prospective participants.

---

[2]*Polymorphisms* are differences in DNA sequences that occur naturally in a population. Single nucleotide substitutions, insertions and deletions of nucleotides, and repetitive sequences (microsatellites) are examples of polymorphisms. A single nucleotide substitution is called a single nucleotide polymorphism or SNP.

[3]Note that the sense of *environmental genomics* used in this chapter pertains to EGP-related initiatives and studies involving human subjects and human populations vis-à-vis factors involving "environmental exposures." There is also another sense of "environmental genomics," however, which is concerned with genomic structures of (non-human) environmental organisms. Environmental genomics studies of this type isolate and analyze genomic fragments directly from the environment. See, for example, the Genomics: GTL (genomics-to-life) and the Microbial Genome Projects, both of which are sponsored by the U.S. Department of Energy.

[4]In particular, Sharp and Barrett identify three kinds of risks associated with environmental genomics: possible discrimination, disrupted relationships between family members, and adverse affects on a participant's self image.

Why, exactly, are certain kinds of risks to research subjects participating in environmental genomics studies so difficult to foresee and to communicate to prospective participants? According to Sharp and Barrett, "uncertainties" involving the very nature of environmental health research[5] complicates the process of informed consent. For example, they note that a particular genomics study may identify a genetic polymorphism that *appears* to play a role in environmental response. However, the authors point out that the extent to which the polymorphism's effects are "mediated by environmental factors" is often unclear. This is further compounded by what they refer to as the problem of the "breadth of the consent" needed from research subjects to conduct environmental genomics studies. Sharp and Barrett point out that as more genes and exposures are considered simultaneously in a given study, it becomes increasingly difficult to anticipate the potential risks and benefits that can result. For example, the authors note that associations between individual alleles[6] and particular environmental exposure are difficult to identify. As a result, researchers are inclined to design studies that concurrently look at "possible associations" between "many different allelic variants and many different exposures." On the one hand, such studies increase the likelihood of identifying "functionally significant polymorphisms." On the other hand, these studies further complicate the consent process for research subjects. So Sharp and Barrett conclude that it is difficult to ensure that research subjects are "fully informed" about the potential risks and benefits of their participation in these studies.[7]

Advocates for research subjects, including Institutional Review Boards (IRBs), worry that without strict oversight of the process currently used to gain the consent of research subjects, the consent process could become a form of "blanket permission" for genomics researchers. Alternatively, however, some environmental genomics researchers are concerned that revisions to policies governing informed consent could become so stringent that they would place "inappropriate restrictions" on their research (Christiani et al., 2001). Sharp and Barrett note that although the present standards for informed consent in genetic research may be appropriate for studies of "highly predictive alleles," these standards may also be overly demanding for studies involving "genetic hypersensitivities to environmental exposures," which generally result in "more limited risks" to individual participants. Thus a key challenge facing environmental genomics research is to establish consent procedures that take into account distinctions of this kind.[8]

---

[5]For example, Sharp and Barrett point out that uncertainties "surrounding the study of genetic hypersensitivities to environmental exposures" make it difficult to identify some of the risks affecting research subjects.

[6]*Alleles* are alternative forms of genes that occupy a corresponding position on each of our two chromosomes, one of which is inherited from each parent.

[7]Because of these kinds of difficulties, Christiani, Sharp, Collman, and Suk (2001, p. 531) recommend that for EGP research, the ELSI framework be extended to include a component that examines "possible" (as well as actual) ethical and social implications. We consider their proposal in the final section of this chapter.

[8]Christiani et al. (p. 528) make a similar suggestion, arguing that a distinction between what they refer to as "'susceptibility genes' (with low-to-moderate penetrance) and relevant 'disease genes' (with high penetrance)" be taken into account in framing "appropriate" informed-consent policies for research subjects.

Another important moral concern that affects environmental genomics research, and which further complicates the informed-consent process, arises because of certain kinds of risks that are posed to *groups*. Because environmental genomics studies typically involve large populations, entire groups and subgroups (of persons) are now vulnerable to discrimination and stigmatism. This point has not gone unnoticed by some environmental genomics researchers, who show why certain groups are especially vulnerable.[9] However, these researchers also tend to limit their discussion of concerns about group-level risks and vulnerabilities to *socially identifiable groups* such as racial minorities.[10] Although these groups are clearly at risk, so too are other groups—e. g., those which may not yet have been identified *as groups* but will conceivably emerge because of the use of computational techniques in environmental genomics research. While authors such as Sharp and Barrett correctly describe some risks for groups that have already been "socially identified," they make no mention of the potential risks to groups that have not yet been identified but that will conceivably emerge as a result of computational techniques used in environmental genomics research. Because these groups also are vulnerable, it would seem that the sphere of the kinds of groups that deserve ethical consideration needs to be expanded.

In the remaining sections of this chapter, our focus on risks to research subjects *as members of groups* will be from the perspective of the controversial "new" groups that can be constructed via a computational technique called data mining. We will see why the construction of these kinds of groups poses some special challenges for environmental genomics research, which go beyond the concerns described by Sharp and Barrett and by Christiani et al. In particular, we will see why these challenges, made possible by data-mining tools, further complicate the informed-consent process required for research subjects participating in genomics studies involving groups and large populations.

2. Data Mining and the Knowledge Discovery Process

The expressions *data mining* and *knowledge discovery in databases (KDD)* are often used interchangeably, even though they refer to different processes. Whereas KDD is the overall process of discovering useful knowledge from data, data mining is a particular step in that process. In its most precise sense, data mining is the application of specific algorithms to "discover" and extract patterns or correlations in data (Tavani, 1999). In this chapter, however, "data mining" is used in a broader sense to include both pattern discovery and data analysis/interpretation, even though the analysis and interpretation of data are, technically, a part of the overall KDD process. We limit our discussion of data mining to applications involving genomic and

---

[9] See both Sharp and Barrett (2000) and Christiani et al. (2001) for descriptions of how these groups are vulnerable.

[10] See, for example, Sharp and Barrett.

genetic information,[11] and our primary concern is with understanding how the use of data-mining technology challenges the principle of informed consent for research subjects in environmental genomics studies. First, however, it is useful to comment on some important contributions that data mining has made to achieving the initial goals and objectives established for the Human Genome Project (HGP). Without data-mining technology, the mapping of the entire human genome would not likely have been completed by 2003, the target date set by HGP's planners. For example, a data-mining technique referred to as the "shotgun method" contributed significantly to the first phase of HGP, where the structure of the human genome was revealed.[12] Data mining tools have also been critical in subsequent phases of HGP, including environmental genomics research, where the emphasis has been on showing functional relationships between genes and diseases.

With the advent of data-mining technology, the nature of discovery has been radically altered in the sense that no hypothesis for a relation between a gene and a certain disease is necessary. Instead, computerized data-mining techniques are used to find correlated genes and gene variations, simply by comparing databases of genetic samples and disease records. Bart Custers (2001) notes that in the past, epidemiological researchers tried to find patterns and relationships in data by forming hypotheses. This was often accomplished through observation, such as when physicians observed certain physical characteristics of individuals they happened to be treating for certain conditions or diseases. For example, Custers notes that a doctor might observe that a patient she is treating for thrombosis has large hands. Noticing this relation in subsequent patients, the doctor might form a hypothesis about a relationship between thrombosis and large hands. A further hypothesis might suggest a linkage to a correlated gene or to a genetic polymorphism. Custers points out that whereas observation and experimentation usually consider only causal relationships, data mining investigates a much wider range of possible relations (including classifications and correlations). Not only are many more patterns and relations revealed in data mining, but the newly disclosed patterns and relations can be recognized more efficiently than in traditional methods where observation and experimentation were primarily used.

The same technology used to discover the patterns and relations that aid geneticists and environmental genomics researchers, however, can also be used to generate group profiles.[13] In certain cases, these profiles can suggest the existence of

---

[11]Anderson and Goodman (2002, p. 114) believe that the future of genomics will depend on the "mining" of data in very large databases and "data warehouses."

[12]See Marturano (2003) for a description of how this technique was used by Celera Genomics to accelerate work on HGP. Marturano's essay is included as Chapter 13 of this book.

[13]It is worth noting that the process of constructing group profiles is not peculiar to data mining. For example, at least some group profiling has been done on the basis of observation. However, Custers notes that group profiling through data mining is very different from group profiling through observation for at least three reasons: (1) the sheer amount (i.e., scale) of profiling that can be accomplished by data mining is substantial; (2) non-causal as well as causal correlations can be more easily revealed; and (3) "trivial" information is much more easily linked with sensitive information. Custers' essay is included as Chapter 9 of this book.

new groups or categories for individuals, some of which also can be considered controversial because of ethical concerns that they raise. Why, exactly, is the discovery/construction[14] of new groups based on group profiles problematic from an ethical perspective? For one thing, individuals who eventually become identified or associated with the newly constructed groups may have no knowledge that the groups to which they can be assigned actually exist. More importantly, however, these individuals can, as a result of their identification with some new groups generated by data mining, be placed at greater risk for discrimination and stigmatization. For the most part, this is due to the fact that personal data at the group level has virtually no legal or normative protection.

2.1 Group Profiles and Personal Data

Groups, or categories of individuals, generated by data-mining technologies are based on the discovery of one or more *profiles* discernable in the data. A *group profile* is a set of properties or characteristics that serve as the basis for identifying an individual with a particular group. Some profiles used in the construction of groups are commonly known and easily identifiable. For example, an individual might be identified as belonging to multiple groups—i.e., the group of males, the group of college professors, the group of homeowners, and so forth. For the most part, people who are assigned to these kinds of groups would not be surprised to learn that they have been identified with such groups. However, group profiles discovered/generated by the use of data-mining tools can also suggest "new facts" about individuals. For example, Custers (2001) notes that a data-mining application might reveal the existence of a group of people who both drive red cars and have colon cancer. The owner of a red car may be very surprised indeed to learn (the "fact") that he belongs to a group of individuals likely to have or to develop colon cancer merely because of a seemingly arbitrary statistical correlation.

Various kinds of group profiles can be derived from personal data after that data has been aggregated. But as Vedder (2004) points out, once an individual's data has become aggregated and generalized, the individual has no say in how that data is further processed because normative requirements for protecting personal data apply only to that data *qua* individual persons.[15] Whereas an individual has a legal right to access and rectify one's personal data as it applies strictly to that individual, one does not enjoy the same rights with respect to personal information that is *derived from* that data. Vedder also notes that information in group profiles is often used *as if* it were personal data (even though it is not), and he points out that the impact on individual persons resulting from the use of the derived (e.g., aggregat-

---

[14] I leave open the metaphysical question of whether these groups are actually *discovered* (and thus already extant in a way that metaphysical realists would assert) or whether they are essentially *constructed* and thus genuinely "new" (in the sense that metaphysical nominalists would suggest). Whereas the patterns in the data can accurately be said to be "discovered" in the data-mining process, the ontological status of the "new" groups revealed as a result of those patterns is less clear.

[15] Vedder (2004) notes that because the kind of personal information contained in many group profiles falls outside the sphere of protection granted to *personal data*, it is not protected by current privacy laws. He argues that because of this, our privacy laws are based on too narrow a conception of "personal data" and thus are inadequate. For example, he notes that personal data is commonly defined and understood as data and information relating to an "identified or identifiable person."

ed) data can sometimes be more severe than in cases involving the use of "real personal data" about that individual. Consider, for example, information about an individual's having a certain probability of manifesting disease X merely because of an association with his or her group profile Y. The use of the information in the group profile, which in this case has been derived from personal data rather than based on the personal data itself, could result in that individual being the victim of stigmatization and discrimination. Vedder believes that the increased production and use of group profiles, especially for groups comprised of data based on what he calls "nondistributional profiles," may adversely affect some individuals.

2.2 Distributional and Nondistributional Group Profiles

We noted in the preceding section that a group profile is defined as a collection of properties pertaining to individuals, and that profiles can serve as the basis for identifying a certain individual with a particular group. We also noted that these profiles can contain information that is already known to many of the individuals identified with (or possibly "assigned to") certain groups. Such is the case for an individual who has been identified with a group based on information in his or her profile having to do with gender, occupation, residential address, and so forth. But some group profiles can contain information about individuals that is not so obvious to them, as in the case of Custers' example of an individual assigned to the group of people who both drive red cars and have colon cancer. Groups of this kind can be constructed on the basis of one or more *nondistributional profiles*. To understand what is meant by this expression, it is useful to contrast it with the notion of "distributional profiles," which employ a concept (used in logic and elsewhere) known as *distribution*.[16]

Vedder (2004) notes that distributional profiles "assign certain properties to a group of persons in such a way that these properties are always and unconditionally manifested by all of the members of that group." For example, every member of the group of female medical workers has both the property of being a female and the property of being a medical worker. Also, each person in this group knows (i.e., possesses self knowledge) that she is a female and a medical worker. Nondistributional group profiles, on the other hand, are constructed in terms of probabilities and averages. They are also based on comparisons of members of the group with each other and/or on the comparison of one particular group with other groups (Vedder, p. 465). Consider, for example, a scenario in which many individuals employed in a certain occupational category, say lab technicians, happen to have a higher-than-average rate of mortgage foreclosures on their homes.

---

[16]Logicians describe a term, *X*, as being *distributed* when we know something about every member of the class designated by the term *X* relative to some other class. So in the case of the proposition "All *X* are *Y*," for example, we know something about all of the members of *X* relative to *Y* (viz., that every member of *X* is also a member of *Y*). Hence we say that *X* is distributed. However, we do not have the same kind of information about the members of *Y* relative to *X*. So in this case, *Y* is undistributed. We should also point out that, when used in the context of statistics and statistical analysis, the term "distribution" often signifies certain relationships involving means, medians, modes, and averages. Vedder seems to have incorporated both senses of "distribution" in his discussion of distributional and nondistributional profiles.

Suppose that a data-mining algorithm discovers this pattern, which in turn generates a new category—viz., an occupational group whose members are likely to lose their homes because of mortgage foreclosures.

Why is the practice of constructing groups on the basis of nondistributional profiles so controversial? According to Vedder, the properties or characteristics used to construct nondistributional profiles can apply to individuals only in so far as they have been identified with the reference group; apart from that group association, these properties have no relevance for the individuals. Also consider that the individuals themselves need not, in reality, exhibit those properties. For example, not every member of the hypothetical new category described above—viz., lab workers likely to default on their home mortgages—necessarily exhibits the property of being a risk for a mortgage loan. As noted in the preceding paragraph, a person may be "discovered" via data mining to be a member of some newly constructed group merely by virtue of an association that she simply *happens to have* with a particular occupation category. Yet, an individual in that group who has been turned down for a mortgage loan because of data based merely on his or her identification with a nondistributional profile may, in fact, have an impeccable credit history. Vedder points out that in these kinds of cases, an individual is being judged and treated on the basis of his or her belonging to the "wrong category of persons," which is made possible because of some statistical correlation that would seem arbitrary. Contrast this with a judgment made about an individual based on a distributional group profile involving female medical workers, where every member of the group is necessarily both a female and a medical worker. Not only are both properties exhibited in *every* member of the group, but each member also knows from (what Vedder calls) an "internal perspective," i.e., from the perspective of her self knowledge, that she possesses the characteristics that apply to the group.

Whereas a member of the occupational group of lab technicians (who is considered a credit risk simply by virtue of her association with that group) will likely know—i.e., have knowledge from an internal perspective—that she is a member of the group of lab technicians, she may have no idea that she is also a member of the group of persons who has a good chance of being turned down for a home mortgage loan merely because of her association with a particular occupational group. From (what Vedder calls) the "external perspective" of others, i.e., from the perspective of the data aggregators who produce and use this kind of group profile, particular associations and inferences about her (as the data subject) have been drawn. To the data subjects themselves, however, such classifications will likely seem to be based on criteria that are arbitrary in the sense that they are non-essential or non-causally linked and thus have no basis in reality.[17] Furthermore, this kind of information about individuals, which can be used to make decisions about them, can easily remain hidden from those individuals (Vedder, p. 466).[18]

---

[17]Vedder notes that the use of these profiles may have a serious impact not only on persons from whom the data was originally taken, but also on those to whom the profiles are eventually applied.

[18]This is because the individuals who comprise the groups based solely on nondistributional profiles can, as Vedder points out, only be identified by those who have defined such groups for one or more specific purposes.

As noted earlier, controversies involving data mining and the use of profiles are not peculiar to genomic research.[19] We should also note that the process of constructing and using nondistributional group profiles is not unique to computerized techniques such as data mining. For example, insurance companies use actuarial tables to determine which socio-demographic groups have a higher incidence of risk when developing criteria for eligibility for insurance coverage. It is well known that in the U.S., teenage boys are placed in a high-risk category for automobile insurance because of the statistical rate of automobile accidents involving male teenage drivers. However, this is a fairly well known correlation; and motorists in the U.S. have generally come to accept insurance premiums based on this particular correlation. More importantly, however, this practice used by insurance companies to determine premiums for young male drivers is *transparent* and thus open to public debate. But many recent correlations discovered by data mining, based on hitherto unnoticed (and oftentimes nonobvious) relationships between characteristics and features of persons, are not transparent. To realize how easily *nontransparent* correlations can be discovered by data-mining tools, recall Custers' hypothetical example of a statistical correlation between people owning red cars and having colon cancer.

Consider that in the future, a data-mining application used by an insurance company might reveal a statistical correlation involving a high incidence of breast cancer among women who read nineteenth-century novels. Thus, a person could conceivably be refused health insurance or life insurance based solely on her association with some non-essential or non-causally-related information[20] that, arguably, is unfair—e.g., the kinds of books one reads (or in the case of Custers' example, the color of one's automobile). Also consider that insurance companies and other organizations could easily conceal the fact that they use certain pieces of information, some of which appear to be altogether irrelevant in making decisions about individuals.[21] So it would seem that the use of data-mining technology, in so far as it facilitates the construction of new groups based on statistical correlations involving nondistributional profiles that unfairly affect some individuals, poses a serious threat to data subjects. And this threat can be especially worrying to research subjects participating in medical and health-related studies in which their genetic data is used.[22]

---

[19] However, the harm resulting to individuals based on inferences made about their genetic predispositions as correlated with non-genetic information in their group profiles can be far more significant than that resulting from inferences based on profiles that include only non-genetic information.

[20] Custers (2001) describes these kinds of associations as ones that use "trivial information."

[21] Vedder points out that insurance companies would be able to select candidates without having to check their health conditions. This practice may even help insurance companies weed out patients in countries where they are forbidden to select on the basis of an applicant's actual health condition.

[22] Because of these kinds of threats, Goodman (1996, p. 226) describes the need for a category of protection that he describes as "collective confidentiality." Along somewhat similar lines, Vedder (2004) believes that we need a new category of privacy protection, which he calls "categorial privacy." Vedder's arguments for "categorial privacy," as well as his distinctions between individual privacy and collective privacy, are briefly examined in the final section of this chapter and are examined in more detail in Tavani (2003).

3. Informed Consent in the Context of Genomics Research

Much has been written about controversies surrounding informed-consent practices, especially as they pertain to human subjects participating in medical research. While there is no need to repeat those discussions here, a few brief remarks about the principle of informed consent would be useful to establish a context for our discussion of ethical concerns that arise because of the consent practices currently used in genomic research. Some authors, such as Sheri Alpert (1998), believe that we need to specify the requirements for "valid" informed consent. Alpert (p. 87) argues that for a person's consent to be valid in the collection of medical data, two conditions are required: (1) individuals must "know and understand the nature of the information being released," and (2) consenting individuals must be made aware of the party or parties to whom the information about them can be released. Conditions similar to those specified by Alpert can be found in the 1993 Office of Technology Assessment (OTA) Report, entitled *Protecting Privacy in Computerized Medical Information*. This report states that (valid) informed consent as it pertains to medical records is possible only when individuals are "familiar with the data contained in their records, so that they understand what they are consenting to disclose." According to the OTA Report (1993, p. 70), patients must: (1) have adequate disclosure of information about the data dissemination process, and (2) be able to fully comprehend what they are being told about the procedure or treatment. Applying OTA's criteria to genomic research, then, both "adequate disclosure" of information to research subjects and "full comprehension" of the process on the part of research subjects are required for valid informed consent.[23]

Why is the principle of informed consent important from an ethical point of view? According to O'Neill (2002), informed consent is "ethically important" because it provides a "tough safeguard" through which individuals can protect themselves against both *coercion* and *deception*. So, in O'Neill's scheme, an ethically adequate system of informed consent requires that an individual's consent be both non-coercive and non-deceptive. That is, the individual's consent must be *voluntary*[24] and the consent process must be *transparent*. O'Neill questions whether these conditions can be satisfied in current genetic testing/research practices. We consider O'Neill's question in detail in a later section of this chapter, where we examine the issue of "opacity" or non-transparency in practices involving informed consent and genomic research. Our purpose in this section has been to identify some requirements essential for "valid informed consent," which we will later apply to issues affecting genomic research. Before doing that, however, we first contrast the principle of informed consent with the notion of "presumed consent," and we contrast "individual consent" with "community consent."

---

[23] OTA also specifies that the patient must be "competent" to consent to the procedure. However, we will not examine this criterion for valid informed consent in this chapter.

[24] The OTA Report also states that the act of consent must be voluntary for the consent process to be valid.

3.1 Informed, Presumed, and Community Consent

Presumed consent, unlike informed consent, places the onus on the research subject to opt out of having his or her personal data collected and used in specific ways. In this scheme, the default practice is that medical information is collected and freely used by researchers in a medical context, unless the subject specifically requests otherwise. Stefansson (2001, p. 53) argues that traditional epidemiological studies have depended on a system of presumed consent. He defends a policy of presumed consent in the collection of medical information on the grounds that it improves the quality of healthcare.

Stefansson also notes that in Iceland, citizens can opt out of having their medical information included in that nation's healthcare database. However, Stefansson's critics point out that individuals who do not explicitly opt out—including people who cannot elect to do so, such as deceased persons, newborns, and the mentally ill—are presumed to have consented to having their data used in whichever ways the researchers decide. But Stefansson (p. 53) responds to these critics by noting that he advocates for a policy of presumed consent only in the collection of medical information that does not involve DNA. He argues that if informed consent had been required for the collection of ordinary medical information in the past, we would have neither the quantity nor the quality of health care information we have today. He also suggests that overly stringent informed-consent policies can diminish the probability that our children will have the same quality of health care in the future that we have enjoyed thus far (Stefansson, pp. 53–54).

Stefansson believes that an additional factor, which he calls *community consent*, must also be taken into consideration in the case of medical research, including the collection of genetic information. He argues that because the use of genetic information has familial and societal implications, we also need to consider those factors when framing an appropriate consent policy for collecting this kind of information.[25] Stefansson (p. 53) claims that if you conduct a major study that involves a large number of people, or perhaps even an entire population (as in the case of Iceland), the notion of community consent is very important. However, he also believes that a policy of *informed* consent should be required in cases where people are asked to contribute DNA samples.[26]

However, even when informed consent is used in the collection of genetic information, protecting the confidentiality of research subjects is still problematic for several reasons. For one thing, we saw how easily that information can be cross-referenced with medical and healthcare data. Moreover, that cross-referenced information can be "mined" for certain patterns. Also, there is an additional worry that is independent of problems associated with cross-referencing and mining

---

[25]In Chapter 1, we discussed some implications that genetic research and testing can have for one's family members, as opposed to individuals themselves, in our examination of the thesis of genetic exceptionalism.

[26]Anderson and Goodman (2002, p. 116) make a similar point when they argue that in the area of genetics, the "weight of informed consent seems greater" than for other kinds of medical information.

personal genetic data. This has to do with the fact that information about a person's genetic data is typically *stored in computer databases*, which, by their very nature, pose a certain kind of challenge for the consent process. We briefly examine this particular challenge.[27]

3.2 The Problem of Protecting Personal Information Stored in Computer Databases: Implications for Informed Consent

What have research subjects consented to when they agree to provide DNA samples that will be stored in computer databases? Have they forfeited all rights to their personal genetic data, including the right to have some say about how that information is used in the future, once that information has been included in a computer database? Consider once again the case of deCODE Genetics (examined in Chapter 1) and its database of DNA records involving 70,000 Icelandic citizens. Suppose that deCODE's genetic database is either sold to or otherwise acquired by some company or organization. Would the confidentiality of the research subjects who contributed their DNA to deCODE still be protected? If not, what does this imply for the principle of informed consent as a meaningful procedure, and particularly as a procedure used to acquire personal genetic information that will be stored in private computerized databases?

Perhaps an analogy will help. Let us draw on some controversies surrounding the status of an electronic database in the commercial sector that was owned and operated by Toysmart.com—a case that does not involve personal genetic data. Customers who dealt with Toysmart were given assurances that any personal information they gave to the online company via electronic transactions would be protected by the company's privacy policy. So "informed" consumers entered into an agreement with Toysmart based on their understanding of the company's privacy policy, viz., that any personal information customers provided during transactions would not be exchanged with or sold to a third party. When Toysmart filed for bankruptcy in the summer of 2000, however, it was required by law to list all of its assets. Among those assets was its database containing customer information, the specific contents of which were presumably protected under Toysmart's privacy policy. However, some of Toysmart's creditors, as well as some companies that were interested in assuming Toysmart's business, indicated that they did not believe that they were bound by Toysmart's privacy policy because of one or both of the following reasons: (1) the data in that database involved a prior contract to which those companies were not signatories, and (2) the database and its contents would now be the sole property of the new owner.

Recall that people who consented to give DNA samples to deCODE did so with the understanding that their personal data would be protected. But if deCODE Genetics were to be sold to, or perhaps acquired (via a "hostile takeover") by, a different company, would the privacy of those who voluntarily provided their DNA samples to deCODE still be protected? The answer to this question is not entirely

---

[27]Although confidentiality issues related to the *storage* of personal information in databases are generally different from confidentiality concerns pertaining to the *mining* of data in those databases, both kinds of concerns raise problems for informed consent in genomic research. Thus we examine both concerns in this chapter.

clear because of the role that the Icelandic government could play in such a case. But we can well imagine a scenario in which an individual's genetic data that is given under one set of conditions for use in a specific context eventually ends up in the hands of some organization that had not been authorized access to that data.[28]

3.3 Informed Consent, Secondary Uses of Information, and the Problem of Opacity

When individuals consent to have their genetic data collected and used by genomic researchers, they agree to have that data used for a specific purpose (e.g., in the context of some particular study), based on assurances that their confidentiality will be protected. But in our discussion of data mining, we saw that when personal data becomes aggregated, it is not clear how the confidentiality of one's personal data can still be protected. Because an individual's data given for use in one context can be aggregated and then used subsequently in secondary applications, we can reasonably ask whether the kind of consent initially granted by that individual was genuinely "informed"—i.e., in the sense of "valid informed consent" described earlier. Recall that for a research subject's consent to qualify as "valid," she must have "adequate disclosure" about how her data will be disseminated and used.

Despite the problems we have examined thus far, many researchers seem to assume that the *principle* of informed consent can still be preserved. But how is it possible to have a system of "valid" informed consent in genomic research when the research subject cannot, in principle, be told how his or her data, once aggregated, might be used in subsequent applications? O'Neill (2002, p. 44) claims that although people speak about the ideal of valid consent, or what she calls "fully informed consent," the full disclosure of information for research subjects is neither definable nor achievable. Part of the reason for this, she argues, has to do with the very nature of the act of consenting, which O'Neill claims is "propositional." It is propositional, she further argues, because consenting is a "cognitive attitude." Like other cognitive attitudes (such as knowing, believing, desiring, or trusting), consenting takes a proposition as its object. O'Neill (p. 43) states:

> ... *informed consent is always given to one or another* description *of a proposal for treatment* ... *it has as its object not a procedure or treatment, but rather one or another proposition containing a description of the intended procedure or treatment. (emphasis O'Neill)*

In effect, then, when a patient or subject gives her consent, she is consenting to a proposition of the form: "I agree to X" (where X is a description of a proposal for treatment). O'Neill points out that in consenting to a proposition, the consenter "may see no further" than the specific descriptions contained in the proposition itself.

---

[28]Among the issues also raised in this scenario are questions about who should have ownership rights to this genetic data and about how long those property rights should apply. We examine issues involving property rights in the context of genetic and genomic data in the chapters included in Section IV of this book.

For example, the consenter may not see that X is either equivalent to Y or that it entails Y. O'Neill states:

> *I might well not be aware of, a fortiori not consent to, the standard and foreseeable consequences of that to which I consent . . . I might consent to a medical procedure described in euphemistic and unthreatening ways, yet not see myself consenting to another more forthright and equivalent description of that treatment . . . I might consent to chemotherapy, and yet when as a result I feel desperately ill and weak may truthfully claim that I never consented to anything that would have this effect—even if these very effects were described among the normal effects of the treatment. I may consent to transplant surgery or to an amputation, and be given extensive information, and yet later feel that I never consented to what has actually been done to me. (pp. 43–44, emphasis O'Neill)*

Thus O'Neill concludes that consent is a much more *opaque* than transparent process.[29] Her claim that consenting is also a cognitive attitude (and thus takes a proposition for its object) is a controversial one, which we will not further examine here. However, O'Neill's remarks about the opacity of consent hold independently of her view that consenting is a cognitive attitude, and thus her remarks about the former can be examined separately from her views about the latter. O'Neill's examples illustrate how easy it is for someone to consent to a particular medical procedure and yet still claim not to have consented to procedures that either are roughly equivalent to, or entailed by, the original consent agreement.

To see how O'Neill's model might be applied to our concern about research subjects in the context of genomics research, consider the following questions. When an individual has consented to give DNA samples for use in one context, does it follow that she has also consented that her DNA can be used in subsequent contexts? For example, does it follow that she has consented that information about her DNA can be used to construct nondistributional group profiles comprised of information that is both sensitive and trivial, in ways that have no causal base in reality? And does it follow that she has also consented to having this information cross-referenced with other kinds of information, which could result in adverse judgments being made about her (that put her at significant risk for discrimination or stigmatization)? We have already seen how data mining can be used to construct new groups via data derived from contexts in which that data was initially given for very different purposes. Thus it would seem that the process of informed consent has become one that, as O'Neill correctly suggests, is far more opaque than transparent. It would also seem that the kind of conditions required for "valid" or

---

[29]O'Neill appeals to the notion of "referential opacity" described by Quine (1953) in her argument to show the nontransparency or "opacity" of informed consent.

"fully informed" consent are extremely difficult, if not impossible, to achieve in cases of genomics studies involving the use of data-mining technology.

4. Responding to the Challenges

Can the informed-consent policies used in genomics research be remedied to meet the challenges posed by data mining? Specifically, can they be revamped in ways that will make the consent process less opaque to research subjects? Is such a goal achievable in light of the potential production and use of nondistributional group profiles generated by data mining? If not, what implications could this have for the future of genomics research? One suggestion for addressing problems associated with group-level protection schemes—i.e., protecting confidentiality at the group or "collective" level—has been put forth by Kenneth Goodman. Goodman describes a category of protection that he refers to sometimes as "group confidentiality" (1998, p.18) and other times as "collective confidentiality" (1996, p. 228).

Noting that it can be difficult to articulate and frame a policy for group-level protection, Goodman considers some conceptual and normative questions that first need to be addressed. For example, he asks whether groups per se have a "right to confidentiality." Assuming that they do, Goodman wonders what such a purported right would amount to—i.e., what it could mean. Additionally, Goodman wonders whether explicit normative policies and laws aimed at protecting group confidentiality would necessarily benefit vulnerable groups and subgroups. He also considers the possibility that, ironically perhaps, these policies might make it even more difficult for members in those groups to avoid certain kinds of group/subgroup stigma. So it would seem that Goodman is correct in pointing out there are some important conceptual and normative questions that need to be answered before we can frame and implement a policy aimed at protecting the confidentiality of groups. But Goodman does not provide us with explicit details for how such a policy would be framed, let alone implemented.

Anton Vedder (2004), who recently introduced a conceptual scheme that explicitly addresses the problem of group confidentiality, proposes a new level of privacy protection that he calls *categorial privacy*. He contrasts this notion of privacy with both "individual privacy" and "collective privacy." In Vedder's scheme, categorial privacy is more closely connected to the former than to the latter kind of privacy. Categorial privacy, like individual privacy, aims at protecting the individuals (in this case, *individuals who are identified as belonging to groups*) rather than protecting groups per se. In this sense, categorial privacy protection would seem to be ideal for protecting individual research subjects who may eventually be assigned to groups that are not yet known or that have not yet been identified.

Vedder also points out that categorial privacy differs from individual privacy in at least one important respect, which makes the former similar to collective privacy. Like collective privacy, categorial privacy takes into consideration what Vedder describes as the "attribution of generalized properties" to members of *groups* being protected. However, collective privacy can only protect groups that are already known—e.g., groups constituted by "typical" distributional profiles. But since groups

Table 1 Three Modes of Privacy Protection Based on the Vedder Model

Individual Privacy	Categorial Privacy	Collective Privacy
Applies to personal data, but *not* after that data has been aggregated.	Applies to personal data in the aggregate, even after it has been abstracted and processed.	Applies to data about the group (of individuals, but not to personal data per se).
Protects individuals *as* personally identifiable individuals, but cannot be extended to protect individuals *as* members of groups.	Protects individuals *in* groups, even when individuals do not know that they have been assigned to a certain group–e.g., a group determined by one or more nondistributional profiles.	Protects groups per se, but only groups that are already known to exist. Does not protect individuals in groups generated from nondistributional profiles.

can also be constructed on the basis of nondistributional profiles, as we saw in our discussion of data mining, the kind of privacy protection needed for people assigned to those groups is more closely tied to the level of individuals, or to what Vedder refers to as "individually identifiable criteria." Hence, Vedder believes that his notion of categorial privacy offers the appropriate level of protection needed for individuals *as members of groups.*

Table 1 illustrates some essential characteristics that differentiate Vedder's three categories of privacy.

Vedder's distinctions involving the three categories of privacy protection are insightful, and they are useful as a conceptual scheme for sorting out the kinds of protection needed at the group as well as at the individual levels. But it is not clear from Vedder's remarks just how his notion of categorial privacy would be implemented in practical ways that would better ensure the confidentiality of research subjects who happened to get assigned to groups based on nondistributional profiles. So, like Goodman's notion of collective confidentiality, Vedder's concept of categorial privacy would seem to be more useful as scheme for identifying and articulating many of the problems involving group confidentiality (inherent in current informed-consent practices) than in resolving these problems. But Vedder does provide us with some important distinctions that can, and arguably should, inform future proposals for informed-consent policies involving genomics research.

Despite the suggestions made by Vedder and Goodman, both of whom do an excellent job of isolating and identifying the kinds of issues that arise in trying to ensure group confidentiality, a solution to the problem they describe does not appear to be imminent. However, if we do not resolve the problematic issues underlying consent and confidentiality concerns for groups (of individuals), which are exacerbated by computerized techniques such as data mining, the future of genomics research involving human populations may be at risk. As we saw in the first section of this chapter, a key challenge facing environmental genomics research is to establish consent procedures that allow individuals to make fully informed choices about their participation in studies. Sharp and Barrett (2000) have suggested that the permissions granted by participants in environmental genomics studies need to be "broad enough to permit diverse research interests, yet specific enough to allow individual participants to assess the possible risks and benefits of their participation." But we have seen that this ideal is challenged by the kinds of concerns generated by the use of computational techniques, such as data mining, in the research process.

At this point, we can only make some very broad or general recommendations about how to proceed. One promising first step would be to expand the existing ELSI requirements[30] for environmental genomics research by including a separate ELSI component or category that specifically acknowledges and addresses problems concerning group confidentiality that *potentially* arise. For example, Christiani et al. (2001, p. 531) have suggested an ELSI category along these lines, which they refer to as "*possible* ethical and social implications."[31] Perhaps this ELSI category, or one similar to it, could explicitly identify the kinds of concerns we have examined in this chapter, thereby directly communicating to the genomics research community some serious group-related risks that prospective research subjects need to understand. In particular, a distinct ELSI "focus area," dedicated to examining risks that are made possible because of new groups constructed from nondistributional profiles, via data-mining tools and analyses, could be established.

Of course, an expanded ELSI framework also needs to address problems involving risks for groups that have already been "socially identified," such as racial and minority groups. We have seen that groups of this type are more easily identified because they are typically based on information involving distributional profiles. In response to concerns involving some of these kinds of groups, Collins et al. (2003) have argued that representatives from ethnic and racial groups, who may be at increased risk of discrimination and stigmatization, need to be included in future discussions regarding ELSI requirements for genomics research projects involving human populations.

Finally, we recommend that three constituencies in the genomics research community—viz., scientists/researchers, Institutional Review Board (IRB) representatives, and ELSI (Program) representatives—work together to draft more robust informed-consent policies for research subjects, which respond to challenges posed by data mining and other computational tools used in genomics research. In this way, the genomic research community can at least begin to engage in a dialog that responds to the kinds of challenges identified and examined in this chapter.

---

[30] In the first section of this chapter, we briefly identified the ELSI (Ethical, Legal, and Social Implications) Research Program. For more information about this program, see Chapter 1.

[31] (Italics added). Their proposal is described in detail in the paper by Christiani et al., which is included as Chapter 5 in this volume. However, Christiani et al. do not seem to have in mind "newly constructed groups" or categories of individuals. Instead, they describe some potential problems for vulnerable groups that have already been socially identified.

5. Concluding Remarks

We have examined some challenges posed to the informed-consent process currently used in environmental genomics research. In particular, we saw how the use of data-mining technology poses some serious threats to the confidentiality of research subjects, both as individuals and as members of groups. This is especially apparent in cases where nondistributional profiles, derived from personal data that is aggregated from both sensitive information (related to DNA) and "trivial" (or non-essential) information, can be used to construct controversial new groups that are based merely on arbitrary statistical correlations. We also examined some ways in which the principle of informed consent is seriously challenged because of the problem of "opacity" in the consent process, which is exacerbated by the use of data-mining tools. And we saw that because of this challenge, the ideals of "valid" and "fully informed" consent are now difficult to achieve. To address these problems, we recommend that the ELSI framework be expanded and that various stakeholder groups in the genomics research community work together to frame a more a robust consent policy that takes into account group-related risks to research subjects who are needed to participate in future environmental genomics research.

Acknowledgments

I am grateful to Ann Backus, Lloyd Carr, Bart Custers, Frances Grodzinsky, Anton Vedder, and John Weckert for some helpful suggestions on an earlier draft of this chapter.

References

Alpert, Sheri A. (1998). "Health Care Information: Access, Confidentiality, and Good Practice." In K. W. Goodman, ed. *Ethics, Computing, and Medicine: Informatics and the Transformation of Healthcare*. New York: Cambridge University Press, pp. 75–101.

Anderson, James G., and Kenneth W. Goodman (2002). "The Challenge of Bioinformatics." In J. G. Anderson and K. W. Goodman, eds. *Ethics and Information Technology: A Case-Based Approach to a Health Care System in Transition*. New York: Springer.

Christiani, David C., Richard R. Sharp, Gwen W. Collman, and William A. Suk. (2001). "Applying Genomic Technologies in Environmental Health Research: Challenges and Opportunities," *Journal of Occupational and Environmental Medicine*. Vol. 43, No. 6, pp. 526–533.

Collins, Francis S., Eric D. Green, Alan E. Guttmacher, and Mark S. Guyer. (2003). "A Vision for the Future of Genomics Research: A Blueprint for the Genome Era," *Nature*, Vol. 42, April. Also available at http://www.genome.gov/11007524. Accessed 12/5/04.

Custers, Bart (2001). "The Risks of Epidemological Data Mining." In R. Chadwick, L. D. Introna, and A. Marturano, eds. *Proceedings of the Fourth International Conference on Computer Ethics—Philosophical Enquiry (CEPE 2001)*. Lancaster University, UK, pp. 61–70.

Goodman, Kenneth W. (1996). "Ethics, Genomics, and Information Retrieval," *Computers in Biology and Medicine*, Vol. 26, No. 3, pp. 223–229.

Goodman, Kenneth W. (1998). "Bioethics and Health Informatics: An Introduction." In K. W. Goodman, ed. *Ethics, Computing, and Medicine: Informatics and the Transformation of Healthcare*. New York: Cambridge University Press, pp. 1–31.

Marturano, Antonio (2003). "Molecular Biologists as Hackers of Human Data: Rethinking IPR for Bioinformatics Research," *Journal of Information, Communication and Ethics in Society*, Vol. 1, No. 4, pp. 207–215.

Office of Technology Assessment. (1993). *Protecting Privacy in Computerized Medical Information*. Washington, DC: U.S. Government Printing Office.

O'Neill, Onora (2002). *Autonomy and Trust in Bioethics*. New York: Cambridge University Press.

Quine, W. V. O. (1953) "Reference and Modality." In W. V. O. Quine, ed. *From a Logical Point of View*. New York: Harper Torchbooks.

Sharp, Richard R., and J. Carl Barrett (2000). "The Environmental Genome Project: Ethical, Legal, and Social Implications," *Environmental Health Perspectives*, Vol. 108, No. 4, pp. 279–281. Also available at: www.mindfully.org/GE/Environmental-Genome-Project.html.

Stefansson, Kari (2001). "Population, Inc." Interview in *Technology Review*, Vol. 104, No. 3, pp. 50–55.

Taubes, Gary (2001). "Your Genetic Destiny for Sale," *Technology Review*, Vol. 104, No. 3, pp. 40–46.

Tavani, Herman T. (1999). "KDD, Data Mining, and the Challenge to Normative Privacy," *Ethics and Information Technology*, Vol. 1, No. 4, pp. 265–273.

Tavani, Herman T. (2003). "Ethical Issues at the Intersection of Population Genomics Research and Information Technology." In A. Backus and H.T. Tavani, eds. *Proceedings of the Conference on Privacy, Informed Consent, and Genomic Research*. Harvard School of Public Health, pp. 1–12.

Tavani, Herman T. (2004). "Genomic Research and Data-Mining Technology: Implications for Personal Privacy and Informed Consent," *Ethics and Information Technology*, Vol. 6, No. 1, pp. 15–28.

Vedder, Anton (2004). "KDD, Privacy, Individuality, and Fairness." In R. A. Spinello and H. T. Tavani, eds. *Readings in CyberEthics*. 2nd ed. Sudbury, MA: Jones and Bartlett, pp. 462–470.

IV: INTELLECTUAL PROPERTY RIGHTS AND GENETIC/GENOMIC INFORMATION

The debate about whether, and to what extent, intellectual property rights should be granted to genetic and genomic data has been especially contentious. Controversies surrounding this debate can be examined from at least two distinct vantage points: (1) issues affecting ownership of personal genetic information that resides in computer databases, and (2) issues involving the patenting of non-personal genetic data (i.e., DNA sequences, genes, and entire genomes). The chapters included in Section IV address either one or both of these concerns. Also examined, in the opening chapter of this section, are some philosophical theories of property that have been used in arguments to justify intellectual property rights in the context of genetic/genomic information.

In Chapter 11, Adam Moore provides an overview of two different kinds of arguments used by proponents of intellectual property rights: the *natural rights* theory of property, and the *utilitarian/incentives* based property theory.[1] Initially, he draws some important distinctions between tangible and intangible "objects," noting that the latter can include medical records, genetic profiles, and gene-enhancement techniques. Moore then examines some arguments that have been used to defend the granting of property rights for intangible objects. He rejects the utilitarian defense, pointing out that utilitarian theories, which have been used in the Anglo-American world to justify the ownership of intangible property, are often based on economic incentives. Next, Moore examines the "Lockean" model of property—a natural rights theory of property (sometimes also referred to as the "labor theory") articulated by John Locke in the 17th century[2]—to analyze the question of ownership rights of intangible objects. Moore defends a version of this model[3] and uses it to show how ownership rights can be justified in the case of "objects" pertaining to medical and genetic information.

---

[1] A third type of property theory, sometimes referred to as the "personality theory of property," has also been put forth by some property theorists. This theory is not examined by Moore in his analysis of intellectual property in Chapter 11. For a discussion of this property theory, see Moore (1997) and Spinello and Tavani (2005).

[2] See Locke (1690).

[3] See Moore (1997, 2001) for a more detailed development of Locke's arguments, as well as Moore's defense of those arguments. An analysis of Locke's theory vis-à-vis property rights for digital information is included in Tavani (2004).

Ownership of Personal Genetic Information in Computer Databases

Deciding who should have ownership rights to, and thus ultimately have control over, personal genetic information that resides in electronic databases will determine who can and cannot access that information. The controversy over access to and control of this information affects three distinct constituencies or groups, whose interests and stakes are vastly different:

(1) *corporations and entrepreneurs*, such as pharmaceutical companies, biotechnology companies, and genetic firms, that claim to have the need for exclusive control over genetic information in their databases as an economic incentive for further research and discovery;

(2) *scientists and researchers in the nonprofit sector*, including government-funded public health organizations, that claim to have a need to access that data to conduct genetic/genomic research and improve the public health; and

(3) *ordinary individuals*, especially research subjects participating in genetic/genomic studies, who believe that they have a need for control over their personal genetic data to protect against the risk of discrimination and stigmatization that could result from others accessing it.

Who should have (legal) ownership rights to the genetic information in electronic databases? Have individuals who consented to supply samples of their DNA necessarily lost all rights to their personal genetic data, including the right to have some say about how that information is used in the future, once it is stored in a computer database? Why are these questions so controversial? Recall our discussion in Chapter 1 of the case of a database containing personal genetic information that is currently owned by deCODE Genetics, Inc., a private genomics company in Iceland. What would happen to the information in that database if deCODE Genetics were to file for bankruptcy? What would happen if deCODE's databases were sold to or otherwise acquired by a different organization?

Consider an analogy involving non-genetic personal data stored in a computer database owned by Toysmart.com, a now defunct online business. This case can help us to understand what could happen to one's personal information that resides in one or more electronic databases.[4] Consumers who dealt with Toysmart were given assurances that any personal information they gave to this e-commerce company would be protected under the company's privacy policy. When Toysmart filed for bankruptcy in 2000, it was required by law to list all of its assets. Included among those assets was its database of customer information, the specific contents of which were presumably protected under Toysmart's privacy policy. Some of Toysmart's creditors, as well as some companies that were interested in assuming Toysmart's business, indicated that they did not believe that they were bound by any prior agreement between Toysmart and its customers.

---

[4] Recall our discussion of the Toysmart case in Chapter 10. There, we examined the Toysmart controversy in terms of its implications for informed-consent issues. In this section, we consider the Toysmart case in light of property-rights issues.

Could what happened at Toysmart.com also happen to a privately owned genetics company in the commercial sector—e.g., deCODE Genetics? Consider that people who voluntarily donated their DNA samples to deCODE consented to do so on the understanding that the confidentiality of their personal data would be protected. But if deCODE Genetics were to be taken over by a different company, would the confidentiality of those who volunteered their DNA samples to deCODE still be protected? Would the Icelandic government intervene in this case? Should it? Because the answers to these questions are not entirely clear, it would seem that we need straightforward intellectual property laws and policies that explicitly address the ownership rights of the highly sensitive personal data stored in genetic databases.

In Chapter 12, Richard Spinello looks at property-rights claims involving the ownership of personal genetic information from the point of view of privacy rights. He considers and rejects an argument advanced by some privacy theorists who maintain that granting proprietary rights to individuals for their personal information is necessary for the realization of individual privacy. However, Spinello also recognizes the importance of safeguarding against the inappropriate release of an individual's genetic information, since that can result in serious harms such as employment discrimination and denial of health insurance coverage. But he argues against the view that individuals should have ownership rights to their genetic information (as well as ownership rights to the source of that information, such as human tissues samples).[5] His argument against what he calls "upstream property rights" is based primarily on utilitarian grounds. Spinello believes that the recognition of such ownership rights would lead to inefficiency, along with the "disutility" of genetic discoveries, and that biomedical research will be hampered if property rights granted for genes and genetic material are too extensive.[6]

If individuals cannot legally own their personal genetic data, then who, if anyone, should have the rights to it? For example, should deCODE Genetics be given ownership rights to the personal genetic information stored in its databases? Spinello rejects this solution as well. Admitting that a middle ground is difficult to locate, he argues that there is a need for a "liberal access policy" to this critical data and that such a policy must also take into account the "necessary role of investment incentives" on the part of entrepreneurs. Although Spinello does not provide a specific proposal for how this "middle ground" can be achieved, his chapter includes a clear and detailed analysis of important issues that need to be resolved with respect to ownership rights involving personal genetic data.

---

[5] With respect to the ownership of human tissue, Spinello also briefly analyzes the *Moore v. Regents of California* case (1990). As Spinello notes, the California Supreme Court rejected Moore's claim that his property rights had been violated when doctors did not share the commercial gains that they had obtained through the use of his spleen cells. The court concluded that Moore did not have a legitimate ownership claim regarding the genetic information coded into his cells. For an excellent discussion of this case, see Boyle (1996).

[6] For example, Spinello notes that although the state of Oregon's Genetic Privacy Act of 1999 initially declared that genetic information was an individual's personal property, the Oregon law was eventually reversed because of pressure from scientists who argued that their ability to conduct genetic research was threatened.

Patenting DNA Sequences, Genes, and Genomes

As already noted, another set of issues involving property rights for genetic information has to do with concerns about laws permitting genes and DNA sequences to be patented. Controversies surrounding the patenting of genetic information have been debated at several levels: scientific, legal, and economic, as well as from the perspectives of philosophy and ethics. When Craig Venter applied for patents on 350 bits of DNA in June 1991 (and later for 2,735 pieces in February 1992), many scientists were outraged because they believed that patents of this type would impede scientific progress. In controversies involving genetic/genomic patents, two important questions arise: (1) *Should* DNA be patented?[7] (2) *Who* will decide the policy for the data's use?

Questions about patenting genetic data have been addressed by the Human Genome Organization (HUGO), an international association that has also taken a policy stance on this issue. HUGO does not oppose the patenting of "useful benefits derived from genetic information," but it strongly opposes the patenting of "short sequences from randomly isolated portions of genes encoding proteins of uncertain functions." For example, HUGO is opposed to granting patents on ESTs (Expressed Sequence Tags) based on their "utility" as "probes to identify specific DNA sequences," because it believes that ESTs constitute research tools. HUGO has also taken the position that the U.S. Patent and Trade Organization (PTO) should rescind earlier decisions to allow patents of ESTs, arguing that it is "untenable to make all subsequent innovation in which EST sequences would be involved in one way or another dependent on such patents."

According to HUGO, DNA molecules and their sequences, be they full-length, genomic, or cDNA, ESTs, SNPs (Single Nucleotide Polymorphisms) or even whole genomes of pathogenic organisms, if of unknown function or utility, should, in principle and as a matter of policy, be viewed as *pre-competitive information*.[8] In this sense, HUGO has welcomed efforts such as the (nonprofit) SNP (TSC) Consortium of industry and academia to map all SNPs and to put them in the public domain. HUGO has also welcomed, in general, the adoption of the European Biology Directive (European Union Directive 98/44/EC) on such issues as "patentable subject matter" and "ethical aspects of patenting in the area of human genomics."

Chapters 13–15 examine concerns surrounding patents for DNA sequences, genes, and entire genomes. In Chapter 13, Antonio Marturano takes up the question of patents and intellectual property rights for DNA by first examining some analogies between molecular geneticists and computer hackers. He uses "hackers" in the benign sense of the term, in which hackers tinker and experiment with software code but do not break into computer systems or engage in malicious activities. Marturano believes that there is a sense in which we can view geneticists as

---

[7]In one sense this question may seem odd, since patents have already been granted for some genes. However, this question is normative in character, asking "what *should be* the case (ethically)?" rather than simply inquiring into what the case *is* or happens to be.

[8]See the *HUGO Statement on Patenting Issues Related to Early Release on Raw Sequence Data* (April 2000), available at http://www.gene.ucl.ac.uk/hugo.

"hackers of code." For example, he points to the use of the "Shotgun Method" by geneticists at Celera Genomics in sequencing DNA, which Marturano suggests can be viewed as a kind of "computer hacking" technique. Continuing with his analogy between computer hackers and contemporary geneticists, especially with respect to the manipulation of programming "code," Marturano examines the "open source code" movement in computer science to see how it might be applied to the intellectual property rights debate involving genetic data. Marturano argues that the "open source model" provides an answer to regulating genetic information—including genetics research as it applies to the pharmaceutical field—that would lead to a "fairer distribution of research opportunities around the globe."[9] He also argues that an open source model of property would shift the debate away from patenting and toward copyright protection for genetic data.

In Chapter 14, law professor Dan Burk continues with the discussion of the "open source theme" and its application to genomic data. However, Burk's suggestions for why we should consider this model in the context of genomic data are different from Marturano's. Like Marturano, however, Burk notes that some similarities can be seen between the so-called "hacker ethic" and the open source movement, both of which advocate for the free flow of information. Burk begins his analysis by carefully distinguishing the open source software movement from the "free software" position. He then describes the kind of licensing agreement that accompanies open source software, in particular the GNU Public License, which allows individuals to make changes to a program's source code. Burk notes that some people refer to this licensing agreement as "copyleft," as opposed to the more standard notion of copyright agreement. Like Marturano, Burk concludes that the open source model could positively influence the debate about ownership rights involving genomic data.[10] But Burk also notes that in the final analysis, the viability of a copyleft scheme, as embodied in the GNU Public License, depends on the threat of lawsuits involving copyright infringement.

In Chapter 15, the closing chapter of Section IV, law professor James Boyle continues with the discussion of the patent theme, and invites us to consider some general points that can be learned from the "squabbles over patents" involving genomic data. According to Boyle, we are currently in a phase that he refers to as the "second enclosure movement of the commons." Whereas the first enclosure movement involved the conversion of the "commons of arable land" into property, Boyle describes the current enclosure movement as one that involves an expansion of property rights over the "intangible commons."[11] The latter, he notes, is "the world of the public domain, the world of expression and invention." Boyle also

---

[9] Also see Chapter 4 where Marturano and Chadwick examine some global implications for sharing genomic data and the relationship that this issue has to the digital divide.

[10] Interested readers may wish to examine Burk's discussion of legal issues involving programming biological code (in Chapter 3) in conjunction with his analyses of open-source code and patent protection in Chapter 14.

[11] Boyle uses the metaphors of the "commons of the mind" and the "grassy commons of Old England" to differentiate the second enclosure movement from the first, even though he acknowledges that the analogies between the two are not perfect.

notes that many of the arguments used to justify the first enclosure are used to defend the second enclosure movement, as well. Perhaps the most prominent of these is the claim that intellectual property is needed to encourage development and investment.[12]

In some of his earlier writings,[13] Boyle focused on economic and utilitarian arguments surrounding the second enclosure movement. In this chapter, however, he has a very different objective—viz., he turns his attention to questions about the "fundamental structure of intellectual property discourse in the context of legal and academic debate." For example, Boyle examines a range of arguments about what it means to "own a gene" to show how this debate has revealed both the "selective focus" and the "selective blindness" of arguments made in the property debate involving genomic patents. So rather than trying to solve the problems introduced by the patenting of genes and genomes, including solutions that might involve the kinds of licensing schemes that Burk describes in Chapter 14, Boyle elects to discuss the debate over gene patents in terms of what he calls a "rhetorical case study." However, the reader can learn much from this "study" through the summary/analysis of the various arguments that Boyle examines in his exposition. For example, Boyle describes some of the arguments that opponents of the second enclosure movement have used against genomic patents, including the argument that our common genome should not be turned into private property because it "belongs to everyone" and thus is the "common heritage of humankind."

Review Questions for Chapters in Section IV
1. Identify three constituencies or "stakeholders" in the property-rights debate involving genetic and genomic data. How are these constituencies differently affected by the outcome of this debate?
2. Describe the two principal kinds of property rights controversies in the current dispute over ownership rights involving genetic and genomic data. Can you think of other categorical distinctions for analyzing the key issues in this particular debate?
3. What is intellectual property? How is this kind of property different from tangible property? What does Adam Moore mean by the expression "intangible object"?
4. According to Moore, what are the two main types of traditional property-rights theories that have been used to defend intellectual property? Briefly describe each.
5. In which ways does Adam Moore believe that Locke's theory of property rights, advanced in the 17th century, can inform the current debate over property rights involving genetic and genomic information? Do you agree with Moore's conclusions? Explain.
6. What arguments and analogies does Richard Spinello use to show that claims involving property rights and privacy rights are closely related? How are claims involving privacy rights and property rights also different?
7. Why does Spinello believe that it would not be appropriate to grant individuals a legal property right to their personal genetic data? Do you agree?

---

[12]See Adam Moore's critique of the utilitarian/incentives based property model in Chapter 11.
[13]See, for example, Boyle (2002, 2003).

8. Which kinds of analogies does Antonio Marturano draw between molecular biologists and computer hackers? Do you agree with Marturano?
9. In which ways does Marturano believe that the "open source movement" in computer software can positively inform the debate involving patents for genomic data?
10. What issues does Dan Burk examine in his discussion of the "open source" philosophy vis-à-vis controversies surrounding the patenting of genomic data? How is Burk's analysis both similar to and different from Marturano's?
11. What does James Boyle mean by the "second enclosure of the commons"? According to Boyle, how is the second enclosure different from the first?
12. Why does Boyle believe that examining the debate about genomic patents can inform us about the "fundamental structure of intellectual property discourse in the context of legal and academic debate?" How does Boyle's analysis also inform us about the substance of the arguments, economic and otherwise, used in the debate about genomic patents?

References

Boyle, James (1996). *Shamans, Software, and Spleens: Law and the Construction of the Information Society*. Cambridge, MA: Harvard University Press.

Boyle, James (2002). "Fencing Off Ideas," *Daedalus*, Vol. 131, No. 2, pp. 13–25.

Boyle, James (2003). "The Second Enclosure Movement and the Construction of the Public Domain," *Law and Contemporary Moral Problems*, Vol. 66, pp. 33–74.

Human Genome Organization. *HUGO Statement on Patenting Issues Related to Early Release on Raw Sequence Data* (April 2000), available at http://www.gene.vcl.ac.uk/hugo/patent2000.html. Accessed 1/26/03.

Locke, John (1690). *Two Treatises of Civil Government*. London: Everyman's Library (1924).

Moore, Adam D. (1997). "Toward a Lockean Theory of Intellectual Property." In A. E. Moore, ed. *Intellectual Property: Moral, Legal, and International Dilemmas*. Lanham, MD: Rowman and Littlefield, pp. 81–103.

Moore, Adam D. (2001). *Intellectual Property and Information Control: Philosophic Foundations and Contemporary Issues*. New Brunswick, NJ: Transaction Publishers.

Spinello, Richard A., and Herman T. Tavani (2005). "Intellectual Property Rights: From Theory to Practical Implementation." In R. A. Spinello and H. T. Tavani, eds. *Intellectual Property Rights in a Networked World: Theory and Practice*. Hershey, PA: Idea Group Publishing, pp. 1–65.

Tavani, Herman T. (2004). "Balancing Intellectual Property Rights and the Intellectual Commons: A Lockean Analysis," *Journal of Information and Communication in Ethics and Society*, Vol. 2, Supplement, pp. S5–S14.

Suggested Further Readings

Aoki, Keith (1998). "Neocolonialism, Anti-Commons Property, and Biopiracy in the (Not So Brave) New World of Intellectual Property Protection," *Indiana Journal of Global Legal Studies*, Vol. 6, pp. 11–58.

Barlow, John Perry (1994). "The Economy of Ideas: A Framework for Copyrights and Patents (Everything You Know About Intellectual Property is Wrong)," *Wired*, Vol. 2, No. 3, March, pp. 47–50.

Benkler, Yochai (2000). "From Consumers to Users: Shifting the Deeper Structures of Regulation," *Federal Communication Law Journal*, Vol. 52, pp. 561–579.

Boyle, James. (2004). "A Politics of Intellectual Property: Environmentalism for the Net." In R. A. Spinello and H. T. Tavani, eds. *Readings in CyberEthics*. 2nd ed. Sudbury, MA: Jones and Bartlett Publishers, pp. 273–293.

Burk, Dan L. (1997). "The Milk Free Zone: Federal and Local Interests in Regulating Recombinant BST," *Columbia Environmental Law Review*, Vol. 22, pp. 227–317.

Burk, Dan L. (2002). "Lex Genetica: The Law and Ethics of Programming Biological Code," *Ethics and Information Technology*, Vol. 4, No. 2, pp. 109–121.

Child, James W. (1997). "The Moral Foundation of Intangible Property." In A. D. Moore, ed. *Intellectual Property: Moral, Legal, and International Dilemmas*. Lanham, MD: Rowman and Littlefield, pp. 57–80.

Eisenberg, Rebecca (1987). "Property Rights and the Norms of Science in Biotechnology Research," *Yale Law Journal*, Vol. 97, pp. 179–231.

Elkin-Koren, Niva (2000). "The Privatization of Information Policy," *Ethics and Information Technology*, Vol. 2, No. 4, pp. 201–209.

Halbert, Deborah J. (1999). *Intellectual Property in the Information Age: The Politics of Expanding Ownership Rights*. Westport, CT: Quorum Books.

Heller, Michael A. (1998). "The Tragedy of the Anticommons: Property in the Transition from Marx to Markets," *Harvard Law Review*, Vol. 111, pp. 621–688.

Heller, Michael A. and Rebecca S. Eisenberg (1998), "Can Patents Deter Innovation? The Anticommons in Biomedical Research," *Science*, May.

Hettinger, Edwin C. (1997). "Justifying Intellectual Property." In A. D. Moore, ed. *Intellectual Property: Moral, Legal, and International Dilemmas*. Lanham, MD: Rowman and Littlefield, pp. 17–38.

Human Genome Organization. "HUGO Statement on Patenting of DNA Sequences—In Particular Response to the European Biotechnology Directive—April 2000. Available at http://www.gene.ucl.ac.uk/hugo/.

Lessig, Lawrence (2002). *The Future of Ideas: The Fate of the Commons in a Connected World*. New York: Random House.

Lipinski, Thomas A., and Johannes J. Britz (2000). "Rethinking the Ownership of Information in the 21st Century: Ethical Implications," *Ethics and Information Technology*, Vol. 2, No. 1, pp. 49–71.

Litman, Jessica (1990). "The Public Domain," *Emory Law Journal*, Vol. 39, pp. 965–1023.

Litman, Jessica (2001). *Digital Copyright*. New York: Prometheus Books.

McFarland, Michael C. (2004). "Intellectual Property, Information, and the Common Good." In R. A. Spinello and H. T. Tavani, eds. *Readings in CyberEthics*. 2nd ed. Sudbury, MA: Jones and Bartlett, pp. 294–3004.

Moore, Adam D. (2000). "Owning Genetic Information and Gene Enhancement Techniques: Why Privacy and Property Rights May Undermine Social Control of the Human Genome," *Bioethics*, Vol. 14, No. 2, pp. 97–119.

Palmer, Tom G. (1997). "Intellectual Property: A Non-Posnerian Law and Economics Approach." In A.D. Moore, ed. *Intellectual Property: Moral, Legal, and International Dilemmas*. Lanham, MD: Rowman and Littlefield, pp. 179–224.

Rai, Arti K. (1999). "Regulating Scientific Research: Intellectual Property Rights and the Norms of Science," *NorthWestern Law Review*, Vol. 94, No. 1, Fall.

Scanlan, Michael J. (2005). "Locke and Intellectual Property Rights." In R. A. Spinello and H. T. Tavani, eds. *Intellectual Property Rights in a Networked World: Theory and Practice*. Hershey, PA: Idea Group Publishing, pp. 83–98.

Stallman, Richard. (1995). "Why Software Should Be Free." In D. Johnson and H. Nissenbaum, eds. *Computing, Ethics & Social Values*, Englewood Cliffs, NJ: Prentice Hall, pp. 199–200.

Tavani, Herman T. (2005). "Recent Copyright Protection Schemes: Implications for Sharing Digital Information." In R. A. Spinello and H. T. Tavani, eds. (2005). *Intellectual Property Rights in a Networked World: Theory and Practice*. Hershey, PA: Idea Group Publishing, pp. 182–204.

11. Intellectual Property, Genetic Information, and Gene Enhancement Techniques

Adam D. Moore

Introduction

Justifying the ownership of intangible works such as books, movies, music, processes of manufacture, credit histories, and purchasing profiles is serious business. Nowhere is this more apparent than in the area of medical technologies. For example, with the completion of the human genome project the next several decades promise profound advancements in genetic enhancement, gene therapy, and cloning. All of this raises important ethical questions. Can intangible works like genetic enhancement techniques be owned like houses, land, and automobiles? If so, how should such moral claims be codified in the law? What limits should be placed on such moral and legal rights?

In this chapter I will argue that intangible works can be owned—that the proper subjects of intangible property claims include medical records, genetic profiles, and gene enhancement techniques. After presenting a few important features of intangible property and a brief analysis and dismissal of the most common justification of intangible property rights, I will consider a new justification—a Lockean model of intangible property rights. The concluding sections will consider several problems with the proposed justification.

Features of Intangible and Intellectual Property

There are a few important differences between intellectual property and tangible or physical property—differences that work to the advantage of the Lockean model that I will defend below. Intellectual property (copyrights, patents, gene

---

This essay appears for the first time in *Ethics, Computing, and Genomics*. Copyright © 2005 by Adam D. Moore. Printed by permission.

enhancement techniques etc.) as well as information, reputation, and the like, is generally characterized as non-physical property where owner's rights surround control of physical manifestations or tokens of some abstract idea or type. Ideas or collections of ideas are readily understood in terms of non-physical types, while the physical manifestations of ideas can be modeled in terms of tokens. Intangible property rights surround control of physical tokens, and this control protects rights to types or abstract ideas.

Intellectual or intangible works, unlike tangible goods, are generally non-rivalrous. Computer programs, genetic enhancement techniques, books, movies, and lists of customers can all be used and consumed by many individuals concurrently. This is generally not the case for cars, computers, and most other tangible goods. Intellectual property, unlike physical property, is also non-zero-sum. In the clearest case, when I eat an apple there is one less apple for everyone else—my plus one and everyone else's minus one sum to zero. With intangible property it is not as if my acquisition leaves one less for everyone else.

Another difference between physical and intangible property concerns what is available for acquisition. There is the domain of ideas yet to be *discovered* (new scientific laws, etc.), the domain of ideas yet to be *created* (the next *Lord of the Rings*, *Star Wars*, etc.), and the domain of intangible works that are privately owned. Since it is possible for individuals to independently invent or create the "same" intangible work and obtain rights, we must include currently owned intangible works as available for acquisition. Only the set of ideas that are in the public domain or those ideas that are a part of the common culture are not available for acquisition and exclusion.

Aspects of these features of intangible property can be found in the case of John Moore. In 1976, John Moore began treatment for cancer at the University of California Medical Center.

> His doctors quickly became aware that some of his blood products and components were potentially of great commercial value. They performed many tests without ever telling him of their commercial interest, and took samples of every conceivable bodily fluid, including sperm, blood, and bone marrow aspirate .... In 1981, a cell line established from Moore's T-lymphocytes was patented by the University of California, with Moore's doctors listed as the inventors. At no time during this process was Moore told anything about the commercial exploitation of his genetic material.[1]

Related to types and tokens, we may ask if Moore actually owned the information found in his T-lymphocytes. As self-owners it may be the case that we each have property rights to our own bodies, capacities, and powers. It does not follow from the notion of "self-ownership" however, that we each own the genetic information found in our cells. Ownership of a token does not entail ownership of a type. In other words, I may own a copy of *The Lord of the Rings* (a token), but this does not mean that I own the intangible work (the plot, characters, theme, and style—or types).

It is also the case that the cell line developed from Moore's T-lymphocytes was non-rivalrous and *created* rather than *discovered*—it is not as if the patent issued eliminated all other ways of producing some outcome.

The Utilitarian Incentives-Based Argument[2]

Anglo-American systems of intellectual property are typically justified on utilitarian grounds.[3] The Constitution grants limited rights to authors and inventors of intellectual property "to promote the progress of science and the useful arts."[4] Beginning with the first Patent Act of 1790 and continuing through the adoption of Berne Convention Standards in 1989, the basis given for Anglo-American systems of intellectual property is utilitarian in nature, and not grounded in the natural rights of the author or inventor.[5] Thomas Jefferson, a central figure in the formation of American systems of intellectual property, expressly rejected any natural rights foundation for granting control to authors and inventors over their intellectual work. "The patent monopoly was not designed to secure the inventor his natural right in his discoveries. Rather, it was a reward, and inducement, to bring forth new knowledge."[6] Society seeks to maximize utility in the form of scientific and cultural progress by granting rights to authors and inventors as an incentive toward such progress. In general, patents, copyrights, and trade secrets are devices, created by statute, to prevent the diffusion of information before the author or inventor has recovered profit adequate to induce such investment. The justification typically given for Anglo-American systems of intellectual property "is that by slowing down the diffusion of information . . . it ensures that there will be more information to diffuse."[7] Moreover, utilitarian-based justifications of intellectual property are elegantly simple. Control is granted to authors and inventors of intellectual property because granting such control provides incentives necessary for social progress. Coupled with the theoretical claim that society ought to maximize social utility, we arrive at a simple yet powerful argument.

Problems for the Utilitarian Argument

The utilitarian incentives-based argument for intellectual property is subject to several serious objections. First, it is not at all clear that adopting Anglo-American institutions of intellectual property maximizes social utility when compared to other models. Building on the work of Michael Polanvyi[8] and Brian Wright,[9] Steven Shavell and Tanguy Van Ypersele offer a compelling case for a reward model.[10] As Shavell and Ypersele note, reward models may be able to provide incentives without affording authors and inventors monopoly control over intellectual works. "Under a reward system innovators are paid for innovation directly by the government (possibly on the basis of sales), and innovations pass immediately into the public domain."[11] If a reward model leads to, or is expected to lead to, greater social utility, then the utilitarian argument for intellectual property rights will have been replaced by a utilitarian argument against current Anglo-American systems of intellectual property.

Recently Raymond Shih Ray Ku has argued that copyright is unnecessary in digital environments. "With respect to the creation of music, [it is argued] that exclusive rights to reproduce and distribute copies provide little if any incentive for creation, and that digital technology make[s] it possible to compensate artists without control."[12] In brief, Ku claims that copyright protects the interests of the publisher—large, up-front distribution costs need to be paid for and copyright does the job. Digital environments, however, eliminate the need for publishers with distribution resources. Artists, who receive little royalty compensation anyway, may distribute their work worldwide with little cost. Incentives to innovate are maintained, as they have been, by touring, exhibitions, and the like. Thus, if Ku is correct the incentives-based argument would lead us away from, not toward, copyright protection for music.

This sort of critique invites us to consider the social benefits that might occur if we weaken intellectual property rights. Perhaps copyrights should lapse after twenty years and patents after five. Consider, for example, the initial profits generated by the sales of certain software packages. The market share guaranteed by initial sales, support services, and the like, may provide adequate incentives. Empirical questions about the costs and benefits of copyright, patent, and trade secret protection are notoriously difficult to determine. Economists who have considered the question indicate that either the jury is out, so to speak, or that other arrangements would be better. George Priest claims that "[t]he ratio of empirical demonstration to assumption in this literature must be very close to zero ... [recently it] has demonstrated quite persuasively that, in the current state of knowledge, economists know almost nothing about the effect on social welfare of the patent system or of other systems of intellectual property."[13] This echoes Clarisa Long's view: "Whether allowing patents on basic research tools results in a net advance or deterrence of innovation is a complex empirical question that remains unanswered."[14] If we cannot appeal to the progress-enhancing features of intellectual property protection, then *the utilitarian can hardly appeal to such progress as justification.*

Finally, it is not at all clear that the theoretical premise in the utilitarian incentives-based justification for intellectual property is secure. This premise holds that we ought to do that act or adopt that policy which maximizes overall social utility. We certainly don't appeal directly to this principle in relation to life, liberty, or physical property rights. Few would suggest that if social utility demanded we should kill an innocent, seize physical property without compensation, or imprison someone. Some rights hold independent of straightforward social utility arguments—perhaps even intellectual property rights. In any event, the theoretical premise stands in need of justification.

A Lockean Model of Intangible Property[15]

Independent of social progress or utility maximization arguments, John Locke offered what has become known as the "labor theory of acquisition." Locke claimed "[f]or this labor being the unquestionable property of the laborer, no man but he can have a right to what that is once joined to, at least where there is *enough and as good left for others.*"[16] As long as the proviso that "enough and as good" is satisfied, an acquisition is of prejudice to no one. Locke argues that "Nobody could think himself injured by the drinking of another man, though he took a good draught, who had a whole river of the same left him to quench his thirst..."[17]

Suppose that mixing one's labor with an unowned object creates a prima facie claim against others not to interfere that can only be overridden by a comparable claim. The role of the proviso is to provide one possible set of conditions where the prima facie claim remains undefeated.[18] Another way of stating this position is that the proviso in addition to X, where X is labor or first occupancy or some other weak claim generating activity, provides a sufficient condition for original appropriation.

Justification for the view that labor or possession may generate prima facie claims against others could proceed along several lines. First, labor, intellectual effort, and creation are generally voluntary activities that can be unpleasant, exhilarating, and everything in between. That we voluntarily do these things as sovereign moral agents may be enough to warrant non-interference claims against others. A second, and possibly related justification, is based on merit. Sometimes individuals who voluntarily do or fail to do certain things deserve some outcome or other. Thus, students may deserve high honor grades and criminals may deserve punishment. When notions of desert are evoked, claims and obligations are made against others—these non-absolute claims and obligations are generated by what individuals do or fail to do. Thus in fairly uncontroversial cases of desert, we are willing to acknowledge that weak claims are generated, and if desert can properly attach to labor or creation, then claims may be generated in these cases as well.

Finally, a justification for the view that labor or possession may generate prima facie claims against others could be grounded in respect for individual autonomy and sovereignty. As sovereign and autonomous agents, especially within the liberal tradition, we are afforded the moral and legal space to order our lives as we see fit. As long as respect for others is maintained we are each free to set the course and direction of our own lives, to choose between various lifelong goals and projects, and to develop our capacities and talents accordingly. Simple respect for individuals would prohibit wresting from their hands an unowned object that they acquired or produced. I hasten to add that at this point we are trying to justify weak non-interference claims, not full blown property rights. Other things being equal, when an individual labors to create an intangible work, then weak presumptive claims of non-interference have been generated on grounds of labor, desert, or autonomy.

The underlying rationale of Locke's proviso is that if no one's situation is worsened, then no one can complain about another individual appropriating part of the commons. If no one is harmed by an acquisition and one person is bettered, then the acquisition ought to be permitted. In fact, it is precisely because no one is harmed that it seems unreasonable to object to what is known as a Pareto-superior move.[19] Thus, the proviso can be understood as a version of a "no harm, no foul" principle.[20]

Bettering, Worsening, and the Baseline Problem

Assuming a just initial position and that Pareto-superior moves are legitimate, there are two questions to consider when examining a Pareto-based proviso. First, what are the terms of being worsened? This is a question of scale, measurement, or value. An individual could be worsened in terms of subjective preference satisfaction, wealth, happiness, freedoms, opportunities, et cetera. Which of these count in determining moral bettering and worsening? Second, once the terms of being worsened have been resolved, which two situations are we going to compare to determine if someone has been worsened? In any question of harm we are comparing two states—for example, "now" after an acquisition compared to "then" or before an acquisition. This is known as the baseline problem.

In principle, the Lockean theory of intangible property being developed is consistent with a wide range of value theories. So long as the preferred value theory has the resources to determine bettering and worsening with reference to acquisitions, then Pareto-superior moves can be made and acquisitions justified on Lockean grounds. For now, assume an Aristotelian Eudaimonist account of value exhibited by the following theses is correct.[21]

1. Human well-being or flourishing is the sole standard of intrinsic value.
2. Human persons are rational project pursuers, and well-being or flourishing is attained through the setting, pursuing, and completion of life goals and projects.
3. The control of physical and intangible objects is valuable. At a specific time each individual has a certain set of things she can freely use and other things she owns, but she also has certain opportunities to use and appropriate things. This complex set of opportunities along with what she can now freely use or has rights over constitutes her position materially—this set constitutes her level of material well-being.

While it is certainly the case that there is more to bettering and worsening than an individual's level of material well being, including opportunity costs, I will not pursue this matter further at present. Needless to say, a full-blown account of value will explicate all the ways in which individuals can be bettered and worsened with reference to acquisition. Moreover, as noted before, it is not crucial to the Lockean model being presented to defend some preferred theory of value against all comers. Whatever value theory that is ultimately correct, if it has the ability to determine bettering and worsening with reference to acquisitions, then Pareto-superior moves can be made and acquisitions justified on Lockean grounds.

Lockeans as well as others who seek to ground rights to property in the proviso generally set the baseline of comparison as the state of nature. The commons, or the state of nature, is characterized as that state where the moral landscape has yet to be changed by formal property relations. For now, assume a state of nature situation where no injustice has occurred and where there are no property relations in terms of use, possession, or rights. All anyone has in this initial state are opportunities to increase her material standing. Suppose Fred creates an intangible work and does not worsen his fellows—alas, all they had were contingent opportunities and Fred's creation and exclusion adequately benefits them in other ways. After the acquisition, Fred's level of material well-being has changed. Now he has a possession that he holds legitimately, as well as all of his previous opportunities. Along comes Ginger who creates her own intangible work and considers whether her exclusion of it will worsen Fred. But what two situations should Ginger compare? Should the effects of Ginger's acquisition be compared to Fred's initial state, where he had not yet legitimately acquired anything, or to his situation immediately before Ginger's taking? If bettering and worsening are to be cashed out in terms of an individual's level of well-being with opportunity costs and this measure changes over time, then the baseline of comparison must also change. In the current case we compare Fred's level of material well-being when Ginger possesses and excludes an intangible work to his level of well-being immediately before Ginger's acquisition.

At this point I would like to clear up a common confusion surrounding the baseline of comparison. What if a perverse inventor creates a genetic enhancement technique that will save lives but decides to keep the technique secret or charge excessive prices for access? Those individuals who had, before the creation, no chance to survive now have a chance and are worsened because of the perverse inventor's refusal to let others use the machine.

The baseline this case implies cannot be correct. On this view, to determine bettering and worsening we are to compare how individuals are before the creation of some value (in this case the genetic enhancement technique) to how they would be if they possessed or consumed that value. But we are all worsened in this respect by any value that is created and held exclusively. I am worsened by your exclusive possession of your car because I would be better off if I exclusively controlled the car—even if I already owned hundreds of cars. Any individual, especially those who have faulty hearts, would be better off if they held title to my heart compared to anyone else's holding the title. I am also worsened when you create a new philosophical theory and claim authorship—I would have been better off (suppose it is a valuable theory) if I had authored the theory, so you have worsened me. Clearly this account of the baseline makes the notions of bettering and worsening too broad.[22]

A slightly different way to put the Lockean argument for intellectual property rights is:

Step One: *The Generation of Prima Facie Claims to Control*—Suppose Ginger creates a new intangible work (a process for changing the genetic structure of mature cells perhaps). Creation, effort, etc., yield her prima facie claims to control (similar to student desert for a grade).

Step Two: *Locke's Proviso*—If the acquisition of an intangible object makes no one (else) worse off in terms of their level of well-being compared to how they were immediately before the acquisition, then the taking is permitted.

Step Three: *From Prima Facie Claims to Property Rights*—When are prima facie claims to control an intangible work undefeated? Answer: when the proviso is satisfied. Alas, no one else has been worsened—who could complain?

Conclusion: So long as no harm is done—the proviso is satisfied—the prima facie claims that labor and effort may generate turn into property claims.

If correct, this account justifies moral claims to control intangible property like genetic enhancement techniques. When an individual creates an intangible work and fixes it in some fashion, then labor and possession create a prima facie claim to the work. Moreover, if the proviso is satisfied the prima facie claim remains undefeated and rights are generated.[23]

Problems for the Lockean Model: The Social Nature of Intangible Works

A common view about the information found in the human genome, one that may undermine property rights, is that this information is publicly owned—thus ownership claims to genetic enhancement techniques may be undermined. Individuals are raised in societies that endow them with knowledge which these individuals then use to create intangible works of all kinds. On this view, the building blocks of intangible works—knowledge—is a social product. Individuals should not have exclusive ownership of the works that they create because these works are built upon the shared knowledge of society. Allowing rights to intangible works would be similar to granting ownership to the individual who placed the last brick in a public works dam. The dam is a social product, built up by the efforts of hundreds, and knowledge, upon which all intangible works are built, is built up in a similar fashion.

Similarly, the benefits of market interaction are social products. The individual who discovers crude oil in their backyard should not obtain the full market value of the find. The inventor who produces the next technology breakthrough does not deserve full market value when such value is actually created through the interactions of individuals within a society. Simply put, the value produced by markets and the building blocks of intangible works are social products.

A mild form of this argument may yield a justification for limiting the ownership rights of authors and inventors—alas, these individuals do not deserve the full value of what they produce given what they produce is, in part, a social product. Maybe rules that limit intangible property rights can be justified as offering a trade-off position between individual effort and social inputs. A more radical form of this argument may lead to the elimination of intangible property rights. If individuals are, in a deep way, social products and market value and knowledge are as well, then the robust property rights found within the Anglo-American tradition are suspect.

This argument, in either version, is deficient for several reasons. First, I doubt that the notion of "society" employed in this view is clear enough to carry the weight that the argument demands. In some vague sense, I know what it means to say that Abraham Lincoln was a member of American *society* or that Aristotle's political views were influenced by ancient Greek *society*. Nevertheless, I think that the notion of "society" is conceptually imprecise—one that it would be dubious to attach ownership or obligation claims to. Those who would defend this view would have to clarify the notions of "society" and "social product" before the argument could be fully analyzed.

But suppose for the sake of argument that supporters of this view come up with a concise notion of "society" and "social product." We may ask further, why think that societies can be *owed* something or that they can *own* or *deserve* something? Notions of *ownership*, *owing*, or *deserving* don't appear to make sense when attached to the concept of "society." If so, and if different societies can *own* knowledge, do they not have the problem of original acquisition?[24] Surely, it does not follow from the claim that X is a social product that society owns X. Likewise, it does not follow from the claim that X is produced by Ginger, that Ginger owns X—what if Ginger produced something out of your property? It is true that interactions between individuals may produce increased market values or add to the common stock of knowledge. What I deny is that these by-products of interaction, market value and shared information, are in some sense owned by society or that society is owed for their use. Why assume this without argument? It is one thing to claim that information and knowledge is a social product—something built up by thousands of individual contributions—but quite another to claim that this knowledge is owned by society or that individuals who use this information owe society something in return.[25]

Suppose that Fred and Ginger, along with numerous others, interact and benefit me in the following way. Their interaction produces knowledge that is then freely shared and allows me to create some new value, V. Upon creation of V, Fred and Ginger demand that they are owed something for their part. But what is the argument from third party benefits to demands of compensation for these benefits—why think that there are "strings" attached to *freely* shared information? If such an argument can be made, then it is plausible as well to maintain that burdens create reverse demands. Suppose that the interaction of Fred and Ginger produces false information that is freely shared. Suppose further that I waste ten years trying to produce some value based, in part, on this false information. Would Fred and Ginger, would society, owe me compensation? The position that "strings" are attached in this case runs parallel to Robert Nozick's benefit "foisting" example. In Nozick's case a benefit is foisted on someone and then payment is demanded. This seems an accurate account of what is going on in this case as well.

> *One cannot, whatever one's purposes, just act so as to give people benefits and then demand (or seize) payment. Nor can a group of persons do this. If you may not charge and collect for benefits you bestow without prior agreement, you certainly may not do so for benefits whose bestowal costs you nothing, and most certainly people need not repay you for costless-to-provide benefits which yet others provided them. So the fact that we partially are "social products" in that we benefit from current patterns and forms created by the multitudinous actions of a long string of long-forgotten people, forms which include institutions, ways of doing things, and language, does not create in us a general free floating debt which the current society can collect and use as it will.*[26]

I would argue that this is also true of market value. The market value of oil, for example, is the synergistic effect of individuals freely interacting. For oil to be worth anything it must be refined and it must be useful—without thousands of prior inventions and advancements oil would be virtually worthless. Moreover, there is no question of desert here. Surely the individual who discovers the oil does not deserve full market value any more than the lottery winner deserves her winnings. Imagine we set up a pure lottery where the payout was merely the entire sum of all the tickets purchased. Upon determining a winner, suppose someone argued that the sum of money was a social product and that society was entitled to a cut of the profit. An adequate reply would be something like "but this was not part of the rules of the game, and if it was, it should have been stated before the investment was made."

On my view common knowledge, market value, and the like, are the synergistic effects of individuals freely interacting. If a thousand of us *freely* give our new and original ideas to all of humankind it would be illicit for us to demand compensation, after the fact, from individuals who have used our ideas to create things of value. It would even be more questionable for individuals ten generations later to demand compensation for the use of, the now very old, ideas that we freely gave.[27]

But once again, suppose for the sake of argument that the defender of this view can justify societal ownership of general pools of knowledge and information. Even in this case we have already paid for the use of this collective wisdom when we pay for education and the like. When a parent pays, through fees or taxation, for a child's education, it would seem that the information—part of society's common pool of knowledge—has been fairly purchased. This extends through all levels of education and even to individuals who no longer attend school.

Finally, it is obviously the case that the information found in the human genome is discovered rather than created. These facts may be discovered by anyone who cares to look hard enough. The genetic enhancement techniques and other technologies that will be built upon this information are created rather than discovered—alas, there may be infinitely many ways to modify human genetic structure. Thus, even if an argument could be marshaled that justified societal ownership of the information found in the human genome this would not automatically yield claims to control every subsequent invention based on this information. Thus if I am correct, the social nature of intellectual works argument will not undermine intangible property rights to creations like genetic enhancement techniques.

The Inequality Argument

One argument commonly given against allowing individual ownership of intangible works is that such technology is expensive and will only impact the rich. Those with the financial resources will genetically engineer their offspring to eliminate defects while the poor will be left what nature gives them by chance. This inequality in health care will lead to further economic and social inequalities. It may also lead to longer, healthier lives for some, ultimately creating a class-based society and discrimination against those who are genetically challenged.

Colin Farrelly puts the inequality argument against intangible property rights the following way.[28] Assume that we are to choose the principles that govern the distribution of social products from an unbiased vantage point—from what John Rawls called the *original position*. In this original position we are behind the veil of ignorance which shields us from knowing our own age, sex, race, goals, projects, biases, and the like.[29] From this position Rawls thinks, and Farrelly agrees, that we would pick the following principle of justice.

> All social values—liberty and opportunity, income and wealth, and the bases of self-respect—are to be distributed equally unless an unequal distribution of any, or all, of these values is to everyone's advantage.[30]

In addition to these "social values" Farrelly thinks that we should add "health and vigor, intelligence and imagination"—i.e. goods that are, in part, controlled by genetic technologies. Farrelly rhetorically asks, "Does one not have a right to a decent distribution of these goods?"[31]

This view is subject to several decisive objections. First, while not included in the original characterization of the inequality argument Farrelly's version relies heavily on the notions of social product and social values. Thus, this view is subject to the kinds of problems that infected the social nature of intangible works argument already discussed. Moreover, it is arguably the case that individuals in the original position would choose a Lockean Pareto-based principle rather than a principle that puts property, projects, and lifelong goals at the mercy of someone's (whose?) notion of "equality" and a "decent distribution of . . . goods."

Initially, almost every medical advancement was available only to the rich. By refining these advancements and techniques, prices dropped, which opened up new markets for those less financially fortunate. In the end, procedures that were once cost prohibitive are now available to everyone. There is no reason to think that genetic enhancement technologies won't follow this same course. In fact, our entire market system seems to necessitate this kind of inequality. Most inventors and companies burn the midnight oil and create or discover new and revolutionary medical procedures in order to make a profit. This process requires large up-front investments that in turn necessitate higher initial prices when a viable commodity does come to market. Nevertheless, sooner or later the "high-priced" market becomes saturated and in order to maintain profits prices are dropped. If this system yields everyone better prospects in the end, the resulting initial inequality of distribution is hardly objectionable.

Moreover, even if gene therapy techniques remain expensive the leveling effect assumed in the inequality argument seems indefensible. Suppose that aspirin-plus is invented and cures headaches and colds. The cost of aspirin plus, however, is very high—suppose $500 per pill. Are we to prohibit the manufacture and administration of aspirin-plus because it is unfair that some will be able to forgo the suffering bought on by colds and headaches while others will not? This sounds like simple envy and mean spiritedness to me—"if I can't have it, then no one can" or "if I have to suffer, then so does everyone else." Let us dispense with the notion that individuals who hold these sentiments are actually concerned with lessening human suffering.

Defenders of the inequality argument may call for government regulation of these technologies rather than prohibition. But regulation, and sometimes public seizure, of property does not undermine the moral claims already established. Eminent domain laws allow for the justified taking of property essential for social utility. Nevertheless these laws are limited in scope and application—they are exercised on a case by case basis and require just and fair compensation with recourse to the courts if necessary. Narrowly construed, I have in principle no objection to eminent domain laws used to provide access to important medical technologies and procedures—so long as just and fair compensation is required in payment for the taking. If correct, these considerations are sufficient to undermine the inequality argument against intangible property rights.

Conclusion

The utilitarian incentives-based model views intangible property rights as state-created entities offered as inducements to innovate. The Lockean account that I have sketched does not—like life rights, and physical property rights, intangible property rights exist prior to and independent of governments and social progress arguments. They are what some theorists call "natural rights."

Consider the simplest of cases. After weeks of effort and numerous failures, suppose I come up with a new combination of chemicals—a drug like aspirin-plus. Would anyone argue that I do not have at least some minimal moral claim to control this new drug? Suppose that you hear of aspirin-plus and desire to purchase the directions and make your own pills. Is there anything morally suspicious with an agreement between us that grants you a limited right to use the intangible work provided that you do not disclose the process? Alas, you didn't have to agree to my terms—you could have just waited out the headache and cold, purchased other drugs, or produced your own concoction of chemicals.

Nevertheless, moral and legal rights to control intangible works are not absolute, and upholding these rights may create social burdens that justify limiting their protection. Eminent domain laws may also warrant certain kinds of takings. Independent of these considerations, however, I have maintained that individuals, corporations, and institutions can justifiably own and control intangible works.

Acknowledgements

Parts of this chapter draw from my article "Owning Genetic Information and Gene Enhancement Techniques: Why Privacy and Property Rights May Undermine Social Control of the Human Genome," *Bioethics* 14 (Spring 2000): 97-119. I would like to thank Bill Kline, Scott Rothwell, Mark Vanhook, Don Hubin, Peter King, and Kimberly Moore for their comments on the initial draft.

Endnotes

1. James Boyle, *Shamans, Software, and Spleens: Law and the Construction of the Information Society* (Cambridge, Mass.: Harvard University Press, 1996), p. 22.
2. For an in-depth analysis of the utilitarian incentives-based argument for intellectual property, see A. Moore "Intellectual Property, Innovation, and Social Progress: The Case Against

Incentives Based Arguments." *The Hamline Law Review*, Vol. 26 (2003) and *Intellectual Property and Information Control: Philosophic Foundations and Contemporary Issues* (Transaction Publishing/Rutgers University, Fall 2004 paperback, Fall 2001 hardback), Chapter 3.

3. See the Committee Report accompanying the 1909 Copyright Act, H.R. Rep. No. 2222, 60th Cong., 2nd Sess. 7 1909. The courts have also reflected this theme: "The copyright law . . . . makes reward to the owner a secondary consideration." *United States v. Paramount Pictures*, 334 U.S. 131, 158 (1948)). "The limited scope of the copyright holder's statutory monopoly, like the limited copyright duration required by the Constitution, reflects a balance of competing claims on the public interest: Creative work is to be encouraged and rewarded, but private motivation must ultimately serve the cause of promoting broad public availability of literature, music, and other arts." (*Twentieth Century Music Corp. v. Aiken*, 422 U.S. 151, 95 S.Ct. 2040, 45 L.Ed.2d (1974)).

4. U.S. CONST., § 8, cl. 8.

5. This view is echoed in the following denials of a common law right to intellectual property. "Wheaton established as a bedrock principle of American copyright law that copyright, with respect to a published work, is a creature of statute and not the product of the common law." See S. Halpern, D. Shipley, H. Abrams, *Copyright: Cases and Materials* (1992) p. 6. The General Court of Massachusetts (1641) adopted the following provision, "There shall be no monopolies granted or allowed among us, but of such new inventions as are profitable to the country, and that for a short time." *See* "Walker on Patents," *Early American Patents* (A. Deller ed., 1964). "The monopoly did not exist at common law, and the rights, therefore, which may be exercised under it cannot be regulated by the rule of common law. It is created by the act of Congress; and no rights can be acquired in it unless authorized by statute, and in the manner the statute prescribes" (Chief Justice Taney, *Gayler et al. v. Wilder*, 1850). See also, *Sony Corp. of America v. Universal Studios Inc.* 464 US 417, 78, L, ED 2d. 574 (1984); *Weaton v. Peters*, 33 US (8 Pet.) 591, 660–1 (1834); and *Graham v. John Deere Co.*, 383 US 1, 9 (1966).

6. Thomas Jefferson quoted in W. Francis and R. Collins, "Cases and Materials On Patent Law: Including Trade Secrets—Copyrights—Trademarks" (4th ed. 1995) p. 92. Prior to the enactment of the U.S. Constitution, a number of states adopted copyright laws that had both a utilitarian component and a natural rights component. A major tuning point away from a natural rights framework for American institutions of intellectual property came with the 1834 decision of *Wheaton v. Peters* 33 US (8 Pet.) 591, 660-1 (1834). See "Copyright Enactments of the United States, 1783-1906," in *Copyright Office Bulletin* 3 (1906), 14. "Unquestionable, the 1834 decision marked an important turning-point, in that it distances American copyright law from the natural law perspectives which were very much in evidence at the end of the eighteenth century." Alain Strowel, "Droit d'auteur and Copyright: Between History and Nature," in *Of Authors and Origins* (Brad Sherman and Alan Strowel ed. 1994) p. 245.

7. See Joan Robinson quoted in D. Nelkin's *Science As Intellectual Property* (1984) p. 15.

8. Michael Polanvyi, "Patent Reform," *Review of Economic Studies* 11 (1943): 61.

9. Brian Wright, "The Economics of Invention Incentives: Patents, Prizes, and Research Contracts," *American Economic Review* 73 (1998): 1137.

10. Michael Kremer offers an auction model where the government would pay inventors the price that it obtains from the public sale of the innovation. See Michael Kremer, "Patent Buyouts: A Mechanism for Encouraging Innovation," *The Quarterly Journal of Economics* 113 (1998): 1137.

11. Steven Shavell and Tanguy Van Ypersele, "Rewards versus Intellectual Property Rights" *Journal of Law and Economics* 44 (Oct. 2001): 525.

12. Raymond Shih Ray Ku, "The Creative Destruction of Copyright: Napster and the New Economics of Digital Technology," *University of Chicago Law Review* 69 (Winter 2002): 263.

13. G. Priest, "What Economists can Tell Lawyers about Intellectual Property," *Research In Law And Economics: The Economics Of Patents And Copyrights* 8 (J. Palmer ed. 1986) p, 21.

14. Clarisa Long, "Patent Law and Policy Symposium: Re-Engineering Patent Law: The Challenge of New Technologies: Part II: Judicial Issues: Patents and Cumulative Innovation," *Washington University of Law and Public Policy* 2 (2000):229. See also Fritz Machlup, *Production and Distribution of Knowledge in the United States* (1962) at 168–169.

15. A more lengthy analysis of intangible property rights appears in *Intellectual Property and Information Control* (Transaction Pub. 2001, 2004), "Intangible Property: Privacy, Power, and Information Control" *American Philosophical Quarterly* 35 (Oct. 1998), and "Toward A Lockean Theory of Intellectual Property" in *Intellectual Property: Moral, Legal, and International Dilemmas*, A. Moore (Rowman & Littlefield, 1997), Cha. 5.

16. John Locke, *The Second Treatise of Government*, § 27 (italics mine).

17. Locke, *Second Treatise*, § 33.

18. This view is summed up nicely by Clark Wolf, "Contemporary Property Rights, Lockean Provisos, and the Interests of Future Generation," *Ethics* (July, 1995): 791–818.

19. One state of the world, S_1, is Pareto-superior to another, S_2, if and only if no one is worse-off in S_1 than in S_2, and at least one person is better-off in S_1 than in S_2. S_1 is *strongly* Pareto-superior to S_2 if everyone is better-off in S_1 than in S_2, and *weakly* Pareto-superior if at least one person is better-off and no one is worse-off. State S_1 is Pareto-optimal if no state is Pareto-superior to S_1: it is *strongly* Pareto-optimal if no state is *weakly* Pareto-superior to it, and *weakly* Pareto-optimal if no state is *strongly* Pareto-superior to it. Throughout this essay I will use Pareto-superiority to stand for *weak* Pareto-superiority. Adapted from G. A. Cohen's "The Pareto Argument For Inequality" in *Social Philosophy & Policy* 12 (Winter 1995): 160.

20. Finally, a virtue of this account is that it proceeds from relatively weak and hopefully widely shared assumptions and builds a case for intellectual property rights. When Colin Farrelly in "Genes and Social Justice: A Reply to Moore" in *Bioethics* 16 (2002): 72–83 replies to an earlier version of this article, his critique rests on numerous controversial assumptions about equality, fairness, and social justice. If one does not agree with Farrelly's assumptions about social justice the entire critique collapses. I hope to avoid this sort of mistake—who could rationally complain about an action or acquisition that does not worsen?

21. For similar views see: Rawls, *A Theory of Justice* (Cambridge: Harvard University Press, 1971), cha. VII.; Aristotle, *Nicomachean Ethics*, bks. I and X; Kant, *The Fundamental Principles of The Metaphysics of Morals*, Academy Edition; Sidgwick, *Methods of Ethics*, 7th ed. (London: Macmillian, 1907); R. B. Perry, *General Theory of Value* (New York: Longmans, Green, 1926); and Loren Lomasky, *Persons, Rights, and the Moral Community* (New York: Oxford University Press, 1987).

22. This sort of baseline confusion infects Colin Farrelly's "Genes and Social Justice: A Reply to Moore" in *Bioethics* 16 (2002): 75. For a similar, yet still mistaken, view of the baseline see

Jeremy Waldron "From Authors to Copiers: Individual Rights and Social Values in Intellectual Property" *Chicago-Kent Law Review* 68 (1993): 866.

23. Note that on this account exclusive patent rights would not be justified—i.e. those who would have independently invented some patented work would be worsened relative to how they would have been prior to the granting of exclusive entitlement.
24. See Nozick, *Anarchy*, p. 178.
25. Lysander Spooner argued that one's culture or society plays almost no role in the production of ideas. "Nothing is, by its own essence and nature, more perfectly susceptible of exclusive appropriation, than a thought. It originates in the mind of a single individual. It can leave his mind only in obedience to his will. It dies with him, if he so elect." Lysander Spooner, *The Law of Intellectual Property*: or An Essay on the Right of Authors and Inventors to a Perpetual Property in Their Ideas, in *The Collected Works of Lysander Spooner*, edited by C. Shively (1971), p. 58.
26. Nozick, *Anarchy*, p. 95.
27. Lysander Spooner puts the point succinctly. "*What* rights society have, in ideas, which they did not produce, and have never purchased, it would probably be very difficult to define; and equally difficult to explain *how* society became possessed of those rights. It certainly requires something more than assertion, to prove that by simply coming to a knowledge of certain ideas—the products of individual labor—society acquires any valid title to them, or, consequently, any *rights* in them." Spooner, *The Law of Intellectual Property*, p. 103.
28. Farrelly, "Genes and Social Justice: A Reply to Moore" in *Bioethics* 16 (2002): 72–83.
29. John Rawls, *A Theory of Justice* (Harvard University Press, 1971), p. 17–22, 136–142.
30. John Rawls, *A Theory of Justice* (Harvard University Press, 1971), p. 62.
31. Farrelly, "Genes and Social Justice: A Reply to Moore" in *Bioethics* 16 (2002): 79.

12. Property Rights in Genetic Information

Richard A. Spinello

Introduction

The mapping of the genome and the advent of genetic testing have triggered a plethora of perplexing ethical conundrums. The most prominent of these involve the interconnected issues of privacy and the ownership of one's "genetic information." That information is broadly defined as information about genes, gene products, or one's inherited characteristics that is derived from a genetic test or a person's DNA sample. A DNA sample refers to any human biological specimen such as human tissue or blood from which DNA can be extracted. Genetic information includes information about any individual's genetic test results or genetically linked diseases. For our purposes, it also includes each person's unique genetic code (or sequence), which can be valuable if there is a genetic variation or mutation in that code that is the cause of a genetic disease.[1]

We will refer to these various forms of genetic information and its bodily source (that is, the DNA sample) as genetic source material. The expansive version of the property argument is that ownership rights should be conferred on genetic source material including the human tissue sample and any information about the genes derived therefrom. There are narrower versions of this argument but in general what we are talking about is the proprietary "rights of sources."

The major rationale for providing these rights is the protection of privacy. Obviously, if this information becomes too widely available or falls into the wrong hands, one's privacy rights are put in the gravest peril. While some maintain that genetic information (in the form of test results) is just an additional component of a person's overall medical record, others support the position of "genetic exceptionalism," that is, "genetic information is sufficiently different from other kinds of health-related information that it needs special protection."[2] Hence, if this position has plausibility, it is imperative that genetic information be regulated more

---

This essay originally appeared in *Ethics and Information Technology*, **6**: 29-42. Copyright © 2004 by Kluwer Academic Publishers. Reprinted by permission.

[1] D. McLochlin. Whose Genetic Information is it Anyway? John Marshall Journal of Computer and Information Law, 19: 609, 2001.

[2] T. Murray. Genetic Exceptionalism and Future Diaries: Is Genetic Information Different from other Medical Information? In M. Rothstein (Ed.) Genetic Secrets: Protecting Privacy and Confidentiality, New Haven: Yale University Press, New Haven, 1997.

stringently than other forms of personal data. Currently, there are no specific federal laws regulating the dissemination of this information, despite the fact that genetic testing (that is, testing of DNA to ascertain an individual's predisposition to a genetic illness or impairment) is becoming more widespread.

The premise of genetic exceptionalism is certainly credible. To begin with, this information is revealing not just about a single individual's medical condition but about the potential condition of his or her relatives as well. Also, while one can change other features or bad habits (e.g., drug addiction) that could lead to discrimination, one's genetic code is fixed and unchangeable. It creates an indelible mark on a person's history, and that "mark" might work against a person's legitimate interests if it becomes known by others. Finally, genetic information is unique in that it doesn't just reveal one's present condition but also future probabilities and predispositions to certain ailments. One's DNA is analogous to a "diary" of a person's future medical conditions. Therefore, unlike other forms of data, it is subject to broad and subjective interpretation, opening up considerable room for bias and manipulation. If an employer administers a drug test the employee either passes or fails that test; there is usually nothing ambiguous about the results. But what if the same employee takes a genetic test and it reveals that he has Gene X which causes diabetes depending upon one's overall genetic background? Perhaps if this mutation were put into a hundred people with different genetic backgrounds thirty of them would contract this disease. What is an employer to make of such information? The 30% probability of diabetes may be enough not to hire or promote this person, but how can one defend the equity of such discriminatory decision making?

Thus, if privacy is a necessary condition of one's security and well-being in our modern, computerized society,[3] preserving the confidentiality of genetic information is of paramount importance. The potential for discrimination and stigmatization is significant and the dignitary harm suffered by the careless dissemination of this predictive data is beyond dispute.

Some of those who support the principle of genetic exceptionalism argue that a property right to one's genetic information would be the most viable means of securing the confidentiality of that information. But if property rights proliferated, and the collection and dissemination of genetic information became too restricted, downstream genetic research would likely suffer. Clearly, an individual's private genetic information when combined with other data or the genetic information of other people can be an extremely valuable research tool. There is well-founded apprehension that an overemphasis on privacy and data protection might preclude or interfere with such research.

Thus, the issue of genetic privacy is deeply intertwined with the question of ownership, and this connection will be the main focus of attention in the following sections of his paper. Who (if anyone) should have a proprietary interest in a person's specific genetic information or the source of that information? Should a person's

---

[3] J. Moor. Towards a Theory of Privacy for the Information Age. In R.A. Spinello and H. Tavani (Eds.), Readings in Cyberethics, pp. 349–359. Jones & Bartlett, Sudbury, MA, 2001; J. Moor. Just Consequentialism and Computing. In R.A. Spinello and H. Tavani (Eds.), Readings in Cyberethics, pp. 98–105. Jones & Bartlett, Sudbury, MA, 2001.

genetic source material belong to that person and be classified as his or her personal property? Should any of that material be regarded as part of the public domain? Is it fair for researchers to receive patents for genes extracted from individuals, especially given that those individuals do not share in the rewards when their genetic material is later commercialized?

deCODE Genetics and the Icelandic Healthcare Database

These contentious issues have come to the surface in the controversy swirling around the Iceland database of genetic information managed by deCODE Genetics, a genomics company located in Iceland. In exchange for its investment and its work to collect and manage this data, the Iceland government agreed to give deCODE Genetics exclusive research access to the genetic data of its citizens.

Iceland's Act on a Health Sector Database[4] authorized the creation of this comprehensive nationwide database that aggregates disparate sources of medical and genetic information. This Icelandic Healthcare Database (IHD) links together separate databases of medical records, genealogical records, and genetic information based on DNA samples received voluntarily from over 10,000 citizens. deCODE's choice of Iceland was based in part upon the genetic homogeneity of its population. Scientists believe that since the Vikings settled here in the ninth century, the genetic code of the Icelanders has changed only marginally.

A major objective of IHD research is that it will provide a better understanding of disease by examining the interaction of the environment and genetics. deCODE has already built pedigrees for patients with certain diseases by scanning the genomes of relatives, searching for a chromosome segment that was inherited from a common ancestor; it then uses these segments to look for an errant gene that might contribute to a disease. According to Wade,[5] "[w]ith this method, deCODE has identified several disease-causing genes, including one for schizophrenia, and has inferred the general location of some others."

Despite its promising results, this ambitious project has attracted trenchant criticism from within and outside the country. The most vigorous resistance within the country has come from the Association of Icelanders for Ethics and Science (Mannvernd) and the Icelandic Medical Association (IMA). According to Mannvernd, "the interests of the company [deCODE] are being put first in order to make money."[6] The IMA's opposition has been based upon "concern about inadequate measures to protect privacy, the lack of access to data among academic researchers, and the belief that individual consent should be required before inclusion of medical records in the data base."[7]

---

[4]Act on a Health Sector Database no. 139, 1998; available at http://brunnur.stjr.is/interpro/htr/htr.nsf/pa-ges/gagngr-log-ensk

[5]N. Wade. Double Helix Leaps from Lab into Real Life. The New York Times, February 25: D6, 2003.

[6]C. Goldwater. Iceland Exploits its Genetic History. BBC News, February 4th, 2000; available at: http://news.-bbc.co.uk/hi/english/sci/tech/newsid630000/630961.stm

[7]G. Annas. Rules for Research on Human Genetic Variation—Lessons from Iceland. New England Journal of Medicine, 342: 1830, 2000.

Although Iceland's Act on the Rights of Patients calls for medical privacy to be safeguarded through the principle of informed consent, such consent was not sought from all those participants in the medical records segment of the deCODE database. Those individuals have a right to opt out, however. According to the Database Act (1998), patients can "request that the information on him/her not be entered into the Health Sector Database." The information in the genetics segment has come only from volunteers who have provided DNA samples and have given their informed consent for inclusion in this database. On the other hand, Icelanders cannot "refuse to allow data on their deceased parents to be entered into the database."[8]

Donors have been assured that privacy safeguards will protect their data. Accordingly, scientists working with this genetic data are blocked from seeing names or other personal information (through encryption) so they are working with anonymous information. Some critics (like the IMA) are skeptical, however, that despite these safeguards there may be inadvertent or even intentional breaches of confidentiality. There are also concerns about the inclusion of genetic information of the deceased in the IHD. In one legal challenge to the Database Act the daughter of a deceased man sought unsuccessfully to exclude her father's information from the IHD. According to Pinto (2002), "the plaintiff in this lawsuit fears that inferences may be drawn about her private life from the data obtained from her father." Since there is a decryption key for this data it is possible that someone will access the information and trace the deceased man's genetic makeup back to this woman. The case has been appealed to the Supreme Court of Iceland.

Even if we concede that privacy rights are adequately protected (and that is far from evident), there are still many questions provoked by the Iceland database project. Is it fair for the government to seize custodial control over this information and then delegate that control to a private company? Should such sensitive information be entrusted with a commercial, profit-seeking enterprise? Will such an entity have the public interest at heart when it makes decisions about the use of this data? And is it morally acceptable for the Iceland Government and deCODE to exploit the genetic information for economic gains without compensating the donors?

The Icelandic Government has contended that health and genetic data is a national resource, so individuals have no proprietary rights to that data. According to Jonatansson,[9] "The Government's rationale is that the normal principles of proprietary rights are not applicable to health data because the information has been gathered by the Government, and therefore with public funding, by virtue of the Government's obligation to provide appropriate medical assistance." In other words, the government's view is that the collection of this data at its expense gives it exclusive stewardship and engenders at least a quasi–property right. Despite the gov-

---

[8] A. M. Pinto. Corporate Genomics: DeCode's Efforts at Disease Mapping in Iceland for the Advancement of Science and Profits. University of Illinois Journal of Law, Technology and Policy, 202: 467, 2002.

[9] H. Jonatansson. Iceland's Health Sector Database: A Significant Head Start in Search of the Biological Grail or an Irreversible Error? American Journal of Law and Medicine, 26: 31, 2000.

ernment's assurances, questions about this approach linger. Can citizens trust the government to exercise responsible stewardship with unwavering consistency so that the confidentiality of this data is never in doubt? Clearly, this information does have financial value but it is by no means obvious how that value can be most equitably appropriated.

Presiding over all this, of course, is the most crucial question—should a person own or have some proprietary right to his or her medical and genetic information (and its source)? Will such an ownership right better protect the confidentiality of this sensitive data and allow for a "just" sharing of the rewards of genetic research? The Iceland Government presumes otherwise, but this question clearly needs more critical attention.

This sort of property issue has been raised before in the debate regarding ownership of personal data (one's name, address, phone number, etc.). It is instructive to review the arguments in that debate to discern their applicability to these concerns regarding the ownership of genetic information.

During the early discussions about data protection many advocated such ownership, primarily as a means of protecting personal privacy. Scholars such as Westin,[10] Miller,[11] and Branscomb[12] have postulated that individuals should have property rights in all personal information about themselves. Personal information includes the simplest data about an individual such as name, address, phone number, and birthday, along with facts about one's financial history or credit background. It is the set of data that describes our unique backgrounds and history.

According to Westin (1967), "personal information thought of as the right of decisions over one's private personality, should be defined as a property right with all the restraints on interference by public and private authorities and due process guarantees that our law of property has been so skillful in devising." Branscomb (1994) made a similar argument, underscoring the tangible benefits that could accrue from such ownership:

> *Our names and addresses and personal transactions are valuable information assets worthy of recognition that we have property rights in them. Unless we assert those rights we will lose them. If such information has economic value, we should receive something of value in return for its use by others.*

Finally, Miller (1969) claimed that the best safeguard for privacy was "a property right vested in the subject of the data and eligible for the full range of constitutional and legal protection that attach to property."

---

[10] A. Westin. Privacy and Freedom. Atheneum, New York, 1967.

[11] Miller. Personal Privacy in the Computer Age: The Challenge of New Technology in an Information-Oriented Society. Michigan Law Review, 67: 1203, 1969.

[12] A. Branscomb. Who Owns Information? From Privacy to Public Access. HarperCollins, New York, 1994.

Once information is regarded as one's private personal property, as an asset with legal attributes, the legal system can begin to develop a set of laws and appropriate regulations. Those nuanced regulations will define when information should be protected from misappropriation, how individuals should be compensated when their information is used by a third party, and when information must be relinquished for the sake of social, public policy, or even technological priorities. Property rights, of course, are not absolute and just as one's physical property may be subject in extreme cases to eminent domain, one's personal information may at times need to be relinquished for the common good.

Such, in general terms, was the rationale for the legal ownership of personal data. This idea was never embraced, however, for a plethora of reasons including strong opposition of direct marketers and other corporate interests which feared higher costs. But do any of the arguments in favor of information property rights constitute a convincing case for a narrower version of this proposal, that is, for the ownership of genetic information and perhaps even the source of that information (DNA samples)? Do they suggest a paradigm for the treatment of this highly sensitive (and sometimes valuable) personal data? Given the validity of genetic exceptionalism, it could be argued that genetic data requires the highest level of data protection, that is, the assignment of proprietary rights.

Legal Support for Privacy-Enhancing Proprietary Rights

While it may not be obvious that proprietary rights are necessary for privacy, it is obvious that the inappropriate release of genetic information can be devastating. As we have seen, it could easily lead to employment discrimination, denial of health insurance coverage, and other forms of dignitary harm or stigmatization due to rash and oversimplified judgements. Hence the concern about genetic privacy is not misplaced or exaggerated and some lawmakers have contemplated the use of a property regime to ensure the protection of genetic privacy.

In the United States there are no federal laws that specifically protect genetic privacy, but the Health Insurance Portability and Accountability Act (HIPAA) does offer protection to all types of medical information. According to HIPAA genetic test results are considered to be "protected health information." But critics argue that HIPAA provides "minimal protection" and does "little to preserve the privacy of all genetic test results."[13]

There have been serious proposals for more specific legislation such as a national Genetic Privacy Act (GPA). Advocates of a GPA see the need for a "unified approach," reassuring the public that genetic information is fully protected.[14] One popular version of this GPA claims that it must "grant a federal property interest in one's own genetic material as well as the right to order the destruction of one's DNA sam-

---

[13] A. Ito. Privacy and Genetics: Protecting Genetic Test Results in Hawaii. Hawaii Law Review, 25: 449, 2003.

[14] J. Weems. A Proposal for a Federal Genetic Privacy Act. Journal of Legal Medicine, 24: 109, 2003.

ples."[15] According to the original draft of this Act, "an individually identifiable DNA sample is the property of the sample source."[16] Advocates of this legislation have consistently argued that a federal property right would protect privacy and enhance clarity about ownership rights in a way that would actually promote commerce by increasing research. They contend that patients would now be incentivized to allow the use of their genetic materials in exchange for a possible monetary reward.

A number of states have enacted legislation to safeguard genetic privacy, and several have sought to use the framework of property rights. New Jersey's genetic privacy statute tried to incorporate a limited property right in genetic tissue samples as a means of allowing individuals to maintain control over their genetic material. Pharmaceutical companies objected, however, since the property right would allow people to demand royalties from products that might be derived from that material. Eventually, New Jersey gave in to these concerns and the property right idea was abandoned (Weems, 2003).

Oregon's Genetic Privacy Act of 1999 boldly declared that genetic information was an individual's personal property. The Oregon legislature saw proprietary rights as the key to securing privacy protection when it enacted that landmark legislation. But researchers and scientists complained that the assignment of a proprietary right to each individual's genetic information (including their genetic code) would be a major obstacle to genetic research. As a result, the law was modified in 2001—the property interest was removed, though criminal penalties would still be imposed for the misappropriation of genetic data. Some privacy advocates, however, were quick to find fault with Oregon's reversal: "By removing the property interest that had been given to individuals with regard to genetic information, Oregon has decreased the strength of privacy associated with their genes" (McLochlin, 2001).

McLochlin suggests that Oregon was on the right track and capitulated too easily to the demands of the scientific community and biotech companies. To be sure, a proprietary right would be a highly efficacious means of preserving one's privacy. As Miller (1969) indicated a "property right vested in the subject" is the surest way to give someone autonomous control over that information, whether it be financial data or one's genetic code. Such a solution also seems to serve the end of distributive justice since people would share in the benefits of research in a way that is proportionate to their contribution. In Branscomb's (1994) terms, if this genetic information is a valuable asset, why shouldn't individuals be compensated for its commercial utilization by a third party? This ownership right and the subsequent negotiations might also engender trust between researchers and donors, especially if the nature of the research on this genetic material became more transparent in that process.

---

[15] M.J. Lin. Conferring a Federal Property Right in Genetic Material: Stepping into the Future with the Genetic Privacy Act. American Journal of Law and Medicine, 22: 109, 1996.

[16] G. Annas, L. Glantz, and P. Roche. Genetic Privacy Act, 1995; available at: http://www.bushp.bu.edu/Depts/Healthlaw

Rule and Hunter,[17] who also advocate property rights in personal data, describe how the granting of such rights will "generate a new balance of power," since owners will now be able to collect royalties for use of their information. According to this classic market-based solution, then, each individual would have exclusive ownership rights over his or her genetic information and source material, and be entitled to compensation for its usage by third parties. That compensation would be determined through "a direct market between buyer and seller with prices based on what people are willing to pay and accept."[18]

Self-Determination and Proprietary Rights

A second argument for property rights in one's genetic information emanates from those concerned about "bio-prospecting," that is, the use of genetic material of indigenous people in order to develop valuable cell lines for profit. Indigenous people have been popular targets of genetic studies, since their isolation may result in unique genetic traits that enable them to resist disease. Ching[19] cites the example of an indigenous group in Panama known as the Guaymi. The U.S. Commerce Department sought to patent the cell line of a Guaymi woman because it believed that this cell line possessed anti-viral qualities. Neither the woman nor her tribe were informed about the patent nor were they asked permission before the patent application was filed. When the patent came to light, the Guaymi tribal president vigorously objected, noting that the patent was "contrary to the Guaymi view of nature," and the patent application was eventually withdrawn. Ching (1997) cites other examples of "molecular colonialism" where genetic samples of indigenous populations have been harvested without proper notice or compensation. In cases such as the one involving the Guaymi, the autonomy rights of indigenous people are being ignored for the sake of profitable research.

A case can be put forth that the basic human right of autonomy and self-determination encompasses the right of all individuals to have maximum control over their genetic information. And many contend that such control is most effectively realized by a legal regime that grants a property right in this data. Accordingly, Ching (1997) calls for the recognition of "indigenous peoples' property right to their genetic material," though such rights might have a different structure than they do in Western cultures. She claims that this right will protect the autonomy of these people and allow them to share in the benefits of the research. In short, a property right will prevent the rampant spread of unabated gene piracy. Ching (1997) recognizes the potential downside for genetic research but argues that "the burden on researchers that would be created by recognizing a property interest in indige-

---

[17] J. Rule and L. Hunter. (1999). Towards Property Rights in Personal Data. In C. Bennet and R. Grant (Eds), Visions of Privacy, pp. 168-181. University of Toronto Press, Toronto, 1999.

[18] L. Andrews. My Body, My Property. Hastings Center Report, October 28, 1986.

[19] K. Ching. Indigenous Self-Determination in an Age of Genetic Patenting: Recognizing an Emerging Human Rights Norm. Fordham Law Review, 62: 687, 1997.

nous peoples' excised cells and any marketable derivative would not have the crippling effect envisioned by those who resist such rights."

This general approach suggested by Ching and other scholars is not confined to indigenous people, but applies to anyone who provides genetic material or undergoes genetic testing. The core argument stipulates the necessity of granting property rights in genetic source material, that is, recognizing self-ownership of tissues and organs and the genetic information derived from those substances. The implication is that the gene or gene sequence derived from this raw material is not eligible for patent protection by the third parties that excise these cells and then isolate and purify the gene. Nor is the genetic data correlated and stored in a database eligible for a proprietary claim by the company or researcher who has aggregated genetic test results. The problem is that these claims contradict the prior "rights" of those sources who provide this material.

According to Boyle,[20] the general rights of sources argument might go something like this: "You can't own this gene because I owned it first. My genetic information is my property. Your gene sequences came originally from a source and source's claims should be recognized, either instead of or as well as, the person seeking the patent." Boyle (2003) notes that this claim often comes from those who have provided genetic material to a research project from which a valuable genetic sequence was derived or from families with a particular genetic disorder who want "to ensure that the development of tests and treatments for the disorder protects the interests of the patients involved."

Flaws with the Property-Rights Approach

In summary, then, property rights are seen as a means of protecting the privacy rights and autonomy of individuals whose genetic information or materials is sought for purposes of research, identification, or as part of a personal profile. Those rights can also ensure that compensation is paid to donors in those situations where it is warranted. With property rights each individual donor will have maximum control. That individual can restrict the flow of genetic information as she sees fit, or even charge licensing fees for the use of such data. By relying on a legally enforceable property claim the person donating the genetic material can demand a share in the revenues of the downstream products derived from that material such as diagnostic tests or treatments.

It is axiomatic, therefore, that a property right will efficiently protect the interests of patients and donors. But, despite the laudable intentions embodied in this property rights approach to genetic source material, there are notable disadvantages and externalities. The major problem with the adoption of a property rights regime for genetic information is economic inefficiency. Fragmented property rights in the genetic data coming from multiple sources would require a substantial integration effort if that data were needed for a particular research project. The

---

[20] J. Boyle. Enclosing the Genome: What the Squabbles over Genetic Patents Could Teach Us, 2003; available at: http://www.creativecommons.org/licenses/by-sa/1.0

higher transaction costs imposed by a property regime would almost certainly constitute an obstacle for biomedical research. A market which recognizes these "upstream" property rights, such as monopolistic patents for genes or proprietary rights in genetic data, would function by licensing this "property" to downstream researchers and biotech firms which are working to develop treatments of genetically based diseases and diagnostic tools. Consider the impediments to that downstream research such as the negotiations with multiple owners required by this property regime, the payment of licensing fees to these owners, the likelihood that some of the owners will act opportunistically and hold up the project. All of this will greatly inhibit research and increase the cost of important end products. Thus, the adverse social and economic effects of recognizing these rights seems beyond dispute. But what does the law have to say about the property rights of sources?

In the most pertinent legal case of Moore v. Regents of California[21] the California Supreme Court rejected Moore's claim that his property right had been violated when doctors did not share the commercial gains they had obtained through the use of his surgically excised spleen cells. A key issue in this case was whether or not Moore owned his human tissue source along with the genetic information coded into his cells, but the court concluded that he did not have a valid ownership claim. The court's rationale was that the bestowal of such a property right would hinder scientific research: "this exchange of scientific materials, which is still relatively free and efficient, will surely be compromised if each cell sample becomes the potential subject matter of a lawsuit" (Moore v. Regents of California, 1990). The court was worried that the nascent biotechnology industry would be irreparably harmed if researchers were forced to "investigate the consensual pedigree of each human cell sample used in research" (Moore v. Regents of California, 1990). In making its decision the Moore court at least implicitly rejected the claim that researchers were bound to share the sometimes ample rewards of that research with those who contribute human tissue (or other samples) like Mr. Moore.

The problem, underscored in the Moore decision, is that society has a critical interest in stimulating biomedical research. And this includes the kind of research that is being conducted by deCODE Genetics. As Harrison[22] observes, progress in biotechnology research will become "unduly burdened by the existence of too many intellectual property rights in basic research tools." Property held in common is subject to a "tragedy of the commons," since individual incentives are often at variance with the collective good. Each individual's marginal exploitation of some common property (such as a fertile tract of land) ultimately destroys that property. But if we effectively remove valuable scientific data from the intellectual commons through the assignment of proprietary rights we get the opposite of a tragedy of the commons, that is, a tragedy of the anti-commons.[23] In this case too few

---

[21] Moore v. Regents of California 793 F 2d 479 [Cal.], 1990; cert denied 111 S. Ct. 1388 (1991).

[22] C. Harrison. Neither Moore nor the Market: Alternative Models for Compensating Contributors of Human Tissue. American Journal of Law and Medicine, 28: 77, 2002.

[23] M.A. Heller and R. Eisenberg. Can Patents Deter Biomedical Research? The Anticommons in Biomedical Research. Science, 280: 698, 1998.

resources are held in common and researchers are blocked from using source material (such as genes or genetic data) that have become privatized, unless they negotiate with a multiplicity of owners. The rights of usage become so fragmented that it is virtually impossible to conduct productive research. The end result is an unfortunate suboptimal underconsumption of human genetic material, and that material is obviously critical for the development of downstream products such as diagnostic tests and gene therapies.

There is considerable risk, therefore, that excessive ownership of information inputs, such as genetic data, and other source material will impose high costs and formidable burdens on the flow of critical scientific information. Biomedical research depends upon the open availability of genetic data resources so long as privacy is ensured. High transaction costs and perverse anti-commons effects, however, will undermine that availability if property rights are granted. Also, while ownership might result in some compensation for those individuals who license their genetic sequences or sell their genetic information, that compensation will be trivial in most cases, and it will be far offset by the social good of better health-care that will be realized by research efforts unencumbered by these transaction costs.

One of the ironies of the Moore case, however, is that the court seemed to contradict itself. It refused to recognize a property right for Moore in the name of research but it concluded that the collectors of genetic data would need such a property right (in the form of a patent) as an incentive. Otherwise, the Court reasoned, "the theory of liability that Moore urges us to endorse threatens to destroy the economic incentive to conduct important medical research" (*Moore v. Regents of California*, 1990). As Boyle[24] points out, the notion that "property rights must be given to those who do the mining [of genetic material]" is indicative of the "doctrinal chaos" of intellectual property law.

DNA material including genes, gene fragments (called expressed sequence tags or ESTs), and related products, are considered to be patentable subject matter. The Patent Act does not cover "the gene as it occurs in nature" (Utility Examination Guidelines),[25] but when a gene has been isolated and purified it is considered to have been modified. This makes it a "new composition of matter" eligible for patent protection. This conclusion seems consistent with the U.S. Congress's apparent intention that the patent statute cover "anything under the sun that is made by man."[26] Three types of patents are possible: structure patents, covering the isolated and purified gene; function patents, covering a new use for the DNA in question (such as a diagnostic test or gene therapy); and process patents, which cover a new method of isolating, purifying, or synthesizing this DNA material.[27]

The patentability of genetic material such as DNA sequences is an intricate and complex issue. Supporters of those patents argue that without the incentive of

---

[24] J. Boyle. Shamans, Software and Spleens: Law and the Construction of the Information Society. Harvard University Press, Cambridge, MA, 1996.

[25] Utility Examination Guidelines. 66 Fed. Reg., 1092, 2001.

[26] Diamond v. Chakrabarty, 447 U.S. 303, 1980.

[27] D. Resnik. DNA Patents and Human Dignity. Journal of Legal and Medical Ethics, 29: 52, 2001.

patents the genome will not be adequately exploited by researchers. Opponents such as Hettinger[28] argue against patents out of respect for life, which should not be the subject of patents. The source or raw material for the gene patents is human tissue, and some ethicists claim that patents should not be given for human material.

The validity and scope of patent protection in the human genome is a question we cannot settle here. But the issue of gene patentability is analogous to the question of ownership of an individual's genetic information and deserves some treatment. What is particularly significant in the Moore decision is the claim that "private ownership of genetic materials could dull the pace of medical innovation."[29] We have argued that this principle should apply to the ultimate sources, that is, patients and donors supplying genetic information and raw material (such as human tissue). But it should also apply to a limited extent to third party researchers and their claims for human gene patents.

As we have observed, the problem is that substances which are upstream in the research cycle must be made easily accessible for downstream research. According to Horn,[30] "This kind of information is considered basic research and provides the data that is necessary for making end products such as drugs, diagnostic tests, and other treatments based on genes and their products." The gene is a basic tool of research, and according to some critics, seeking an exclusive right to a gene by means of a patent is like "trying to gain ownership of the alphabet."[31] When DNA is abstracted from human tissue or blood the goal is to produce a gene that functions exactly as it would in the human body. The lab that does this work can get a patent for this gene or DNA sequence even though the utility of this "invention" is sometimes vague or trivial and the gene in question may be critical for future research projects conducted by other scientists. We now know, for example, that there are many diseases caused by defects in multiple genes. According to Pinto (2002), research has shifted from Mendelian diseases (involving a single gene) to "polygenic disorders" involving several genes. But under the current system researchers will need to locate the patentees of these genes and pay royalty fees for doing research. As Horn (2002) points out, "if the licensing and transaction costs are too high, these valuable downstream innovations [such as therapeutic and diagnostic products] will never take place."

It is a mistake, therefore, to award proprietary rights too far upstream in the research and development value chain for biotech products. This includes patents for genes and gene fragments and it also includes providing property rights in the ultimate source material: human tissue and the genetic information that it contains. As we have seen, the upshot of granting such rights is a tragedy of the anti-

---

[28] N. Hettinger. Patenting Life: Biotechnology, Intellectual Property and Environmental Ethics. Boston College Environmental Affairs Law Review, 22: 267, 1995.

[29] R. A. Epstein. Steady the Course: Property Rights in Genetic Material. The Chicago Working Paper Series, 2003; available at: http://www.law.uchicago.edu/Lawecon

[30] M. E. Horn. DNA Patenting and Access to Health-Care: Achieving the Balance among Competing Interests. Cleveland State Law Review, 50: 253, 2002.

[31] A. Pollack. U.S. Hopes to Stem Rush Toward Patenting Genes. The Patriot Ledger, June 28: 18, 2002.

commons: resources will be underutilized and downstream research and commercialization efforts will be hampered.

In addition, loosely awarding proprietary rights too far upstream in the value chain will most likely raise the cost of downstream innovations. According to Pollack (2002), about 14% of the cost of gene therapies is attributable to the royalties that must be paid to gene patent holders. Those costs will surely increase even further if donors of DNA samples or participants in genetic studies also demand royalties for therapeutic products based on their genetic information.

It is also worth considering whether or not gene patents can be justified on a normative basis. For example, are they justifiable when viewed from the perspective of Locke's labor-desert theory? Stripped of its subtleties, that theory claims that labor engenders a property right. According to Locke, people engage in labor not for its own sake but to reap its benefits; as a result, it would be unjust not to let people have these benefits they take such pains to procure. In short, property rights are required as a return for the laborers' painful and strenuous work. As Locke[32] maintains, one who takes the laborer's property "desire[s] the benefit of another's pains, which he has no right to." In the case of gene fragments the labor desert view provides no support since little labor is involved in sequencing these fragments. On the other hand, sequencing the entire gene is more labor intensive so a patent for a full length gene appears to be on firmer ground.[33]

But Locke calls for limits on the acquisition of property even when a property right seems to be commensurate with the labor performed. According to Locke, the bestowal of a property right should be denied unless there is "enough, and as good left for others" (Locke, 1952). This proviso, when applied to intellectual property, implies that the granting of a property right should not harm the intellectual commons. In this context, awarding an exclusive entitlement will interfere with the research activities of other scientists who will find it difficult to pursue related or similar research trajectories. As Shaw[34] indicates, there is compelling evidence that gene patents are having ill-effects on research: "With genetic patents staking private claims to huge chunks of [genetic] code, researchers and clinicians are finding their genetic research and diagnostic efforts thwarted by various restrictions imposed by commercial, and in some instances, academic, patent holders." Given that the commons is impaired by the removal of a gene that has been patented, a Lockean justification for genetic patents seems dubious. Arguably, a modified property right that requires access at a reasonable cost (for example, some form of compulsory licensing) will resolve this problem and be more consistent with Locke's liberal philosophy of property rights.

Given the high social costs of strong and broad patent rights, they are also hard to justify from a utilitarian perspective. There is a case to be made, however, on utilitarian grounds for some types of gene patents as a basis for stimulating genetic research. The American Medical Association advocates patents on processes

---

[32] J. Locke. The Second Treatise of Government. Bobbs-Merrill, Indianapolis, 1952. (Original work published 1690).

[33] M. Holman and S. Munzer. Intellectual Property Rights in Genes and Gene Fragments. Iowa Law Review, 85: 735, 2000.

[34] G. Shaw. Does the Gene Stampede Threaten Science? AAMC Reporter, February, 2000.

used to isolate and purify gene sequences, substance patents on purified proteins, and gene patents "only if the inventor has demonstrated a practical, real world, specific and substantial use (credible utility) for the [gene] sequence."[35] Thus, according to this policy a gene patent would not be awarded as a tool for scientific research but only in cases where a practical use has been demonstrated. The AMA policy also calls for access to gene patents by all "certified laboratories at a reasonable cost" so that other researchers will be able to build upon these innovations (Horn, 2002). These limited genetic patents would preserve incentives but minimize interference with downstream research initiatives.

But what about the protection of privacy and autonomy rights? Are we sacrificing these basic human goods for the sake of biomedical research? Absolutely not. We have made a tenable case on utilitarian grounds that property rights in genetic material are unsound since the social costs are disproportionate to the benefits received. Those costs include the inhibition of research and higher prices for therapeutic products and genetic screenings. Society benefits tremendously from the enhancement of human health through biomedical research and when that research is constrained the social welfare loss (such as undeveloped treatments or genetic screening products) is quite substantial. It may be that privacy and autonomy are marginally safer with a property right, but, as we will demonstrate in the next section, sound privacy legislation based on informed consent can also be an effective means of ensuring genetic privacy and autonomy.

Some might still argue that social welfare concerns such as innovation in biotechnology should not decisively trump an ownership right in one's genetic material and information. But there is no evidence that a property right is a necessary condition for protecting the privacy of genetic information or the autonomy of genetic data subjects. It is certainly possible to develop an alternative means for safeguarding genetic privacy. Thus, the enhancement of biomedical research and the protection of privacy are not mutually incompatible goals. The principle of informed consent can go a long way to protect basic human rights without the need for an exclusive entitlement.

Privacy and Informed Consent

In order to address the matter of privacy we can turn for some guidance to normative frameworks such as Moor's (2001b) notion of just consequentialism. As Moor and others have opined, it is problematic to resolve ethical questions such as this one purely on the basis of utility or social welfare concerns. While the "good" or end does justify the means to some extent, it should not justify the use of unjust or unfair means. One of the problems with consequentialism has been its lack of sensitivity to ethical principles such as justice in the name of pragmatically optimizing consequences.

---

[35]AMA Policy (2000). Report 9 of the Council of Scientific Affairs (1-00); available at: http://ama-assn.org/ama/pub/article2036-3063.html

In the case of genetic information, the moral imperative is to guarantee each individual's privacy and autonomy by keeping this information strictly confidential and respecting the person's right to determine how that information will be utilized. As we have demonstrated, genetic information is highly sensitive and its unauthorized distribution has the potential to cause significant harm. This need to safeguard confidentiality for the sake of privacy and autonomy is demanded by the principles of justice and fairness. An impartial observer, behind the Rawlsian veil of ignorance would not want to risk the loss of control over his or her genetic information. That person would not want such information to be made available to others without permission, given the potential for harm. Rather, she would want to be able to restrict access and determine how that information is utilized by third parties.

At the same time, the means of achieving this end, the protection of genetic privacy, are open and variable. The choice of these means should depend to some extent upon factors such as efficiency and social welfare. If there are several equivalent options for how we can achieve this moral imperative we should select the option that optimizes social welfare, which should be understood as the aggregate well-being of members of society. As long as we protect the basic human goods of privacy and autonomy, we should not adopt needlessly inefficient methods that might waste important opportunities such as the enhancement of human health. Prudence dictates that we attend to the consequences of our decisions and policies so long as we foster human well-being and do not interfere with one's integral participation in human goods such as privacy and autonomy. The ideal should be a solution that most fairly respects the rights of vulnerable stakeholders (donors of genetic material and genetic data subjects) while ensuring that the negative externalities associated with protecting those rights are minimized.

As we have been at pains to insist in the previous analysis, the upstream propertization of genetic material is most likely to hamper research by increasing transaction costs. Privacy and autonomy can also be protected by mechanisms that do not involve property rights, such as informed consent. We cannot review the nuances and details of a sound genetic privacy statute but we can suggest its general requirements. Well-crafted litigation should guarantee that "the individual has the right to be informed and to control subsequent use of biological material from the individual's body" (Ito, 2003). This legislation should prohibit the covert collection of DNA, "require an individual's voluntary consent for collection and analysis of DNA, and...require that the DNA collectors obtain a written, informed consent before dispersing any genetic information to anyone else" (McLochlin, 2001). The law should also prohibit genetic discrimination. Where applicable, the law should require hospitals and researchers to inform patients of their intention to use tissue or other genetic source material for commercial applications.

The overriding objective is to prevent misappropriation and misuse of this genetic data, and this goal can be accomplished by implementing these strictly enforced controls such as informed consent and the requirement of strict confidentiality. There is no empirical evidence that we cannot achieve this objective by means of this alternative as opposed to the granting of property rights. As a general principle, we should avoid treating information as property unless it is absolutely necessary. As Samuelson (1991)[36] has noted, "A world in which all information is...property under all circumstances is unthinkable."

Therefore, given that informed consent and legally mandated confidentiality requirements do not impose the same level of transaction costs as property rights, a presumption should be given to the use of informed consent as a means of assuring that privacy and autonomy rights will be safeguarded. The law should also be strictly enforced with stiff penalties for violators. It may be that the Iceland's Act on a Health Sector Database does not adequately protect privacy rights given the nature of the sensitive data in the IHD. But the remedy for this problem is not to succumb to the temptation to grant property rights in the genetic material of Iceland's citizens. Rather, the solution is to avoid presumed consent for the living and deceased and to require that each data subject's freely given informed consent be given before the inclusion of their medical and genetic data in the IHD.

Similarly, international law needs refinement and enhancement so that it will adequately protect the autonomy of indigenous people whose genetic material is attractive to foreign researchers. A discussion of how to avert gene piracy is beyond the scope of this essay, but Ching (1997) describes how international law could evolve to ensure that the right of self-determination is better safeguarded. Of course, the same basic principle of full informed consent would be a basic part of this solution. In addition, some developing countries are already requiring that those applying for intellectual property rights prove that they have the consent of native citizens.[37]

Finally, it should be pointed out that although the vision articulated by Miller (1969) and Branscomb (1994) of personal information as property has never been realized, privacy rights have not eroded throughout the world. While the U.S. track record on privacy protection is not exemplary, privacy rights are reasonably well protected in Europe thanks to the European Union Directive on Privacy. This directive requires that any "identifiable person" must be guaranteed fundamental privacy rights. It provides a model statement of principles (such as informed consent) that has been translated into regulatory systems which adequately safeguard the privacy rights of most European citizens.[38] The European experience demonstrates that property rights are not a necessary condition for the preservation of privacy and suggests that the protection of genetic privacy can be accomplished by means of a clear and consistent genetic privacy directive.

Access and Equity Issues Reconsidered

One last related issue deserves some brief elaboration. What about access to the DHA in Iceland? As we have seen, deCODE has argued that without exclusive access rights there is no incentive to invest, and there is some merit to this claim. Intellectual

---

[36]P. Samuelson. Is Information Property? Communications of the ACM, 34 (3): 15–18, 1991.

[37]The Right to Good Ideas. The Economist, June 23: 21–23, 2001.

[38]D. Elgesem. The Structure of Rights in Directive 95/46/EC on the Protection of Individuals with Regard to the Processing of Personal Data and the Free Movement of Such Data. In R.A. Spinello and H. Tavani (Eds.), Readings in Cyberethics, pp. 360–377. Jones & Bartlett, Sudbury, MA, 2001.

property rights or liability rules that dictate the terms of access generally provide that incentive. What's necessary in this case is a careful balance between a reward structure that acknowledges those incentives and broad access to this information for the sake of scientific research and the common good. While de-CODE's reward seems sufficient, more should be done to ensure greater access to this resource. One of the major criticisms of the IHD was lack of access for scientists who were not employed by deCODE. We must object to this if we consistently follow the principles suggested by our analysis that emphasize limited genetic patents and broader access to genetic data. If we deny that Iceland's citizens have a proprietary right in their genetic source material for the greater good of scientific research, it seems contradictory to assert that deCODE Genetics should have a de facto property right by virtue of its exclusive access to this information and its ability to reap the lion's share of the rewards. Neither the Iceland Government nor deCODE Genetics should be vested with a quasi–property interest that gives them exclusive, custodial control over this research data.

Currently, the Iceland Act on a Health Sector Database (1998) provides for a committee that will make all decisions about access, but one of the three members of that committee is a deCODE employee. The committee is expected to allow access unless there is an "adverse effect" upon deCODE's commercial interests. There seem to be two problems with this provision. First, this criterion is too general and one-sided; other factors need to be considered such as the merits and expected results of the proposed research. Second, it is reasonable to assume that the deCODE representative will take a very broad view of the term "adverse effects" and as a result legitimate access will be denied. According to Jonatansson (2000), this cozy arrangement implies that "scientific access to data is...not accorded on scientific grounds."

What's necessary is a better balance between regard for deCODE's financial incentives and the need for less-fettered access to this data. One possibility is a compulsory licensing scheme that would avoid the potential for biased judgments by this committee. This scheme, if properly and fairly implemented, will provide deCODE with compensation for its investment without proprietary controls. Such an arrangement seems to be a more equitable means of ensuring the availability of this data while avoiding exclusive access to this data, which is tantamount to a quasi–property right for deCODE Genetics. Another alternative for providing liberal access to this data on objective grounds is to redefine the criteria for access. As we have suggested, those criteria should take into account the scientific merits of proposed research along with "adverse effects" on deCODE's investment. Efforts should also be made to ensure that the committee is impartial so that the public interest is fully represented. The committee should make its determinations about access to and licensing of this data based purely on the merits of each case (Jonatansson, 2000).

Finally, what about the issue of compensation for Iceland's DNA donors and others who contribute their genetic material? Justice demands a fair distribution of the burdens and benefits of research. While donors should be fully informed about the profit potential and incentives of deCODE Genetics, if they willingly and altruistically provide their genetic material it may be hard to see any inequity. In some cases, however, a donor may deserve to share more directly in the tangible benefits of successful research.

Informed consent, which must include informing donors about the potential financial value of their donated genetic material (particularly diseased material), is a good starting point for protecting donors' interests. The presumption, however, is that any cell line for downstream products (such as a test for a genetic mutation) would be developed from an aggregation of many donor cells. As Epstein (2003) observes, "if genetic materials gathered from a large cohort of treated individuals is used to concoct some new genetic compound…no compensation…seems appropriate." In the vast majority of cases the rewards for each Icelander's participant would be minimal. On the other hand, if there is a situation where a cell line is developed exclusively from one donor's cells and that cell line had great therapeutic value, the donor deserves to share in the benefits of this research as a matter of fairness. We must acknowledge the valuable nature of this raw biological material before it is donated for research along with the donor's entitlement to some compensation for providing that material. As a general principle, companies and research labs that refuse to compensate under these circumstances are acting unjustly and should be subjected to moral and social pressures if they fail to live up to their obligations. These situations are rare, however, and should not require a new regime of property rights in human tissue. Beyond any doubt, there is a need to better allocate the financial benefits of genetic research that takes into account the donor's interests when that donor makes a contribution that has material value. But this should be accomplished through revised public policy rather than the introduction of new property rights.

Conclusions

This paper has scrutinized in a cursory way deCODE's unique agreement with the government of Iceland, which gives the company an exclusive license to construct a genetics database using medical records and genetic information from patients. This type of genetic research is being emulated by other private companies engaged in "bio-prospecting," that is, the development of genetic databases of small homogenous populations. The purpose of this bio-prospecting is to locate key genes with potential medical significance. The collection of this data raises troubling privacy issues and the deCODE case allows us to see how positions regarding privacy rights and access privileges can be polarized. Proprietary rights in source material such as cells and genetic information will better protect privacy and ensure some compensation for donors. However, recognizing these "rights of sources" will preclude unfettered access and hinder genetic research. A regime of exclusive ownership also entails the risk of opportunistic hold-up. On the other hand, if this genetic data is simply considered to be part of the public domain, research opportunities will be greatly enhanced. But privacy rights might be in some jeopardy and data subjects will have no ability to recover any benefits from the commercial exploitation of their genetic data. There is a compelling need, therefore, to find some reasonable middle course of action, a proper balance between data protection and access.

While that middle course may be hard to discern, we must resist the impulse to propertize genetic information and the DNA samples from which it is derived, despite legitimate concerns about the need to safeguard privacy rights. A property regime for genetic information is an overreaction and a misguided solution to this problem. As we have insisted, genetic data should be unencumbered by proprietary rights, as the State of Oregon has belatedly realized. Otherwise we will end up with a tragedy of the anti-commons. The *Moore v. Regents of California* case has also confirmed this conclusion. Upstream property rights are not conducive to the advancement of scientific research and must be replaced by an ethos of information sharing. Those rights include patents for genes and gene fragments, which should be awarded on a more limited basis in accordance with the AMA's policy. Instead of relying on property rights, privacy and autonomy should be safeguarded by strongly enforced laws that protect genetic information by informed consent and tight regulations governing the disclosure of such information (including genetic test results).

It remains somewhat unclear whether the Icelandic Government or deCODE Genetics fully accepts that philosophy, since they rely in some cases on presumed consent. Informed consent should be implemented as the universal standard for inclusion. Also, as we have argued, if data subjects must forego proprietary rights in their genetic information for the sake of facilitating research, then we must conclude that a private company or the government should not have proprietary entitlements to data, since such an entitlement might also impede genetic research among other scientists throughout the world. Instead, there is a need for a liberal access policy to this critical genetic data that takes into account the necessary role of investment incentives. Compulsory licensing to certified researchers at a reasonable cost is one way to achieve this tenuous balance.

References

Act on a Health Sector Database no. 139, 1998, available at http://brunnur.stjr.is/interpro/htr/htr.nsf/pages/gagngr-log-ensk

AMA Policy. Report 9 of the Council of Scientific Affairs (1-00) 2000, available at: http://ama-assn.org/ama/pub/ article2036-3063.html

L. Andrews. My Body, My Property. Hastings Center Report, October 28, 1986.

G. Annas. Rules for Research on Human Genetic Variation—Lessons from Iceland. New England Journal of Medicine, 342: 1830, 2000.

G. Annas, L. Glantz and P. Roche. Genetic Privacy Act, 1995; available at: http://www.bushp.bu.edu/Depts/ Healthlaw

J. Boyle Shamans, Software and Spleens: Law and the Construction of the Information Society. Harvard University Press, Cambridge, MA, 1996.

J. Boyle. Enclosing the Genome: What the Squabbles over Genetic Patents Could Teach Us, 2003, available at: http://www.creativecommons.org/licenses/by-sa/1.0

A. Branscomb. Who Owns Information? From Privacy to Public Access. HarperCollins, New York, 1994.

K. Ching. Indigenous Self-Determination in an Age of Genetic Patenting: Recognizing an Emerging Human Rights Norm. Fordham Law Review, 62: 687, 1997.

Diamond v. Chakrabarty, 447 U.S. 303, 1980.

D. Elgesem. The Structure of Rights in Directive 95/46/EC on the Protection of Individuals with Regard to the Processing of Personal Data and the Free Movement of Such Data. In R.A. Spinello and H. Tavani, editors, Readings in Cyberethics, pp. 360–377. Jones & Bartlett, Sudbury, MA, 2001.

R.A. Epstein. Steady the Course: Property Rights in Genetic Material. The Chicago Working Paper Series, 2003, available at: http://www.law.uchicago.edu/Lawecon

C. Goldwater. Iceland Exploits its Genetic History. BBC News, February 4th, 2000, available at: http://news.-bbc.co.uk/hi/english/sci/tech/newsid630000/630961.stm

C. Harrison. Neither Moore nor the Market: Alternative Models for Compensating Contributors of Human Tissue. American Journal of Law and Medicine, 28: 77, 2002.

M.A. Heller and R. Eisenberg. Can Patents Deter Biomedical Research? The Anticommons in Biomedical Research. Science, 280: 698, 1998.

N. Hettinger. Patenting Life: Biotechnology, Intellectual Property and Environmental Ethics. Boston College Environmental Affairs Law Review, 22: 267, 1995.

M. Holman and S. Munzer. Intellectual Property Rights in Genes and Gene Fragments. Iowa Law Review, 85: 735, 2000.

M.E. Horn. DNA Patenting and Access to Healthcare: Achieving the Balance among Competing Interests. Cleveland State Law Review, 50: 253, 2002.

A. Ito. Privacy and Genetics: Protecting Genetic Test Results in Hawaii. Hawaii Law Review, 25: 449, 2003.

H. Jonatansson. Iceland's Health Sector Database: A Significant Head Start in Search of the Biological Grail or an Irreversible Error. American Journal of Law and Medicine, 26: 31, 2000.

W. Klug and M. Cummings. Essentials of Genetics 2nd ed. Prentice-Hall, Upper Saddle River, NJ, 1996.

M.J. Lin. Conferring a Federal Property Right in Genetic Material: Stepping into the Future with the Genetic Privacy Act. American Journal of Law and Medicine, 22:109,1996.

J. Locke. The Second Treatise of Government. Bobbs-Merrill, Indianapolis, 1952.

D. McLochlin. Whose Genetic Information is it Anyway? John Marshall Journal of Computer and Information Law, 19: 609, 2001.

A. Miller. Personal Privacy in the Computer Age: The Challenge of New Technology in an Information-Oriented Society. Michigan Law Review, 67: 1203, 1969.

J. Moor. Towards a Theory of Privacy for the Information Age. In R.A. Spinello and H. Tavani, editors, Readings in Cyberethics, pp. 349–359. Jones & Bartlett, Sudbury, MA, 2001a.

J. Moor. Just Consequentialism and Computing. In R.A. Spinello and H. Tavani, editors, Readings in Cyberethics, pp. 98–105. Jones & Bartlett, Sudbury, MA, 2001b.

Moore v. Regents of California 793 F 2d 479 [Cal.], 1990; cert denied 111 S. Ct. 1388, 1991.

T. Murray. Genetic Exceptionalism and Future Diaries: Is Genetic Information Different from other Medical Information? In M. Rothstein, editor, Genetic Secrets: Protecting Privacy and Confidentiality, Yale University Press, New Haven, 1997.

A.M. Pinto. Corporate Genomics: DeCode's Efforts at Disease Mapping in Iceland for the Advancement of Science and Profits. University of Illinois Journal of Law, Technology and Policy, 202: 467, 2002.

A. Pollack. U.S. Hopes to Stem Rush Toward Patenting Genes. The Patriot Ledger, June 28: 18, 2002.

D. Resnik. DNA Patents and Human Dignity. Journal of Legal and Medical Ethics, 29: 52, 2001.

The Right to Good Ideas. The Economist, June 23: 21–23, 2001.

J. Rule and L. Hunter. Towards Property Rights in Personal Data. In C. Bennet and R. Grant, editors, Visions of Privacy, pp. 168–181. University of Toronto Press, Toronto, 1999.

P. Samuelson. Is Information Property? Communications of the ACM, 34(3): 15–18, 1991.

G. Shaw. Does the Gene Stampede Threaten Science? AAMC Reporter, February, 2000. Utility Examination Guidelines. 66 Fed. Reg., 1092, 2001.

N. Wade. Double Helix Leaps from Lab into Real Life. The New York Times, February 25: D6, 2003.

A. Westin. Privacy and Freedom. Atheneum, New York, 1967.

J. Weems. A Proposal for a Federal Genetic Privacy Act. Journal of Legal Medicine, 24: 109, 2003.

13. Molecular Biologists as Hackers of Human Data: Rethinking Intellectual Property Rights for Bioinformatics Research

Antonio Marturano

> "In order to combat the huge agricultural corporations such as Monsanto and their ridiculous claims of intellectual property rights on genetic code, we need a movement similar to the Open Source movement of the computer industry.
>
> Groups of geneticists and farmers can work together to produce enhanced products without having to cater to these huge corporations.
>
> However, I will still just continue to buy organic to make sure I get the real things as it was intended."
>
> Evoke the Tiger, July/26/2001

1. Introduction

According to Gezelter (1999), modern science relies to a very large extent on computer simulations, computer models, and computational analyses of large data sets. To what extent is the new genetics dependent upon computing/information technology? Castells (2001: 264) argues that "[w]ithout the massive computing power and the simulation capacity provided by advanced software, the Human Genome Project would not have been completed—nor would scientists be able to identify specific functions and the locations of specific genes." Noting also that "there is theoretical convergence between the two [information and genetics] technological fields," Castells believes that these technologies are "obviously information technologies, since they are focused on the decoding and eventually reprogramming of DNA, the information code of living matter" (id.). Given this interpretation, some philosophical issues arise with respect to whether we can properly regard notions such as DNA and the Human Genome as pure informational concepts.

This essay originally appeared in the *Journal of Information, Communication and Ethics in Society*, **1**: 207-215. Copyright © 2003 by Troubador Publishing, Ltd. Reprinted by permission.

2. The Philosophical Framework

New genetics regards the Human Genome and DNA as if they were concrete and real objects of nature as a table or a sand of grain are. An alternative view, according to the orthodox philosophy of science (as expressed in the so-called Popper-Hempel model), is to understand such kinds of entities (such as the Genome or the DNA), on the contrary, as theoretical. Theoretical entities are, according to this *non-realist* interpretation, unobservable entities that are theory-dependent. We don't have empirical evidence they actually exist; rather, we have an indirect (weak) sign, or, so to speak, circumstantial evidence (by the theory which presupposes it) of their existence. On the contrary, we understand observable terms such as those terms that describe things directly observable.

The main tool for observing DNA is the X-rays diffraction technique (Watson, 1991: p. 24), which is a kind of indirect method of observation. It is indeed possible to achieve some diagrams of diffraction with a beam of X-rays. Those diagrams will allow us to *deduce* partly or completely structures of nucleic acids. Particularly, diagrams of nucleic acids are quite poor in details, both in disuniformity of the constituency units in the chain and in the lacking of co-ordination within the adjacent molecules. Nonetheless such diagrams are good enough to individuate some of the fundamental parameters in the structure (De Petris, 1960).

Some have no particular problems with the presupposition that theories are correct and unchangeable (in this way, we could interpret Lewontin's claims that new genetics has embedded a kind of "astrological" or religious vision of its background theories—see Lewontin, 2000, p. 137). But theories are subject to continuing and stressing revisions, even in the actual practice of molecular biologists' studies. Actual genomic theories could, in the future radically change or significantly change in accord with the "falsificationist" position.

The challenge to the position in genetics, in my opinion, is actually arising from the second step of the Human Genome Project, that is the understanding of the meaning and function of genes (Trivedi, 2000). The "Shotgun" and other methods of genetic sequencing have led to rough drafts of the human genome. However, according to Lander (one of the scientists at Celera), the work is not complete: "We have the structure but no meaning yet". Scientists, in fact, have successfully identified the sequence of nucleic acids, but not the location of all the genes or their precise functions. Adding to the complications is the fact that 99% of the human genome is "junk" DNA and does not code for genes. Whether these nucleic acids even have a purpose is still unknown. In other words, the ongoing research about the "meaning" of the Human Genome can revolutionise our ideas and concepts of human biology in an unexpected way, which can revise our previous conceptions.

According to Lewontin (1993: p. 66), the actual reason in devising causal information from DNA messages is that the same "words" (that is the amino acid triplette) have different meanings in different contexts and have multiple functions in a given context, as in any complex language. Lewontin (2000: p. 152) claims that we do not know how the cell decides among the possible interpretations. In working out the interpretative rules (that is, rules which decide whether code sequences are meaningful), it would certainly help to have a very large numbers of different gene sequences, and Lewontin himself suspects that the claimed significance of the genome sequencing project for human health is an elaborate cover story for an interest in the hermeneutics (I would say a computerised one) of biological scripture (Lewontin, cit., p. 153).

The "realist" way molecular biologists use to talk about DNA and the Human Genome is, in my opinion, rather misleading and, from a point of view of the scientific studies, suspicious. The scientific community, in fact, should be open to any discordant opinion at all times. In the case of the theoretical representation of the DNA in genetics there is quite a monolithical agreement despite the history of the DNA structure, which has its roots, as Watson and Crick acknowledged, in the theory of communication by Shannon and Weaver (computers again!).[1]

3. The Human Genome Project as a Bio-Informatics Project

"I thought Open Source Genetics would be about university days, when you tried to share your DNA with as many people as possible."

UnaBubba, Aug/01/2001.

Geneticists and molecular biologists are confounded by the claim raised by several scholars that they are not working on the human genome "in vivo" or "in vitro" (as in the traditional experimental research), but are working instead "in silico" (Kaku, 1998, p. 158); that is, they are working with one or more representations of actual entities (such as the genome or the DNA) which are being sequenced using computers. In other words, the idea is that the Human Genome Project was not a biomedical project but a bio-informatics one (OSCE Forum Megascience, 1995: p. 16).

3.1 The Shotgun Method

A well known example of the inter-relatedness between computer technology and molecular biology for the studies of the Human Genome is the so-called "Shotgun" sequencing method (which is indeed one of the methods, even the most famous, used for sequencing the genome). The Shotgun method is a mathematical algorithm—that is, computer software. The Shotgun method involves randomly sequencing tiny cloned sections of the genome, with no foreknowledge of where on a chromosome the section originally came from. The partial sequences obtained are then reassembled to a complete sequence by use of computers. The advantage with this method is that it eliminates the need for time-consuming mapping because of increased computer speeds to solve such complicated algorithms (Trivedi, 2000). In fact, other methods of sequencing such as BAC to BAC sequencing needed to create a crude physical map of the whole genome before sequencing the DNA. In the BAC to BAC sequencing computers become important in the final step of the protocol when the sequences collected in the so-called MI3 libraries are fed into a computer program called PHRAP, which looks for common sequences that join two fragments together (Trivedi, 2000).

In the case of the Shotgun method we cannot see geneticists as working on living material (although perhaps there is some work on living material in the first phase of the sequencing, see Trivedi, 2000). Rather, they are working on representations of DNA sequences. Moreover, the Shotgun method is able to shift two steps in the sequencing, because of its abstractness. Geneticists' work, therefore, is similar to the software developers who are working on complicated mathematical functions in order to create useful software.

Genetics, according to Karp, is turning into an information science: "Many biologists consider the acquisition of sequencing to be boring. But from a computer science point of view, these are first-rate and challenging algorithmic questions" (cited in Kaku, cit. p. 157). According to Lewontin, "Many of the founders of molecular biology began as physicists, steeped in the lore of the quantum mechanical revolution of the 1920s. The Rousseau of molecular biology was Erwin Schrödinger, the inventor of the quantum wave equation, whose *What is Life?* (1944) was the ideological manifesto of the new biology" (Lewontin, 2000: p. 136).[2] Today, "The intrusion of computers into molecular biology shifted power into the hands of those with mathematical aptitudes and computer savvy" (Kaku, p. 158).

4. The Open Source Philosophy

According to Barr (2001), the Open Source philosophy is often confused with the Free Software movement.[3] They share, indeed, some practical projects but have different views and outcomes. Basically they are connected because their polemical target is proprietary software.

Open Source doesn't just mean access to the source code but, accordingly to the definition provided by the Open Source Initiative (2003), it must comply with the following criteria:

1. Free Redistribution: the license shall not require a royalty or other fee for such sale;
2. Source Code: the program must include source code, and must allow distribution in source code as well as compiled form. The source code must be readily available;
3. Derived Works: the license must allow modifications and derived works, and must allow them to be redistributed under the same terms as the license of the original software;
4. Integrity of the Author's Source Code: users have a right to know who is responsible for the software they are using. Authors and maintainers have reciprocal right to know what they're being asked to support and protect their reputations;
5. No Discrimination Against Persons or Groups;[4]
6. No Discrimination Against Fields of Endeavour;
7. Distribution of License: the program may be redistributed without the need for execution of an additional license by those parties;
8. The License Must Not Be Specific to a Product;
9. The License Must Not Restrict Other Software: distributors of Open Source software have the rights to make their own choices about their own software;
10. The License Must Be Technology-Neutral.

It is instructive to make some observations about the rationale for licensing. The first criterion alludes to the potential universal distribution of the licensed product. Although these ten criteria are talking about software, the last protocol (number 10) could suggest that the concept of Open Source "may be predicated on any individual technology or style of interface" (see Open Source Initiative, 2003) and, in particular, it cites explicitly genetics research. The Open Source philosophy therefore could be used as a model to be applied to other spheres of technology because of its universality and neutrality. The Open Source philosophy, finally, does not exclude business or commercial enterprises: "We want commercial users to join our community, not feel excluded from it" (ibid.).

The main limits of the Open Source definition is that it does not provide any restrictive clause in case of breach. It is very difficult to understand which kinds of punishment the infringer could incur besides an implicit threat to be cast out from the Open Source "Heaven". In other words, the Open Source definition accepts principles of distributive,[5] contributive,[6] and commutative[7] justice, but there is a gap in terms of retributive[8] justice.

Finally, the Open Source definition, because of its application to Big Science, which involves research centres from all over the world, seems to be a sort of explicit and updated ethical code. It is reminiscent of Merton's four rules for scientific practice. Such an ethical code may seem utopian because of its lack of coercion and its need to emphasise the notion of "hacker attitude".

4.1 The Notion of Hacker

The notion of "hackers" proposed here is not the one used in the ordinary sense of the term. That concept sees hackers (mainly adolescent males) who get a kick out of breaking into computers and phreaking the phone system. "Real hackers" call these people "crackers" and want nothing to do with them. Real hackers also tend to think crackers are lazy, irresponsible, and not very bright. They also object that being able to break security doesn't make you a hacker any more than being able to hotwire cars makes you an automotive engineer. Unfortunately, many journalists and writers have used the word "hacker" to describe crackers, and this irritates real hackers.

When I refer to "hackers," I mean those people who solve problems and build things; they believe in freedom and voluntary mutual help. To be accepted as a hacker, you have to behave as though you have this kind of attitude yourself. And to behave as though you have the attitude, you have to really believe the attitude.[9] The philosophy of Open Source explains this idea better: when programmers can read, redistribute, and modify the source code for a piece of software, the software evolves. People improve it, people adapt it, and people fix bugs. And this can happen at a speed that seems astonishing, especially if one is accustomed to the slow pace of conventional software development. The open source community has learned that this rapid evolutionary process produces better software than the traditional closed model, in which only a very few programmers can see the source and everybody else must blindly use an opaque block of bits.

5. Adopting the Open Source Philosophy in the Genetics Field: Challenges for Intellectual Property Rights

"'I claim this land in the name of the Queen.'
'But...excuse me...em...we already live here.'
'Do you have a flag?'
'No.'
'Sorry, no flag, no country'."
etgrrl, Aug/01/2001.

In my view, the ideas discussed above are relevant for a rethinking of the Intellectual Property Rights (IPR) in the field of Genetics. My proposal is to regulate genetics discoveries not in term of patents but in terms of copyrights. Patenting is used for a type of industrial property, which is protected primarily to stimulate the innovation, design and creation of the technology. The social purpose is to provide protection for the results of investment in the development of new technology, thus giving the incentive and means to finance research and development activities. A functioning intellectual property regime should also facilitate the transfer of technology in the form of foreign direct investment, joint ventures and licensing. The protection is usually given for a finite term (typically 20 years in the case of patents). On the other hand, the rights of authors of literary and artistic works (such as books and other writings, musical compositions, paintings, sculpture, *computer programs* and films) are protected by copyright, for a minimum period of 50 years after the death of the author. The main social purpose of protection of copyright and related rights is to encourage and reward creative work.

In the genetics field one value of patenting a gene sequence, according to Lewontin (2000), "lies in its importance in the production of targeted drugs, either to make up for the deficient production from a defective gene or to counteract the erroneous or excessive production of an unwanted protein". Alternatively, Lewontin continues, "the cell's production of a protein code by a particular gene, or the physiological effect of the genetically encoded protein, could be affected by some molecule synthesized in an industrial process and sold as a drug. The original design of this drug and its ultimate patent protection will depend upon having rights to the DNA sequence that specified the protein on which the drug acts" (Lewontin, 2000: p. 181). Lewontin concludes, "Were the patent rights to the sequence in the hand of a public agency like the NIH, a drug designer and manufacturer would have to be licensed by that agency to use the sequence in its drug research, and even if no payment were required the commercial user would not have a monopoly, but would face possible competition from other producers" (Lewontin, cit. p. 182). But, in my opinion, an international competition would be difficult, because, as Lewontin claims, it wants "the NIH to patent the human genome to prevent private entrepreneurs, and especially foreign capital, from controlling what has been created with American funding" (Lewontin, 1991: p. 75). Following the successes of Celera Genomics, lead by Craig Venter, it could be argued that the actual patenting strategy has focused on protecting the interest of the few corporations working in this field. As Sir W. Bodmer says, "the issue of ownership is at the heart of everything we do" in the field of genomics (id.).[10]

6. Conclusion

"GNU genome licensing—yeah! I'll volunteer my DNA."

Dog Ed, Jul/26/2001.

Finally, another important feature that the new genetics shares with the computer is the low technological costs. According to Kaku, "With only a modest $10,000 investment, one can conduct biotech experiments in one's living room and begin to manipulate the genome of plants and animals. With a few million dollars, one can create a fledging biotech industry. The low initial investment, high return, and potential for feeding its people are some reasons why a poor nation such as Cuba has decided to jump into biotechnology" (Kaku, cit.: p. 244). Information technology has similar or even lower costs, and even similar potential economic returns, which make both technologies a resource for underdeveloped nations that wish to invest in technological innovation.

The New Genetics is, in my opinion, a field quite similar to Computer Science."[11] Genetics theories are more like a work of literature (that is, a patient and scrupulous coherent reconstruction of our genetics knowledge) and partly like computer software, which helps us in this scientific reconstruction. Another advantage from the change of paradigm in the regulation of IPR in the genetics field is to adopt an "open-source" philosophy (which resembles in its basic nature the old and original way of scientific research) in the genetics research. An involvement of a larger number of researchers (even if for small subprojects) and a probable increased speed of the research are the benefits. On the other hand, we will witness the decreasing power of big pharmaceutical multinationals and a reduction of secrecy in software corporations, which prevent unauthorised copying. Stallman (1994) claims that such practices resemble those used in the former Soviet Union; but the motive for information control in the Soviet Union was political, while in the USA the motive is profit. Stallman concludes, "Any attempt to block the sharing of information, no matter why, leads to the same methods and the same harshness" (ibid.).

Obviously, regulating genetic research using the Open Source paradigm (that is, using licenses rather than patenting) will yield flexible and less rigid regulations. In my view, this kind of option would help to shorten the gap between developing and developed nations, at least in the field of genetics and pharmaceutical research.

I have defended the view that genetics, when interpreted in a realist (as opposed to "non-realist") way, becomes the science of biological *information*; it involves management, processing and modification of genetic information by mean of computers. This also requires the resources of an "information rich" country.

The global economic production structure is rapidly changing from a manufactured-based to an information-based one (accentuated by globalisation) which, in turn, is stimulated mainly by the phenomenon called advanced capitalism (Rifkin, 1995: p. 236). On the contrary, the Open Source philosophy promises to shift, according to Raymond, to a "gift economy," where status among peers is achieved by giving away things that are useful to the community. In particular, social aspects of science work in a similar way; activities such as publishing papers, giving talks, and sharing results can help scientists to obtain status among scientific peers. Science, in this sense, is a sort of gift economy of ideas (see Gezelter, cit.); the Open Source model thus gets at the basic nature of the old and originary way of scientific research.

The Open Source philosophy as a model for scientific enquiry fits Merton's four related norms of scientific practice; these norms, according to Merton, were guidelines for the practice of scientific enquiry in order to ensure the growth of certified knowledge. Universalism means that scientific truths are of impersonal kinds, that is valid *erga omnes*, independently from the scientist and the place of discovery. Communitarism means that science is basically a social practice, based on past efforts influencing the future ones. Disinterest is about a scientist's commitment to truth as his/her first motivation. Organised scepticism means the valuation of possible truths by means of open debate, peer review, and experimental replicability. According to the analysis of Rabinow, the most important reward for a scientist is appreciation by their community and professional prestige. According to Merton, the trick in the system stems from the fact that scientists, while working for their interest, are on the other hand reinforcing collectively the public good (Rabinow, 1996: p. 22). This model indeed has been recently revised, especially under the change of paradigm in the biotechnology industry under the so-called patenting-and-publish strategy (Rabinow, cit., p. 37), which entails the merging between scientific research and business. Although debate and discussions continue, a large biotechnology industry funded by a massive infusion of venture capital and an equally significant amount of capital from large, often multinational, pharmaceutical companies has become an established force (Rabinow, 1999: p. 3).

The most appealing feature of the Open Source philosophy for genetics and software research is the fact that it is possible to create a research network based on the model that the source code can be given away and other researchers can fix and improve that software. Open Source projects also tend to have much stronger communities. The entire premise is one based on sharing and the enjoyment of creation for the good of the community (Torvalds, 2001). As systems become more complex, calculations that are verifiable in principle, such as those used in genetics, are no longer verifiable in practice without a public access to the code; it is therefore pivotal for sceptical (or neutral) scientific inquiry that software used for simulating complex systems be available in source-code form (Gezelter, 1999). Therefore the Open Source philosophy could bypass the problem that in underdeveloped countries there is no proper or adequate "info-structure" based on ICT and knowledge about the access and use thereof, which will limit a society's participation in the virtual economy and will hamper global trade. This also would avoid cutting poorer countries off from the higher growth sectors (such as the biotechnology) of the global economy, and would enable everyone an equal opportunity to take part in the different economic processes including the creation, processing and use of information (Britz and Blignaut, 1999). The redistribution of resources allowed by Open Source would be led, (paradoxically perhaps) not directly by the ICT (information and computing technology) corporations but from the more sensitive field of bioinformatics which makes an abundant use of ITCs.

Acknowledgments

My work on this project began during my first research visit at the Institute of Cultural Research, Lancaster University (1998). I would like to thank Paolo Palladino and Simon Rogerson for their helpful discussions and their support on this project, and I would like to thank Rick Spinello for his helpful comments on an earlier draft of this chapter.

Notes

All mottoes are from http://www.halfbakery.com/idea/Open_20Source_20Genetics;
   1. There is no general agreement about the paradigm used by the new genetics: Castells is indicating in the work of Fritjof Capra the main genetics paradigm (*op.* and *loc. cit.*) while Watson (1999),

and in a recent interview on *Scientific American*, states his debt to E. Schrödinger's work *What Is Life?* (1944), where Schrödinger boldly asserted that living things could be understood by the quantum theory of atoms and that life was governed by a "genetic code" (a sentence he coined) locked in the arrangement of our molecules.

2. Watson tells us that a major factor in his colleague F. Crick leaving physics and developing an interest in biology was the reading of Schrödinger's book. Watson comments on that book as follows: "This book very elegantly propounded the belief that genes were the key components of living cells and that, to understand what life is, we must know how genes act. When Schrödinger wrote his book (1944), there was general acceptance that genes were special types of protein molecules" (Watson, 1977: p. 23).

3. For a brief history of the Open Source movement, see Grodzinsky, Miller and Wolf (2003).

4. This criteria seems to be based on Rawls' principle that "Each person is to have an equal right to the most extensive basic liberty with a similar liberty for others (Rawls 1973: p. 60). This principle is based on the anthropological viewpoint that human beings are all fundamentally equal and have equal intrinsic human rights. Applied to information poverty it must imply that everybody, irrespective of colour, race, gender or religion has a fundamental right of access to information needed to satisfy all basic human rights (Britz and Blignaut, 1999).

5. In its broad sense, distributive justice is concerned with the fair allocation of the benefits of a particular society (i.e. income, wealth, power and status) to its members. According to Blignaut and Britz (*op. cit.*), because developing countries often use knowledge produced in industrial countries, they have a responsibility to disseminate it to these developing countries. This would imply the finding of new innovative ways of maintaining incentives to create knowledge (intellectual property) while ensuring its broad diffusion.

6. Contributive justice implies that an individual has an obligation to be active in the society (individual responsibility), and that society itself has a duty to facilitate participation and productivity without impairing individual freedom and dignity. One of these duties would therefore be the creation of equal opportunities.

7. Commutative justice calls for fundamental fairness in all agreements and exchanges between individuals or social groups. In its economic application, it calls for equality in transactions.

8. It refers to the fair and just punishment of the guilty as well as the punishment of the guilty.

9. The notion of "believing the attitude" is very similar to the notion of "ethical internalism". The notion of ethical internalism has been recently defended by Bernard Williams where, in order for something to be a genuine reason for action, it must be related to what Williams terms the agent's "subjective motivational set" or "thus making reasons in some sense necessarily relative to each individual agent's psychological make-up. Many important critics of Williams' views, however, raise at least two standard objections: first, that internalism leads to a radical agent-relativity about reasons, thereby impugning the alleged normativity of practical judgments; and second, that internalism unjustifiably excludes *a priori* what may be described as a more "realist" viewpoint according to which reasons for action can exist totally independent of any particular agent's contingent motivations.

Secondly, a basic worry arises here: Does this internalist/externalist debate amount to a mere terminological dispute? In general, it appears that internalists simply choose to define "reasons" in such a way that they must be necessarily associated with what an agent has motivation to do, whereas externalists instead define "reasons" as referring to certain alleged "practical truths" that may exist entirely apart from an agent's motivations (Williams, 1981).

10. Another important feature in the genetic patenting quarrel is the distinction between what is "natural" (which cannot be patented) and what is not natural (which can be patented). Isolated genes are not natural and as such they can be patented even though the organism from which they are taken may be. An analogous quarrel was in the field of information technology with computer programs: are they a *sui generis* formalised reasoning? If they are, they cannot be patenting or copyrighted. If, on the contrary, they are similar to a new chemical compound or something producing something new, they can be patenting or copyrighted. It could be argued that some software (such as an operating system or an Internet browser) are public goods: we cannot indeed use a computer or search the Internet without such tools. Such software therefore is neither divisible nor excludable and should be free.

11. The analogy between hackers and molecular biologists, however, has been voluntarily presented in a questionnaire to molecular biologists in an ambiguous way, as one of the research aims was to explore if there was room for a molecular biologist actually to act as a hacker (in the second sense) for commercial purposes. The result was that I sent the questionnaire to about 100 personality figures in the world of biology, computer science, law and bioethics. Only a very small number of them replied to my questionnaire (mainly from computer science). Molecular biologists avoided replying, likely because, as I understood during a private conversation with one of them, they simply refused to be paralleled to hackers, and rejected the questionnaire.

A general observation could be drawn from this experience: is the molecular biology community an "open-to-criticism" community?

References

Barr, J. (2001) Why "Free Software" is better than "Open Source", in Stallman, R. (2002) http:1 lwww.gnu.org./philosophy/free-software-for-freedom.html (accessed 03/0412003).

Britz, J.J. and Blignaut, J. N. (1999) An Ethical Perspective on Information Poverty and Proposed Solutions, in D'Atri, A., Marturano, A., Rogerson, S. and Bynum, T. (eds.) (1999).

Castells, M. (2001) Informationalism and the Network Society, in Himanen P. (2001), pp. 155–178.

D'Atri, A., Marturano, A., Rogerson, S. and Bynum, T. (eds.) (1999) *Proceedings of Ethicomp 99, Look to the Future of Information Society*, Rome, Luiss CeRSIL, ISBN 88-900396-0-4.

De Petris, S. Diffrazione dei raggi X (X-rays diffraction) (1964), in *McGraw-Hill Encyclopedia of Science and Technology*, New York, McGraw-Hill, 2nd Italian edition, pp. 49–54.

Gezelter, D. (1999) Catalyzing Open Source development in science: The OpenScience Project;http://www.openscience.org/talks/bnl/imgo.htm (accessed 03/04/2003).

Grodzinsky, F., Miller, K. and Wolf, M. (2003) Ethical Issues in Open Source Software, *Journal of Information and Communication in Ethics and Society,* vol. 1, no. 4, 2003, pp. 193–205.

Hempel, C. (1966) *Aspects of Scientific Explanation*, New York, Free Press.

Himanen, P. (2001) *The Hacker Ethic and the Spirit of the Information Age*, London, Vintage.

Kaku, M. (1998) *Visions. How Science Will Revolutionize the Twenty-First Century*, Oxford, Oxford University Press.

Lewontin, R. (1993) *The Doctrine of DNA, Biology as Ideology*, London, Penguin.

Lewontin, R. (2000) *It Ain't Necessarily So*, London, Granta.

OCSE Forum Megascience (1995) *Le Grand Programme sur le Génome Humain*, Paris, Les Éditions de l'OCDE.

Open Source Initiative (OSI) (2003) *The Open Source Definition*, version 1.9; http://www._opensource.org/docs/definition.php (accessed 03/04/2003).

Popper, K. (1959) *The Logic of Scientific Discovery*, London, Routledge.

Rabinow, P. (1996) *Making PCR: A Story of Biotechnology*, Chicago, Chicago University Press.

Rabinow, P. (1999) *French DNA. Trouble in Purgatory*, Chicago, Chicago University Press.

Rifkin, J. (1995) *The End of Work. The Decline of the Global Labor Force and the dawn of the Post-Market Era*. G.P. New York: Putman's Sons.

Schrödinger, E. (1944) *What Is Life? The Physical Aspect of the Living Cell*, Cambridge, Cambridge UP.

Stallman, R. (1994) Why Software Should Not Have Owners, in Stallman, R. (2002), http://www.gnu.org./philosophy/whyfree.html (accessed 03/04/2003).

Stallman, R. (2002) *Free Software, Free Society*, New York, GNU Press.

Torvalds, L. (2001) What makes Hackers Tick? a.k.a. Linus's Law, in Himanen, P. (2001) XIII–XVIII.

Trivedi, B. (2000) Sequencing the Genome, *Genome News Network*, June 2; http://gnn.tigr.org/_articles/06_00/sequence_primer.shtml, (access 04/04/2003).

Watson, J.D. (1999) *The Double Helix*, 2nd edit., London, Penguin Books.

Williams, B. (1981) Internal and External Reasons, reprinted in *Moral Luck*, Cambridge, Cambridge University Press, 1981, pp. 101–113.

14. Bioinformatics Lessons from the Open Source Movement

Dan L. Burk

As biotechnology begins to move into the bioinformatics area and as it becomes a tool and an issue, we are seeing the merger of the two areas that tend to get talked about as being "high technology" innovation: the biotechnology area and the informatics or digital revolution. Those who have watched each of these areas know that they have moved at very different paces for quite some time. The pace of innovation and change in biotechnology has been relatively slow, in part because of the long development times and the product cycle in biotechnology. In the area of cyberspace or the digital revolution, things have been moving very rapidly.

The pace of change in informatics has been very rapid, and as biotechnology and informatics become married, we will see the acceleration of the pace of change in biotechnology. The question I want to ask is, can we learn some things in the biotechnology and bioinformatics area from the scholarship and discussion that has been taking place in the rapidly evolving area of cyberlaw and digital technology? I am going to suggest is that the answer is "yes." There are some interesting parallels and contrasts, and some of the scholarship that has been developed in one area—the so-called "open source movement"—can be carried over into the other.

The open source movement is sometimes called the free software movement. Those terms are sometimes used interchangeably, sometimes they refer to slightly different communities, but they are generally used to refer to communities of programmers who are committed to certain principles in terms of the way they work and the software they produce and the way that software is used. The phrase that Richard Stallman uses is, "Free software is not like free beer." The term "free software" does not refer to software that you get for free or at no cost. Rather, it refers to the ability to manipulate and to change programs that are written, to use them in certain ways, and to make certain that people further downstream can continue to use them that way as well. One of the tenets of this movement is that the

---

This essay originally appeared in the *Boston University Journal of Science and Technology Law*, **8**:254-271. Copyright © 2002 by Dan L. Burk. Reprinted by permission.

source code that will allow you to understand and change and manipulate the software should be freely available, which is typically not the case with most commercial packages, which are distributed as object code.

Typically the software that is produced by open-source programmers or free software programmers is accompanied by some sort of licensing agreement that is designed to keep this freely available to other users downstream. The most famous of these is the so-called GNU Public License. GNU is an acronym for "Gnu's Not UNIX," which is the type of operating system that is produced using this open source methodology. There is a little recursive joke to it. It is accompanied by a public license, and there are other versions of these licenses, but they tend to have in common certain features. Some people refer to this as a *copyleft*, as opposed to copyright, license. The features of copyleft are that you are allowed to take the program and its source code and modify it under the condition that you allow other people further "downstream" to continue to modify it and that you make your source code available as well. That is why they refer to it somewhat jokingly as copyleft. Rather than trying to assert proprietary rights to keep that code in one particular form and keep it away from other people who might modify it, copyleft uses proprietary rights for the purpose of keeping the program open and available and free to people who might modify it.

There is a very striking sort of parallel when you think about the tenets of this open source movement that is committed to the free flow of information, and the tenets of the scientific research community. Many of open source programmers work as volunteers providing their work as a public service. We have seen some of the same tenets of this movement talked about in the literature with respect to the Human Genome Project and similar types of scientific research projects. Both share this sort of distributed production model in which you have many small units of production that are not, as far as is immediately apparent, centrally coordinated. Each of them has a normative structure that is committed to this idea of the free flow of information and the commons and sharing the output communally. Rebecca Eisenberg has written about this in the scientific research area.[1] Arti Rai has also written about this a little bit,[2] and they show that some of this behavior in the genomics area is captured in the work of Robert Merton.[3] Merton has been criticized for not getting everything right, but I think anyone who has worked in a biotechnology laboratory still feels some of those norms that Merton talked about: that you are supposed to share data freely, not to work so much for commercial rewards as for reputational rewards, etc.

One sees many similar statements about the norms of this open source programming community, the so-called "hacker ethic," committed to the free flow of information and sharing their output, and working, again, for reputational rewards rather than for monetary rewards. Both communities seem to be very concerned

---

[1] *See* Rebecca Eisenberg, *Property Rights and the Norms of Science in Biotechnology Research*, 97 Yale LJ 177 (1987).

[2] *See* Arti K. Rai, *Regulating Scientific Research: Intellectual Property Rights and the Norms of Science*, 94 Nw L. Rev. 77 (1999).

[3] *See* Robert K. Merton, *The Normative Structure of Science*, in The Sociology of Science: Theoretical Empirical Investigations (Norman W. Storer, ed., University of Chicago Press 1985).

about the capture or commercial conversion of their output. Articles in *Science* and *Nature* and so on indicate the concern of academic researchers and the scientific community that Celera or other types of commercial firms might capture this information that they are generating. There have been attempts to try and keep that from happening by creating prior art so that patents cannot be filed. You see the same sort of concern in the open source community's norm against commercialization, against proprietary capture of the software that is produced, particularly concern that Microsoft is somehow going to capture this code and turn it into yet another part of Bill Gates' dominion over the free world.

Thus, there are some very similar sorts of norms and concerns in each area. The "public domain capture problem" has been discussed a good deal in the digital rights literature. First Yochai Benkler,[4] and now Larry Lessig[5] have talked about different levels at which you might have the kind of openness that the open source movement is committed to, or conversely different levels at which technology that is available to the public might be captured and made proprietary. They talk about the physical layer, the logical layer, and then the content layer. If you think about the telephone network or the Internet, the way they function now, you can quickly see these different layers. Information flows over physical wires that are owned by somebody, frequently by AT&T or your cable carrier or someone like that. There is a logical layer that routes and directs the information, and in the case of the Internet, that is open and available to everyone; the Internet protocol is not proprietary. Then, riding on top of that, you have content that is sent over those wires using those protocols, and the content may be copyrighted and proprietary or it may be public domain and nonproprietary. Thus, you could have any of those layers either open or proprietary, and to the extent that any of those layers is captured or made proprietary, that particular layer, at least, would be unavailable to the public.

There is a great deal of concern at each of those levels about the possibility of capture of the Internet by Microsoft or AT&T or whomever. The same is true if we think about the bioinformatics realm. There is going to be some hardware level, and there is also going to be some logical level to that bioinformatics information. There already has been some discussion of nomenclatures, of normalization, of standardization, of indexing, etc. Finally, there is the data, the content itself. There could be proprietary rights at any of those levels. With respect to each of these levels, we are going to see what economics talks about as network effects. The telephone system is the paradigm example of this: once you have the telephone system in place, it does not cost much to add another user, but when you do, the value of the network most certainly goes up to the people who are already on the telephone system. If you only have one or two people on the telephone network, it is not very useful or valuable, but if you have hundreds and thousands of people joining it, each new user that joins increases the value of it. The same is true of other types of real and virtual networks, and the Internet and certainly bioinformatics would fall into that category.

---

[4] Yochai Benchler, *From Consumers to Users: Shifting the Deep Structures of Regulation*, 52 Fed. Comm. L.J. 561 (2000).

[5] Lawrence Lessig, The Future of Ideas: The Fate of the Commons in a Connected World (Random House 2001).

There are problems involving standardization and there is the problem that Dennis Karjala describes about trying to put different protocols or types of databases together.[6] If those are standardized, interaction will be much easier, and the information will be much more valuable. The problem that we know from the economic literature is that if you believe there is such a thing as a network effect, if the standard that becomes settled upon is proprietary, you get a kind of lock-in that may be undesirable. There is this also the problem with tipping, which is what the *Microsoft* case[7] was all about: can people with intellectual property rights influence or push the standardization process in a direction that benefits them, and if they do that, is that desirable? Those are things we might want to be worried about in the bioinformatics area. Certainly there has been concern about it in the digital technology area. If there are certain standards or conventions that are settled upon for bioinformatics, either in terms of the content or the indexing or the logical layer, would we want those to be proprietary standards? Would we be concerned if people pushed things in a direction so that their proprietary standard became the one that was generally adopted?

There has been an evolution of the literature in the open-source and digital rights area, talking about this movement in terms of the theory of the firm. This is related to the concern that we might have in bioinformatics about proprietary standards and the sorts of norms and concerns that each of these two communities share. The question here is simply, when you have these communities like the open source community or the human genome sequencing project that seem to be very atomistic and distributed and lacking in central coordination, how does the work get done and why does it get done? Would it be better done in a traditional commercial sort of firm? Certainly Celera seems to think that that is the case. In the software area, Microsoft or even Red Hat would seem to think that that is the case, and yet a lot of work does get done. How can that be the case?

Coase has discussed the concept of the firm.[8] This is one of his major contributions to economics and to our understanding of the law. He described the firm as an area of economic activity that is organized in a certain way, hierarchically rather than as an open market with market transactions, directed by some central authority, an entrepreneur, where these interactions are not negotiated but they are much like command and control. Other people took that a little further and said, well, the size of the firm and the scope of the firm will be determined by, first of all, the transaction costs. It is costly to use markets, and if it is sufficiently costly, it might be better to have this area of activity that is organized hierarchically, but it is also going to be bounded by the ability to coordinate efforts within that firm and also by agency costs. You have many different people acting within the firm, and some of them might have interests that are not well aligned with the interests of the firm.

---

[6] See Dennis Karjala, *Data Protection Statutes and Bioinformatic* Databases, 8 Boston University Journal of Science & Technology Law 171 (2002).

[7] *United States v. Microsoft Corp*, 253 F.3d 34 (DC. Cir. 2001).

[8] Ronald H. Coase, *The Nature of the Firm*, in The Firm, The Market, and the Law 33 (University of Chicago Press, 1990).

You can also think about the firm as a nexus of contracts. We say it is hierarchically organized, but part of that is by contracts that could be employment contracts or other kinds of production contracts, depending on how vertically or even horizontally integrated that firm might be. These contracts define the kinds of things that happen within that economic space, and as I said, agency costs are going to be a problem. You may have employees who have interests that differ from that of the entrepreneur who is directing things in the firm. How do you get those employees to fall into line? Well, the employment contract is one thing. You do not pay them or do not give them certain rewards if they step out of line. There may also be other kinds of bonding or coordinating mechanisms to cut down opportunism. In the software area, we have seen the use of stock options. If the firm does well, the employees are richer, and if the market tanks like it has, the employees are poorer but we hope they will pull together to make those stock options worth more than they would have otherwise.

How about intellectual property rights? Can those be used in these ways to make the firm operate better? Arti Rai[9] has discussed some of Ed Kitch's work in the so-called prospect theory of intellectual property rights,[10] and one way to think about that is that it is a type of coordinating mechanism, sometimes between firms, but even within a firm. We give an intellectual property right to one particular entrepreneur who then coordinates and directs the development of that particular resource, and that was how Kitch thought we should view intellectual property rights. There are also intellectual property rights that are bonding mechanisms.[11] The classic example here might be something like trade secrecy. In order to keep my employees headed down the same path I want to go down, I might put some legal restrictions on how they use certain information and what information they can take with them when they leave the firm. That raises the question, then, if intellectual property can be used to coordinate development of certain resources, do we see that happening in these communities that do not look very much like traditional firms? In the open source community writing software, and with the Human Genome Project trying to put together a picture of what the human genome looks like, we see intellectual property being used to try and help coordinate those efforts.

Now, a lot of the coordination in that community is clearly normative. There is a certain expectation that you will write code if you are part of this community, you will share that code with other people, and you will not work on projects that someone else has staked out as their territory. There is still concern, however, that someone will defect, will try and commercialize one of these software packages and will try and take it away from the commons. This GNU public license that I mentioned, this idea of copyleft, is used to prevent that from happening and also

---

[9] Arti K. Rai, The Proper Scope of IP Rights in the Post-Genomics Era 8 Boston University Journal of Science & Technology Law 233 (2002).

[10] See Edmund Kitch, *The Nature of the Patent System*, 20 J.L. & Econ. 265 (1977).

[11] Dan L. Burk, *Intellectual Property and the Firm*, 71 U. Chi. L. Rev 3 (2004).

to coordinate the development of these software packages. The most famous example of such a package that is accompanied by one of these licenses is Linux. The copyleft license that accompanies it is supposed to keep people from defecting and turning a modified version of the software into a proprietary package.

Now curiously enough, over in the human genomics side, there has been discussion of using intellectual property rights, but that has been either in a very negative sense or it is been largely rejected. Rebecca Eisenberg has discussed the NIH patents that were filed on all these expressed sequence tags when Craig Venter was still part of that effort.[12] Reid Adler, who was the technology transfer guy who did that filing, said the reason they filed these patents was to prevent people from staking out claims or capturing information. They wanted the government to do it and essentially used the patents to keep the information open and free. The NIH EST patents caused an enormous outcry among the scientific research community, and eventually the NIH stopped trying to prosecute those patents. There has been a fair amount of pressure not to go back there, and the patents have been filed instead by the commercial firms, with some degree of disdain from the rest of the research community.

Whereas the open source software coding community has used the copyright and the license of that copyright to prevent "capture," patent control was an option that was rejected in the genomics area. That means that people who do want to move into this area and commercialize this information are operating in very different licensing environments. A company like Red Hat, which wants to commercialize and have a more traditional business model using the Linux software package produced through this open source effort, is unable to do certain things with Linux software. In fact, in the disclosure they filed with the Securities and Exchange Commission, they talked about the fact that they do not really have control over the product they are selling since it is accompanied by this license which requires them to make public the source code of any changes that they made to Linux. Red Hat's product is subject to this copyleft or GPL, and they are somewhat constrained in what they do in their business model. With a company like Celera, however, you see something very different. There has been an effort to try to prevent commercial patenting of gene sequences by throwing information into the public domain as quickly as possible. Of course, Celera and Incyte and other pharmaceutical firms that specialize in providing these databases sucked that public information right up and made it part of the product they already have. As a result, they are relatively unconstrained by intellectual property rights that people might have in the way they want to use the information.

This difference in approach to using intellectual property rights to try and coordinate developments has resulted in two very, very different environments for commercialization. We may think that is a good thing or a bad thing depending how you view commercialization of information, but it clearly has led to different outcomes. A bit of different take on this can be found in the literature that has grown

---

[12] Rebecca Eisenberg, *Molecules vs. Information: Should Patents Protect Both?* 8 Boston University Journal of Science & Technology Law 190 (2002).

up in the open source area on so-called actor-network theory.[13] I mentioned the work that Robert Merton had done on the normative structure of the scientific community. Things have advanced since then, and Merton's work has been criticized, and people have advanced other sorts of theories as to how science really operates.

One view principally put forward by Bruno Latour postulates that really anything in society, but science in particular, operates as a series of relational networks,[14] and this parallels, in some sense, Coase's theory of the firm that I mentioned a moment ago. These networks define areas of effort and contain both human and technological components that people call *actants*. They are not really actors because some of them are bits of technology rather than humans. But a scientist, for example, is only a scientist because she has a laboratory and centrifuges and test tubes and Eppendorf tubes, etc., that really define that particular role as a scientist. Part of this theory of these relational networks discusses something called blackboxing, which is sort of a social shorthand. We talk about someone as a scientist, but we do not try to describe or look behind what that means. We are sort of vaguely aware that there is this network of training and universities and government funding and centrifuges and workbenches and so on back there, but we just talk about it as a "scientist," and that is sort of a black box into which we do not look.

This is true of other types of entities as well. We talk about Microsoft as a firm, and there are really lots of employees and machinery and all sorts of inputs there within that particular black box. We have talked today about the open source movement, which is a sort of black box, or the Human Genome Project, which is a black box full of all sorts of *actants* and relationships. When you blackbox something like that, Latour says you suddenly have a control point for whatever is in that box, depending on what the dimensions of that box are and how you define that particular label that you attach to it. One of the ways we might think about the idea of capturing this information that would otherwise be out in the public domain is as blackboxing it. A software package or genomic information could be incorporated it into one of these other black boxes or one of these other relational areas that can then be marketed and sold. This blackboxing process involves control of some of these relationships that are there.

Intellectual property has been discussed in the open source area, and it is an important part of that blackboxing phenomenon. We talk about someone being subject to a patent or a work that is copyrighted, and that is part of what is going on in the genomics area just as it goes on in the open source area. One can take information, put it into that black box or put a label on it, and then you have a type of control over it because of the copyright or the patent or the mechanism of blackboxing that you would not have had previously. This is an area where it seems there is some fruitful cross-fertilization potential between software and the open-source movement and bioinformatics.

---

[13] See Ilkka Tuomi, *Internet, Innovation, and Open Source: Actors in the Network*, First Monday, at http://www.firstmonday.org/issues/issue6_1/tuomi/index.html (Jan. 8 2001).

[14] See Bruno Latour, Science in Action: How to Follow Scientists and Engineers Through Society (Harvard University Press 1998).

This chapter probably leaves more questions hanging than it answers. At least two questions seem to deserve some further thought: (1) Why has there been or is it a good thing for there to have been this divergence in approach between these two communities with regard to intellectual property rights? (2) Would it make sense for there to have been something like copyleft or GNU Public Licensing for information that is put out into the public domain or the bioinformatics area? There are probably a number of additional questions worth contemplating as well, as we consider the application of open source philosophy to the genomics realm.

15. Enclosing the Genome: What the Squabbles over Genetic Patents Could Teach Us

James Boyle[1]

I Introduction

In other writing,[2] I have argued that we are in the middle of a second enclosure movement. The first enclosure movement involved the conversion of the "commons" of arable land into private property. The second enclosure movement involves an expansion of property rights over the intangible commons, the world of the public domain, the world of expression and invention. Quite frequently it has involved introducing property rights over subject matter—such as unoriginal compilations of facts, ideas about doing business, or gene sequences—that were previously said to be outside the property system, uncommodifiable, "essentially common," or part of the common heritage of mankind.

The justifications given for the first enclosure movement were often, though not always, centered on the need for single-entity private property rights over land to encourage development and investment, prevent over- and under-use, and in general to avoid the phenomena which we refer to today as "the tragedy of the commons." Enclosure's defenders argue that it helped increase agricultural production

---

This essay originally appeared in *Perspectives on Properties of the Human Genome Project* (F. Scott Kieff ed., 2003), pp 97-124 Copyright © 2003 by James Boyle. Reprinted by permission.

[1] A version of this chapter is made available by the author under a Creative Commons License http://www.creativecommons.org/licenses/by-sa/1.0 The final version is to be published in *Advances in Genetics*. Thanks to Lauren Dame, Bob Cook-Deegan and Alex Rosenberg for their comments and criticisms. Bob, in particular, should not be held responsible for the results. His comments helped me greatly by making clear out how *alien* it is to the world of pragmatic intellectual property policy to take the rhetorical structure of a discipline seriously, rather than making arguments within that structure. To the extent that it is now clear that this chapter is not an attack on gene patents, nor an embrace of the critics' arguments, but rather an examination of what the gene patenting debate can teach us about intellectual property scholarship generally, he bears much of the credit.

[2] James Boyle, *Fencing Off Ideas* DAEDALUS 13 (Spring 2002); *The Second Enclosure Movement and the Construction of the Public Domain* 66 LAW AND CONTEMP. PROBLEMS 33 (2003). The Introduction to this chapter draws heavily on the latter article.

and, in the long run, to generate an agricultural surplus sorely needed by a society whose population had been depleted by the mass deaths of the sixteenth century. Private property saved lives. Though "overuse" is rare in the intellectual commons, the rest of the arguments are exactly the ones used to support the second enclosure movement. Intellectual property is needed to encourage development and investment. This argument is made in defenses of drug patents from compulsory licensing claims, in the debates over the creation of new intellectual property rights over data, over business method patents, in the rhetoric of support for the Digital Millennium Copyright Act.[3]

But as a cursory study of the newspapers demonstrates, it is the idea that the genome has been turned over to private ownership that has fueled real public attention. Again, the supporters of enclosure have argued that the state was right to step in and extend the reach of property rights; that only this way can we guarantee the kind of investment of time, ingenuity and capital necessary to produce new drugs and gene therapies. To the question, "should there be patents over human genes?" the supporters of enclosure would answer again, "private property saves lives." Again, the opponents of enclosure have claimed that our common genome "belongs to everyone" that it is "the common heritage of humankind," that it should not and perhaps in some sense *cannot* be owned and that the consequences of turning over the human genome to private property rights will be dreadful, as market logic invades areas which should be the farthest from the market. What damaged view of the self, what distorted relationship between human beings and the environment, flows from a world in which our genetic code itself is fenced off, made both alienable and alien at the same moment?

The analogy is not perfect, of course; the commons of the mind has many different characteristics from the grassy commons of Old England. Some would say that we never had the same *traditional* claims over the genetic commons, that the victims of the first enclosure movement had over theirs; this is more like newly discovered frontier land, or perhaps even privately drained marshland, than it is like well-known common land that all have traditionally used. In this case, the enclosers can claim (though their claims are hotly disputed) that they discovered or perhaps simply made usable the territory they seek to own. The opponents of gene patenting turn more frequently than the farmers of the 17th century to religious and ethical arguments about the sanctity of life and the incompatibility of property with living systems. Importantly, too, the genome—like most of the subjects of the second enclosure movement—software, data, digital music and text—is non-rival; unlike the "commons" enclosed during the first enclosure movement, my use does not interfere with yours. I can work it and you can work it too. All of these differences might be relevant, or not, depending on the underlying normative framework we chose to use to assess their property claims. But for the moment, think of the critics and proponents of enclosure—locked in battle; hurling at each other incommensurable claims about progress, efficiency, traditional values, the boundaries of the market, the saving of lives, the disruption of a *modus vivendi* with

---

[3]For details see Boyle, *The Second Enclosure Movement, supra* note 2 at 33–44.

nature. We have done this before and perhaps we can learn something from the process.[4]

In my earlier work,[5] I concentrated on the economic and utilitarian arguments for the second enclosure movement. In sum, I concluded that for a variety of reasons the embrace of the logic of enclosure to justify the expansion of intellectual property across so many fields and so many dimensions was probably a mistake. I argued that while the basic argument for intellectual property protection remains as strong as ever, we have adopted an asymmetric analytic framework which a.) overestimates the applicability of the general logic of enclosure to the special case of intellectual material b.) undervalues the importance of the public domain and the commons to intellectual production thus c.) focusing only on the (very real) arguments in favor of private property while neglecting the role of the raw materials out of which future innovation is constructed. The resulting policies also d.) overestimate the potential threats and underestimate the potential benefits of the technologies of cheap copying to existing intellectual property rights and e.) fail to take seriously enough the important potential for various types of distributed production, which require a rather different intellectual property environment in order to flourish. Some of those conclusions, a.) and b.) in particular, seem to me strongly applicable to fundamental gene patents—particularly those which may operate to block or channel future research. Others, for example, have a possible but more dubious relevance.

In this chapter, however, I turn my attention to the other arguments over enclosure, the ones that economic historians—and intellectual property scholars for that matter—tend to note briefly and somewhat dismissively before turning their attention to the real meat of the economic incentives set up by property systems. My question, then, is **not** "will licensing solve the potential monopolistic problems threatened by fundamental gene patents?" Or "will this particular bit of surgery on the Bayh-Dole Act cure the incentive system so as to encourage fundamental research to yield real products while discouraging 'blocking patents' across the arteries of research?" Those questions are vital ones on which important work has already been done, some of it by my wonderful colleagues at Duke.[6] But I want to ask a different question, one about the fundamental structure of intellectual property discourse. What can the debate over gene patents teach us about the structure of our discipline, about our pattern of inquiry? What does that debate reveal

---

[4] The analogy to the enclosure movement has been too succulent to resist. To my knowledge, Ben Kaplan, Pamela Samuelson, Yochai Benkler, David Lange, Christopher May and Keith Aoki have all employed the trope, as I have myself on previous occasions. For a particularly thoughtful and careful development of the parallelism see Hannibal Travis, "Pirates of the Information Infrastructure: Blackstonian Copyright and the First Amendment," 15 Berkeley Tech. L.J. 777 (2000). It is also worth noting that Jeremy Rifkin, one of the most outspoken critics of gene patents, has recently turned to this analogy as well. See JEREMY RIFKIN, THE BIOTECH CENTURY (1998).

[5] *See* note 2 *supra*.

[6] John Walsh, Ashish Arora, & Wesley Cohen, *The Patenting of Research Tools and Biomedical Innovation*, draft (on file with author); Arti K. Rai & Rebecca S. Eisenberg, *Bayh-Dole Reform and the Progress of Biomedicine*, 66 LAW & CONTEMP. PROBS. 289, 289 (Winter/Spring 2003). For the seminal study of the early days of the genome project and the conflicts over gene patenting, which goes beyond economic considerations to consider some of the ethical objections made, see ROBERT COOK-DEEGAN, THE GENE WARS: SCIENCE, POLITICS AND THE HUMAN GENOME (1994).

about both the selective focus and the selective blindness in the way that intellectual property scholars analyze the question? In short, this article treats the debate over gene patents as a rhetorical case study, a place to reflect on the limits of our discipline. It does not attempt to enter that debate on one side or the other.

II You Can't Own a Gene

Like the debate over the first enclosure movement, the debate over the enclosure of the genome is not one debate but many. The commonly heard phrase "you can't own a gene" turns out to cover a remarkably broad range of arguments and beliefs. These arguments are not necessarily consistent, of course, and intellectual property lawyers would disclaim many of them as being beyond their purview. In addition, the arguments vary importantly depending on the subject matter of the patent; the debate is not merely, or even most interestingly, about patents on genes. Claims to patents on different types of genetic sequences raise very different issues. What's more, I mean here to catalogue types of objections, rather than to investigate them in any detail, which requires painting with a distressingly broad brush. Sweeping all of these qualifications aside for a moment, let me attempt a breathless summary of the most commonly heard objections to genetic patents. My hope is that, for the cognoscenti, it may be useful to review territory so well-known it is seldom examined. For the newcomer, this may serve as a rough guide to the territory, to be deserted later in favor of more detailed maps.

The Sacred: "You can't own a gene because it would be against the tenets of my faith. God wrote those sequences of C, G, A & T. It is heresy, or at least plagiarism, for you to claim to do so." In a more modest version, the accusation is not heresy or plagiarism, but simply ludicrous *hubris*. Who are *you* to claim a patent on a *human* genetic sequence, even a "purified" version of a genetic sequence with the introns spliced out?[7] (A version of this argument will reappear as an argument about whether genetic sequences fit the statutory and constitutional requirements of patent law.)

The Uncommodifiable: "You can't own a gene because it would be immoral; some things should be outside of the property system—babies, votes, kidneys; if they become commodified the market seeps into aspects of our lives that should be free

---

[7] "In the 19th century, another great debate ensued with abolitionists arguing that every human being has intrinsic value and "God-given rights" and cannot be made the personal commercial property of another human being. The abolitionists' argument ultimately prevailed, and legally sanctioned human slavery was abolished in every country in the world where it was still being practiced...Now, still another grand battle is unfolding on the eve of the Biotech century. This time of the question of patenting life—a struggle whose outcome is likely to be as important to the next era in history as the debate over usury, slavery, and involuntary indenture have been to the era just passing. In May of 1995, coalition of more than 200 religious leaders, including the titular heads of virtually every major Protestant denomination, more than 100 Catholic bishops, and Jewish, Muslim, Buddhist, and Hindu leaders, announced their opposition to the granting of patents on animal and human genes, organs, tissues, and organisms.... The coalition, the largest assemblage of U.S. religious leaders to come together on an issue of mutual interest in the 20th century, said that the patenting of life marked the serious challenge to the notion of God's creation in history. How can life be defined as an invention to be profited from by scientists and corporations when it is freely given as a gift of God, asked the theologians? Either life is God's creation or a human invention, but it can't be both. " JEREMY RIFKIN, THE BIOTECH CENTURY 64 (1998).

from market logic."[8] Sometimes the proponents of this argument rely on a teleological view of human flourishing, at other times on a theory of "spheres of justice" within a liberal state, at still others they conjure up a dystopian future in which transgenic sub-human creatures are traded as chattels, our reverence for thinking beings having been undermined by the twin expansions of genetic science and market logic. This view in turn, segues into…

The Environmental Ethic: "You can't own a gene; to do so is to embrace a system in which nature, even our own nature, is to be manipulated, traded and commodified." While the environmentalist critics of gene patenting generally have concerns that stretch beyond the human genome alone, they see it as a particularly revealing example. Many argue that the frame of mind which would permit such property rights is one which displays a breathtaking willingness to tinker with delicate environmental systems.[9] Others claim that the availability of such juicy state-granted monopolies establishes a set of incentives which positively invite corruption of the legislative process and the undermining of rational environmental policy.

The Common Heritage of Mankind: "You (*alone*) can't own a gene because the genome belongs to all of us. Like the deep sea bed, the history of the human species, or outer space the genome is 'the common heritage of mankind,' one of the quintessential resources that must be held in common."[10] Sometimes this view rests

---

[8] "Now, the most intimate Commons of all is being inclosed and reduced to private commercial property that can be bought and sold on the global market. The international effort to convert the genetic blueprints of millions of the years of evolution to privately held intellectual property represents both the completion of a half millennium of commercial history closing of the last remaining frontier of the natural world." Rifkin at 41. Not all agree. "The reductive objectification of life has enlarged understanding of vital processes, but it does not relegate living creatures to the category of objects. We understand that animals are physics and chemistry. But we can realize that they are sentient creatures and grant them rights that they do not inherently possess. We can realize that people perform physical and chemical functions, but we need not treat them as physical and chemical machines." Daniel Kevles, *Vital Essences and Human Wholeness: The Social Readings of Biological Information* 65 S. CAL L. REV 255, 277.

[9] "Biotechnology is not simply another mechanical or chemical procedure aimed at making the world better for us. With biotechnology, we are not reshaping matter, but harnessing life. We take a 3,500 million year old process that shaped our existence and the existence of every other organism on the planet and restructure it for our benefit. We need a more thoughtful conceptualization of this technology and more careful control over its development and use than is allowed by gung ho biopatent policies. Biotechnology does offer promise and hope for bettering human life and perhaps other life as well. Opposing biopatents does not entail opposing biotechnology. Organism and gene patents should be resisted not because biotechnology should be resisted, but rather because these biopatents are a morally dangerous and inappropriate way of thinking about and encouraging biotechnology." Ned Hettinger, *Patenting Life: Biotechnology, Intellectual Property, and Environmental Ethics*, 22 B.C. ENVTL. AFF. L. REV. 267, 304 (1995).

[10] "The application of the Common Heritage concept to the genome would balance the interests of developed and developing countries, thus holding true to the traditional purpose. As stated previously, most developing countries are opposed to patenting genes, and consider it a form of neo-colonialism. Because the genome technically belongs to the citizens of LDCs to the same extent as citizens of the developed countries, the Common Heritage Principle is necessary to balance the property rights of both LDCs and developed country citizens despite their different interests." Melissa L. Sturges, Who Should Hold Property Rights to the Human Genome? An Application of the Common Heritage of Humankind, 13 Am. U. Int'l L. Rev. 219 251-2 (1997); See also J.M. Spectar, *The Fruit of the Human Genome Tree: Cautionary Tales About Technology, Investment, and the Heritage of Humankind*, 23 LOY. L.A. INT'L & COMP. L. REV. 1. On both sides of these arguments, the background assumptions are often hard to parse. "From the perspective that genes are our common, universal possession symbolizing humankind's collective heritage, genes seem an inappropriate substance in which to grant individual intellectual property rights. Yet, in light of patent availability in most other scientific research fields, denying gene patents appears to be inequitable." Patricia A. Lacy, *Gene Patenting: Universal Heritage vs. Reward for Human Effort*, 77 OR. L. REV. 783.

on some Teilhard de Chardin-esque teleology of the human species, while at other times it represents a straightforward appeal to international distributive justice. If the technologically advanced countries can secure property rights over resources that only advanced technology will reach, goes the argument, then patent rights over the genome are a kind of second colonial expansion. The moon or the deep sea bed or the genome should not become the next place to be claimed for Queen Isabella of Spain.

The Rights of Sources: "You can't own *this* gene because I owned it first. My genetic information is my property. Your gene sequences came originally from a source and the source's claims should be recognized, either instead of or as well as, the person seeking the patent." This objection is sometimes raised on behalf of those who unwittingly provided genetic material to a research project from which some desirable genetic sequence was derived.[11] At other times, the claim comes from the families of those with a particular genetic disorder. Here the genetic material was donated knowingly but the families are using property and contract claims to ensure that the development of tests and treatments for the disorder protects the interests of the patients involved.[12] Other variants include attempts by particular groups or populations whose genetic information is somehow of interest, to prevent what they see as "gene piracy." [13]

Patentable Subject Matter: "You can't own a gene because you can't patent it; it doesn't satisfy the basic requirements of the patent law and the constitution." The challenges differ depending on the type of genetic material involved. The main objections are as follows: **Novelty**: Many genetic patents cover material that is not "novel" because they are in fact patents on naturally occurring products; if patents are being given for laborious transcription of a naturally occurring substance, that hardly meets the novelty standard. Even if the sequences are "purified," for example by removing non-coding segments, critics argue that this should not meet the standard for novelty, and might also run into problems because of the "obviousness" of the process. **Non-obviousness**: In general, the standard methods of genetic sequencing are such that each step in the process of cell

---

[11] The classic case is that of John Moore. See JAMES BOYLE, SHAMANS SOFTWARE AND SPLEENS; LAW AND THE CONSTRUCTION OF THE INFORMATION SOCIETY 97-107 (1996).

[12] "When the Terry family of Sharon, Massachusetts, realized that disease researchers didn't have the family's interests at heart, they took control of their bodies intellectual property. The result? Do-it-yourself patenting was born. Boyd (who hasn't been funded by PXE International) agreed with Terry's morally driven, but practical, view that the laws of this market left patients no choice but to seek control of the intellectual property. "This was a way to ensure that that the test isn't going to cost an arm and a leg," notes the scientist.... By laying legal claim to their bodies, the Terrys have advanced research enough that doctors genuinely may be able to arrest the progress of PXE before their children go blind, which often happens to patients in their late 30s. Ian Terry, now 11, already has initial signs of eye trouble. At the very least, the parents' organizing has inspired Elizabeth Terry, now 13, to say she'll become a patent lawyer—though her mother says she's also flirting with criminal law. No doubt, Elizabeth should pursue what she wants: By the time she graduates from law school, the patient advocacy movement should have plenty of patent attorneys on its side." Matt Fleischer, Patent Thyself AMERICAN LAWYER June 2001.

[13] Kara H. Ching, *Indigenous Self-Determination in an Age of Genetic Patenting: Recognizing an Emerging Human Rights Norm*, 66 FORDHAM L. REV. 687.

line isolation, purification and sequencing may be obvious from the prior step in the process. One of the larger concerns being expressed here, as in the discussion of novelty, is that intellectual property rights are being given in raw data, based on the labor required to produce it. Such "sweat of the brow" claims are supposed to be anathema to the law of copyright and patent, while the subject matter here—genetic sequences—seem to be straightforward arrangements of factual data; a subject matter that the Constitutional patent and copyright power does not reach. **Utility**: Some of the genetic patents have been shotgun claims over large number of sequences, without clear knowledge as to what function these sequences actually have. Others cover sequences the main function of which is as probes in research so that their "utility" is as research intermediaries rather than useful end products.

As is often (and appropriately) the case in patent law, debates about the meaning of the statutory and constitutional requirements for patents frequently devolve into questions about the effects on innovation that a particular interpretation of the subject matter requirements would have. Since the patent system is, in its constitutional and theoretical origins, guided explicitly by utilitarian concerns, almost all would concede that the ultimate benchmark for any interpretation of the subject matter requirements should be its effect on the encouragement and promotion of the progress of the useful arts.[14] In the case of patents on genetic sequences, this means defining the requirements of patentable subject matter with the assumption that those requirements exist so as to encourage both abstract research and final practical deployment of some therapy or test. Thus discussions of patentable subject matter tend to segue into the free-standing debate that already exists about the effects of genetic sequence patents on innovation.

Innovation Policy: "You can't own a gene because if you did so, it would actually hurt research and innovation, the very things you are trying to encourage." Here the arguments are particularly subtle and nuanced. The basic difference between the genetic commons and the earthy commons of old England, at least from an innovation policy point of view, is its non-rivalrous quality. The danger to be sought here is not unrestrained overuse, but underinvestment in research and product

---

[14]Unfortunately, the Court of Appeals for the Federal Circuit (CAFC), which handles patent appeals nationwide, does not appear to share this point of view. To this external observer it seems that the CAFC lurches from formalism to utilitarian analysis and back again, guided by some muse of its own. [For a related critique see Arti K. Rai *Specialized Trial Courts: Concentrating Expertise on Fact* 17 Berkeley Tech. L.J. 877 (2002).] The court also appears gratifyingly indifferent to academic opinion; many distinguished patent scholars appeared to think the State Street bank opinion [*State St. Bank & Trust Co. v. Signature Fin. Group,* 149 F.3d 1368, 1373 (D.C. Cir. 1998)] was the jewel in the crown of bad patent decisions,[See eg Robert P. Merges, *As Many as Six Impossible Patents Before Breakfast: Property Rights for Business Concepts and Patent System Reform* 14 Berkeley Tech. L.J. (1999).] Yet despite the storm of academic protest, one judge on the court recently remarked in clear self-satisfaction, that everyone was clearly now comfortable with the court, so that perhaps its jurisdiction should now be expanded to cover copyright as well! Being ignored like this is clearly good for the otherwise non-existent humility of legal academics, but since the court appears instead to take its lead from the patent bar, which has a something of a self-interest in maximalist protection, one wonders whether it is also a prescription for a good patent law jurisprudence.

development. What's more, as is generally the case, intellectual property deals with a subject matter in which future products are generally assembled from the pot-shards of prior efforts. Each property right tacked onto your stream of intellectual outputs increases the cost of my stream of intellectual inputs. Even if, in principle, I could either "invent around" prior protected innovation,[15] or secure licenses on those chunks which it proved impossible to bypass, the transaction costs involved in the process of identification, weighing the relative costs of inventing around, negotiating and so on, might undercut a large number of innovations at the margin.[16] It is the work of intellectual property scholars to work at this point of delicate equipoise; the balance between, on the one hand, failing to guarantee sufficient protection (or government funding) in order to encourage both research and commercialization, and on the other, slowing down the process of research and commercialization with a thicket of property rights. The best work on the subject refines this point of balance; pointing out counter-intuitive processes at work within it,[17] expanding on institutional dynamics, bringing in the insights of other disciplines, and of empirical research on the innovation process.[18]

III The Limits of Intellectual Property Policy

As promised, this has been a somewhat breathless summary of the arguments about the patents over human genetic sequences. It is far from complete. In particular, I have not expanded on the many thoughtful counterarguments that can be raised to each of these objections, because I do not mean to enter the debate so much as to point out something about it. It is my hope that the summary I have given helps us to see that the academic intellectual property debate has tended to be much narrower than the popular or policy debate. First, patent scholars and patent lawyers have largely viewed all but the last two sets of criticisms as outside their purview;[19] the legislature decides what is and is not patentable, they argue, so to address arguments about the immorality, or environmentally hurtful quality of gene patents to the Patent and Trademark Office, to the patent bar or even to patent scholars is to mistake the institutional function of those entities. The role of patent spe-

---

[15] "Inventing around" is normally a safety valve in intellectual property policy, the possibility of which operates to reduce the drag on progress caused by prior rights, and to create downward pressure on the price of licenses. One particular (though disputed) concern with the human *genetic* patents is the extent to which there may be no feasible way to innovate around a particular patent, precisely because of the fundamental quality of the subject matter.

[16] Michael A. Heller & Rebecca S. Eisenberg, *Can Patents Deter Innovation? The Anticommons in Biomedical Research*, SCIENCE, May 1, 1998, at 698.

[17] See eg Pamela Samuelson & Suzanne Scotchmer, *The Law and Economics of Reverse Engineering*, 111 YALE L.J. 1575, 1608-13 (2002); F. Scott Kieff, *Property Rights and Property Rules for Commercializing Inventions*, 85 MINN. L. REV. 697 (2001).

[18] See eg Wesley Cohen et al., *Protecting Their Intellectual Assets: Appropriability Conditions and Why U.S. Manufacturing Firms Patent (or Not)*, National Bureau of Economic Research, Working Paper No. 7552 (2000); Mark Lemley & John Allison, *Who's Patenting What? An Empirical Exploration of Patent Prosecution*, 53 VANDERBILT LAW REVIEW 2099 (2000).

[19] I would happily acknowledge that there are salutary exceptions to this claim. See eg Molly A. Holman & Stephen R. Munzer, *Intellectual Property Rights in Genes and Gene Fragments: A Registration Solution for Expressed Sequence Tags*, 85 IOWA L. REV. 735 (2000). I would argue, however, that the general claim holds true.

cialists is to make positive claims about what the law is, and restricted normative claims about how current or proposed patent rules will further the goals of the patent system, which in turn are understood in means-ends terms as a utilitarian calculation of the best way to foster technologically usable "progress." Thus patent scholars do not eschew normative analysis altogether; most of them simply restrict it to the consensus utilitarian goals of the system.

Patent scholars will happily debate the claim that current PTO utility guidelines, or the PTO's practices in reviewing patents on Expressed Sequence Tags, are moving intellectual property rights too far "upstream" in the research cycle in a way that will actually discourage future research and innovation. But for most of them, claims about the effect of commodification on attitudes towards the environment, about distributive justice and the common heritage of mankind or the moral boundaries of the market are simply outside the field. (The claims about the relationship of the genome to the realm of the sacred, of course, are even more so.) Ironically, some of those who in other areas of scholarship might be more inclined to criticize the utilitarian, means-ends metric as partial, or as a "thin and unsatisfactory epistemology" through which to understand the world[20] are, in the patent field, particularly wedded to it. A brief digression is necessary to explain why.

The Bipolar Disorders of Intellectual Property Policy

At its crudest, the basic division in the intellectual property field is between maximalists or high protectionists, on the one hand, and minimalists, or those with a heightened concern about the public domain, on the other. While actual positions are considerably more nuanced, if one had to construct an "ideal type" of these two positions, it would run something like this. The maximalists favor expansive intellectual property rights. They tend to view exemptions and privileges on the part of users or future creators as a tax on rights holders and have sympathy for thinly disguised "sweat of the brow" claims. They exhibit a kind of economic bipolar disorder: being deeply pessimistic about market functioning around potential public goods problems in the **absence** of intellectual property rights, and yet strikingly, even manically, optimistic about our ability to avoid transaction costs and strategic behavior "anti-commons effects" that might be caused by the **presence** of intellectual property rights.

The minimalists have exactly the opposite set of attitudes. They start from the presumption that the baseline of American law is that "the noblest productions of the human mind are upon voluntary disclosure to others, free as the air to common use." Thus intellectual property rights, understood as "monopolies," look like dangerous state granted subsidies, which should be confined in amount and extent to the minimum demonstrably necessary. Minimalists exhibit their own economic bipolarity; they are often deeply optimistic about the ability of creators and innovators to adapt business methods so as to gain returns on innovation without recourse to legal monopolies. (For example, by being "first to market," relying on

---

[20] This description of the philosophical shortcomings of economic analysis is, amusingly enough, Richard Posner's. RICHARD POSNER, THE PROBLEMS OF JURISPRUDENCE xiv (1990) (citing the words of Paul Bator).

the provision of tied services to generate revenue, or by adopting physical and virtual methods of exclusion in order to avoid the underlying public goods problem.) They are also relatively optimistic about the impact of technology on the innovation marketplace; they see new technologies that lower the cost of copying, such as the internet or PCR, as providing benefits as well as costs to rights holders, so that every reduction in the cost of copying need not be met with a corresponding increase in the level of protection. On the other hand, the minimalists are deeply, almost tragically, pessimistic about the ability of rights holders to bargain around the potential inefficiencies and transaction costs that their rights introduce into the innovation process. Here, private action to avert potential market failure seems much harder, for some reason, than it did when we were discussing the adoption of possible business methods to gain a return on innovation.

The maximalists' arguments are, in general, weaker, because the minimalist account is more faithful to thoughtful innovation economics and to the historical and constitutional traditions of American intellectual property law. More importantly, years of relentless expansion in intellectual property rights mean that the disputed frontier is now very far out indeed; maximalists must now defend rights that are, frankly, silly. However, what the maximalists lack in argumentative power, they more than make up in political power; their views are extremely hospitable to a variety of film, recording, software and pharmaceutical companies who can supplement dubious economic *analysis* with indubitable economic *payments*—in many cases, payments direct to legislators. More interestingly, there are also some philosophical reasons why we tend to undervalue the public domain,[21] and why strong intellectual property rights breed even stronger intellectual property rights.[22]

To be sure, the minimalist and maximalist I have just described are exaggerated ideal types, seldom met in their pure form. (And certainly no such caricatured scholars are present in this volume I am sure, indeed a number of the scholars here have written important pieces that point out the dangers of both under and over protection.[23]) Still the distinction catches something important. Now notice the irony. Minimalists might seem like the natural allies of those who are skeptical about gene patents for religious, environmental, moral or distributive reasons. Yet inside the world of intellectual property policy, *the minimalists are often the ones who are put in the position of insisting on the need to confine analysis to the economic need for intellectual property rights as incentives to progress.* Normally, this happens when a minimalist confronts some intellectual property claim built on Lockean labor principles, or Hegelian ideas of personality, or even just based on the fecklessly unattractive behavior of some alleged "infringer" who is scarfing up a work created with the investment of much sweat and capital.

---

[21]See eg, JAMES BOYLE, SHAMANS SOFTWARE AND SPLEENS; LAW AND THE CONSTRUCTION OF THE INFORMATION SOCIETY(1996). James Boyle, *A Politics of Intellectual Property: Environmentalism for the Net*. James Boyle, *A Politics of Intellectual Property: Environmentalism for the Net?*, 47 DUKE L.J. 87 (1997).

[22]James Boyle, *Cruel, Mean or Lavish?: Economic Analysis, Price Discrimination and Digital Intellectual Property* 536 VANDERBILT LAW REVIEW 2007 (2000).

[23]Compare Rebecca Eisenberg, *Proprietary Rights and the Norms of Science in Biotechnology Research,* 97 Yale L.J. 177 (1987) & Heller & Eisenberg, *supra* note 15.

In the face of these appealing arguments, the minimalists insist sternly that the constitution and the American tradition of intellectual property law forbid us such normative appeals. We must confine ourselves to incentives, no matter how hardworking or morally attractive the potential recipient of intellectual property rights, only the encouragement of the *next* creator is important. Thus when the issue turns to gene patents, the minimalists are happy to point out the dangers in moving intellectual property rights too far upstream, or the need to keep some teeth behind the utility, non-obviousness and novelty requirements. Yet their relentlessly utilitarian framework makes it much harder for them to consider arguments that fuel much of the general popular, journalistic, and policy debate, *even when those arguments are not deployed to argue for the extension of intellectual property rights, but for their circumscription*. Having insisted on the necessity of a means-ends analysis focused on producing incentives to progress it is much harder to engage the claims about the moral limits of the market in a liberal democracy, or the argument that the growth economics of a consumer society so permeates our assumptions as to make it impossible to consider the environmental consequences of our property systems. To be sure, there is excellent scholarly work that avoids this potential pitfall, but nevertheless the ambit of discussion in the intellectual property debate is a relatively narrow one, and I believe this is at least one of the reasons why.

Reasons To Be Narrow: Take 3

So far I have argued that the debate over gene patents allows us to reflect on a curious fact: a number of factors have come together to produce a remarkably narrow disciplinary self-conception in the world of intellectual property scholarship. Thus, for example, I argued that our concentration on the clash between maximalist and minimalist visions of intellectual property has produced as an unintended sideeffect a curious methodological tunnel vision. The critical scholars most likely to question the ambit of new rights are, paradoxically, firmly wedded to the notion that the only legitimate rubric for intellectual property policy is the maximization of innovation. All other normative criteria are to be exiled beyond the pale of the discipline.

But is narrowness such a bad thing? First, are these debates *worth* participating in? For example, one does not have to be a logical positivist to believe that policy debates based on religious faith tend to collapse fairly quickly into mutually incompatible thunderous denunciations. There *might* be a principled deal to be brokered between the Darwinians and the theists on the status of the human genome, but the odds are against it. Does this caution hold true for most of the objections to genetic patents?

Second, even to the extent that there are fruitful intellectual debates to engage with here, do intellectual property scholars have anything to add to them? What do *we* know of environmental ethics, or the moral limits of market under liberal capitalism, or of the distributional or internationalist commitments raised earlier? Shouldn't the cobbler stick to his last?

Third, isn't this whole question moot? We may believe that there are reasons, both inside and outside the legal system, why genetic patents should not have been granted, but the genie is now out of the bottle, at least in the United States. A multi-billion dollar industry has been erected on the premise that quite fundamental genetic patents are available. In practice, neither the Supreme Court nor Congress nor the PTO is going to upset that premise, even if some have their doubts about whether it was ever correct either as a matter of law or as a matter of desirable innovation policy.

These are weighty objections, but I think that there is a response—at least a partial response—to each of them.

First, let me concede that the religious objections present the hardest case for the benefits of disciplinary openness (and are thus frequently used as a proxy to suggest that none of the noneconomic arguments are worth considering, a conclusion that hardly follows). Let me note, though, the fact that debates turn on matters of non-falsifiable religious faith is hardly anything new to the world of legal scholarship. Indeed, it is often the case that such debates—abortion is the obvious example—attract an astounding amount of scholarly attention, some of it actually *premised* on the claim that the controversy is faith-based; some scholars have argued the fact that a policy is faith-based, or that the moral claims raised are incommensurable, has concrete legal or constitutional ramifications.[24] In other words, the religious components of fundamental policy debate have just as frequently in the past been reasons *for* scholarly intervention—wise or unwise though that intervention might be. One might also note that it is hard to draw a clean line between religious faith-based systems and secular moralities which developed out of them or away from them, and which also rest on *a priori* claims. And if it is hard to justify our conceptual apartheid even in the strongest case, straightforward claims to religious faith, then surely it must be very hard indeed to justify exclusion of the environmental, distributional and other normative concerns, which certainly form an important part of more conventional property scholarship. Put differently, if one reverses the question and asks, "Is it appropriate for intellectual property scholarship to exile all forms of morality other than wealth-maximizing utilitarianism?" then the boundary lines of our discipline look considerably less justifiable. But even if the frankly religious component of the debates drops out—a result which would not displease me, I confess—or is admitted only through some "purifying" secular filter of abstract moral philosophy, there is still much of substance that remains, much that fits within the ambit of the constitutional basis of American intellectual property law, yet that lies beyond the current bounds of intellectual property scholarship, policy and debate.

Think for a moment of more conventional property scholarship and policy debates. Imagine that we were debating whether or not to allow the commodification of non-reproducible organs, such as kidneys or corneas. Or imagine a debate about the sale of contracts of employment, or the question of ownership of Indian relics. Think of a discussion of the possible partition and sale of America's national parks. Or, for that matter, of the deep sea bed, the surface of the moon, or of licenses to pollute. There are those, of course, who would pursue these questions solely through a narrow utilitarianism, in some framework of wealth maximization or theoretical Pareto superiority. We all have such colleagues and I am sure they are

---

[24]See eg RONALD DWORKIN, FREEDOM's LAW: THE MORAL READING OF THE AMERICAN CONSTITUTION 29 (1996).

very dear to us. But most of us would find such an approach a mark of a sadly crabbed legal and philosophical imagination, would we not? This is not to say that one would *neglect* the economic questions, in fact they would often play a vital role, but they would surely do so in the context of a much wider analysis. Such analyses would in all likelihood be marked by explicit consideration of the effects of the property regime on human dignity, and on the environment, by consideration of the legitimacy of traditional property claims, by discussion of the requirements of international distributive justice. Even within the utilitarian analysis, they would be marked by a skepticism about the normative force of extrapolations from the existing distribution of wealth, and would in all likelihood involve a frank choice among multiple possible measures of efficiency. *Is a discussion of patent rights over the genome so different that it warrants about a completely different set of methodological assumptions?* Does being an intellectual property scholar offer a license to be parochial?

What of the second objection, that modesty should cause intellectual property experts to stay away from the environmental, moral or distributive objections to genetic patents? Modesty is a great virtue to be sure, but one must note that it is altogether an uncharacteristic one for *legal* scholars. In fact, following their discipline's longstanding imperialistic urges, legal scholars have exercised eminent domain over a remarkable range of questions and specialties—from Rawlsian moral theory through game theoretic analysis, to literary theory, neo-classical economics, deconstruction, queer theory, and chaos theory. Why stop here? It may be that intellectual property is the only place where one can find a modest legal scholar, though that would not be the descriptive term that immediately leaps to my mind to characterize the group. But we surely need some explanation of why the arguments for methodological modesty are so dominating here that they win *almost without discussion*, and yet so weak elsewhere.

More particularly, as I just pointed out, if one looks at parallels in the scholarship over real and personal property, one finds no such reticence. Debates over commodification and the limits of the market are commonplace,[25] as are discussions of the rights of indigenous peoples, or the extent to which property has a fundamental status within our political tradition that demands greater constitutional protection.[26] If they can do it, why can't we? It is worth pointing out that the link between environmental policy and property law has been particularly fruitful, both in terms of practical explorations of the intersection, in the law of nuisance for example, and theoretical explorations of the juncture in the context of such issues as tradeable emissions rights, ecosystem services or "the substitutes for a land ethic."[27] There is no reason to believe that the same number of insights, provocations and engagements could not be found in the fight over the enclosure of the intangible genetic commons, the intersection between intellectual property policy and environmental policy or the search for a Leopoldian "substitutes for a genome ethic."[28]

---

[25] For an uncommon example, see MARGARET JANE RADIN, CONTESTED COMMODITIES (1996).

[26] See eg RICHARD A. EPSTEIN, TAKINGS: PRIVATE PROPERTY AND THE POWER OF EMINENT DOMAIN (1985).

[27] ALDO LEOPOLD, A SAND COUNTY ALMANAC 210-211 (1949).

[28] See infra note 33.

The third objection is that the entire question is moot because a multi-billion dollar industry has been built on the premise of gene patents. Even if those were initially of questionable validity, and I think there is a respectable argument in law and policy to say that this is the case, no court is now going to upset the financial arrangements that have already been built. This is an important and legitimate pragmatic consideration. Still, scholars are not supposed to give up their inquiries even if there are strong practical reasons why the social arrangements they are challenging look particularly stable. This point is, or at least should be, axiomatic. Moreover, the insights that come from engaging the broad range of arguments for and against genetic enclosure are not limited to some up or down vote on gene patents. The same issues are presented at every level of the analysis, from the debate over the scope of stem cell patents, the PTO's utility guidelines, the exercise of governmental march-in rights, the negotiation of the next round of the TRIP's, the revision of Bayh-Dole, the debate over database protection and so on and so on. Those issues continue to arise and it will ill-serve us if we have a double-decker structure of policy debate about them—a lower deck of tumultuous popular and non-legal arguments about everything from the environment to the limits of the market, and a calm upper-deck which is once again fine-tuning the input-output table of the innovation process. We need a fusion of the two—or at least a staircase between them.

IV Reconstructing Scholarship

Let me assume optimistically that a couple of the arguments I have made here cause you at least a moment's thought before they are rejected. Perhaps you find it hard to explain the disparity in methodological breadth between property scholarship and intellectual property scholarship when it comes to questions of commodification. Alternatively, perhaps you are struck by the fact that even much of the most "critical" intellectual property scholarship tends to be relentlessly utilitarian, whether the question being debated is the expansion of rights or their circumscription. It could even be that you notice how seldom intellectual property scholarship deals explicitly with questions of distributive justice, or even with more expansive definitions of economic efficiency. Those questions, after all, are meat and drink to scholars of tort law, property law, taxation and so on, even those operating in a largely economistic framework. Can they be irrelevant here? Let us say, in other words, that you are willing to entertain the *possibility* that the methodological boundaries of the discipline might be a little narrow and that the gene patenting controversy could illustrate some of that narrowness. I would count this chapter a success if it raised even this much of a doubt.

The next, entirely reasonable, question is, "so what would we be looking at if we agreed with you?" What kinds of potentially productive study are suggested if we take seriously some of the points I have tried to developed here? It is a daunting task first to imagine and then defend an entire body of scholarship. Nevertheless, here are two lines of inquiry that strike me as particularly fruitful and, while they go far beyond the debate over gene patents, seem to me provocatively to be suggested by some of the arguments we find there.

1.) **Questioning and refining the ideal of perpetual innovation**: Like most scholars of intellectual property, I use innovation as my touchstone. For both constitutional and economic reasons, the ideal of continued innovation in science, technology and culture lies at the heart of the discipline. We worship at the church of innovation. We take it as an *a priori* good. But is the concept as unproblematic as our work would make it seem? In other contexts, environmentally inspired economics has raised powerful concerns about the different but related ideal of perpetual growth; is our ideal to build a market that makes more and more *stuff*, and in the process makes us (unequally) richer so we can afford to buy more *stuff*, fueling yet another cycle of productive expansion and relentless consumption? The environmentalists ask us whether such a conception of economics is environmentally sustainable, or suited to promote human flourishing. It would be hard to disagree that our conceptions of economics and growth policy are stronger as a result of their objections. Are there a similar set of questions about the ideal of perpetual innovation?

To put it more precisely, the environmental critique of perpetual growth depends on **1.) an empirical challenge** to the assumption of infinite resources, **2.) an epistemological challenge** to the notion that human welfare is reducible to consumer surplus, and **3.) a deontological or teleological challenge** to the "form of life" implied by consumer materialism. But for most scholars, the response to these thoughtful challenges is not to *reject* growth as a goal. It is, rather, to talk of sustainable development, to disaggregate the notion of growth and talk about those forms of growth that are desirable. Can we, should we, do the same thing with innovation, seeking to disaggregate it and fine-tune our policies to achieve the types of innovation we wish, whether to preserve the environment or to promote "human flourishing," or honor the logic of the original position, or to take account of the diminishing marginal utility of wealth? Is our discipline's goal of innovation really reducible to the attempt to "develop new stuff that sells?"

For those who practice the economics of the Chicago school, current revealed consumer preferences (*based on the existing distribution of wealth and the pattern of ability and willingness to pay which that distribution yields*) have an almost totemic power. This is an absolutely central point. It is not simply preferences in the abstract that generate our measure of value: doubtless parents of children with sleeping sickness in developing countries have an overwhelming "preference" for the development of a drug that would offer a cure. But preferences in the Chicago model are measured only at the conjunction of ability and willingness to pay: the distribution of wealth thus dramatically effects the "value" of the goods desired. If one ignores the question of a justification for the existing distribution of wealth, one can assume that it is only our current pattern of preferences (based on our current distribution of wealth) that gives rise to "neutral" or "scientific" descriptions of human welfare. Attempts to interfere with the pattern of innovation called for by those preferences, (for example, by shifting research away from male pattern baldness and towards malaria and sleeping sickness) represent a "political choice" or a preference for "equity" over "efficiency," and thus possibly a market distortion and resulting welfare loss. The decision blindly to follow the demand dictated by the current distribution of rights and wealth does not represent such a political choice.

(To which one can only respond, "Humbug!") Thus the Chicago school economist would effectively answer that innovation is reducible to the attempt to develop new stuff that sells. This is a neutral account of innovation and an appropriate guide to intellectual property policy. Any other notion, for example, seeking those forms of innovation that might be called for by a more equal distribution of wealth, or those forms which hold the best prospect of preserving the biosphere, represents a contentious political choice and is an inappropriate focus for scholarly inquiry. *Scholars should not be choosing among or between preferable types of innovation, but simply seeking to promote* **innovation**—*understood as a black box which takes its preference inputs from the existing distribution of wealth, and its constraints from the existing pattern of rights and regulation.*

But the Chicago school account is, to use a precise term of art vouchsafed to me by my children, "bogus." There is no neutral account of innovation, or of efficiency. Note well, please, that I am **not** saying we must "supplement" our analysis of neutral efficiency with warm, fuzzy distributional concerns. Our conceptions of efficiency (and thus of innovation) themselves **already** contain strong, and frequently unannounced or justified, distributional assumptions. We must, and we already do, pick between different visions of that concept. For example, we choose between those visions of efficiency which do and those which do not rely on ability and willingness to pay as the sole measure of value. Some measures of valuation take "asking price" rather than offering price as their criterion; if you had the right to have sea otters floating off Alaska, how much would I have to pay you to give up that right? Wealth thus becomes less important. The decision of whether it is "efficient" to engage in development that threatens the sea otters changes dramatically—not because we have added some marginalised notion of distributional equity as an annex to our efficiency analysis, but because all efficiency analyses implicitly contain such assumptions already. The same point applies to more technical efficiency criteria. Do we pick Kaldor-Hicks efficiency with its notion that a change is economically efficient if the winners could compensate the losers and still be better off, while not actually requiring that they *do* engage in compensation? Or do we apply Pareto superiority, where a change is only "efficient" if the winners could and do *actually* compensate the losers?

Now apply this critique to the world of intellectual property. Different notions of efficiency and market demand for particular innovations will obviously yield very different patterns of efficient innovation policy. More complex methodological assumptions about the scope of the analysis too, will influence the outcome. For example, our analysis of innovation is influenced by market structures and ideals of progress that are partly caused by the very rules we are analyzing. We must choose whether to hold those structures constant, or to put them, too, under the microscope of analysis. Consider, for example, the disparate treatment of investment in compiling factual data on the one hand and investment in invention or expression on the other. Are these differences to be taken as given, as exogenously determined rights which structure the market we study, or as a part of the field to be questioned? Our work is influenced by implicit selections *between* types of innovation already prefigured in the legal doctrine. Consider, for example, the treatment of pure research tools under the utility guidelines, or the different treatment of innovative parody and satire in copyright law. Some types of innovation are to be hailed, others not. My point is that, *implicitly*, scholars are already selecting among types of innovation.[29] We hail certain market structures as more innovative, when what we really mean is that they promote more desirable forms of innovation when judged against some other metric—based on a particular and contentious conception of efficiency, distributional justice, or what have you. These other metrics, however, are generally assumed rather than defended.

The result of these patterns of thought is particularly unfortunate. To the extent that we discuss distribution at all, we relegate it to a second class status, after the *real* scientific analysis, devoted to maximising innovation, has taken place. But this makes no sense. We know—or we should know—from the cognate debates over the meaning of efficiency in economic analysis, that our definitions of innovation *already* incorporate contentious distributional assumptions. Whether something is "innovation," like whether something is efficient, will depend on complex judgements about how to incorporate the impact of existing distributions of wealth, including distributions of prior legal rights, on the "valuations" that the analysis uses as criteria. We know this when we pick between Kaldor-Hicks and Pareto superior measures of "efficiency," when we incorporate or do not incorporate the diminishing marginal utility of wealth, or the endowment effects of behavioral economics, into our tax policy. We know it when we decline to find that it is efficient to move the resource of a glass of water to the mildly overheated rich man rather than the pauper dying of thirst. We know it when we use contingent valuations of "what it would cost to make you give up X right, or drug" rather than "what could you afford to pay for X right, or X drug" as our measure. We know it when we decide whether the pollution driven destruction of wildlife for which there is no market is inefficient, which in turn depends on what types of contingent valuation we use to measure that loss. Why then do we *not* know it in the world of intellectual property scholarship?

If the gene controversy, where all these issues are forcibly thrust to the surface of the debate, forced us to disaggregate innovation, and then to articulate and defend the precise incarnation of it that we believe to be justified—whether on environmental, distributional or deontological grounds—then it would have done us a favor. It is striking to see the uncritical way that appeals to big "P" Progress and big "E" Efficiency are accepted in our discipline, where in other areas of legal scholarship there is much more of an attempt to clarify which type of efficiency or which

---

[29] I take it that some economists are perfectly prepared to agree to certain versions of this argument. See eg Brett Frischmann *Innovation and Institutions: Rethinking the Economics of U.S. Science and Technology Policy* 24 VT. L. REV. 347 (2000). Yochai Benkler's work comes closest towards unpacking the meanings of innovation and selecting among and between them based on an analysis that is informed by economics, but goes beyond the narrowly economic. Yochai Benkler, *A Political Economy of the Public Domain: Markets in Information Goods vs. The Marketplace of Ideas* in EXPANDING THE BOUNDS OF INTELLECTUAL PROPERTY: INNOVATION POLICY FOR THE KNOWLEDGE SOCIETY (R. Dreyfuss, D. Zimmerman, H. First eds.) (2000); *Free as the Air to Common Use: First Amendment Constraints on Enclosure of the Public Domain* 74 NYU Law Review 354 (1999).

type of progress is being appealed to, an attempt which must reach outside the charmed circle of economic analysis in order to be meaningful. To be sure, early Chicago school law and economics scholarship talked as if there were one uncontestable vision of efficiency, and all other visions of the concept could be exiled to the touchy-feely world of "equity claims"; claims which were always to be pursued *somewhere else* than in the current analysis. But it would be hard to find many contemporary scholars who echo those claims, or who treat wealth maximization within the current distribution of wealth as an unproblematic definition of efficiency. Is that not, though, what we do all too often in intellectual property scholarship—at least when we worship at the church of innovation without clarifying what implicit choices about types of desirable innovation are being concealed by the very structure of our analysis? To say this is not automatically to nominate any particular successor notion of innovation. My own preference would be for considerably more scholarship that measures our system of current innovation against one that, in questions of basic human need, such as access to essential medicines, stipulated a certain minimum valuation to human life, even among the global poor. There are many possible solutions to the failures *and inefficiencies* this analysis would reveal in our current intellectual property system—ranging from supplementing the patent system with government bounties or prizes, to offering dual zone patents, to directly subsidizing research. Intellectual property scholars have an extremely valuable role to play in working out the bugs in such solutions and they are less likely to play that role while they are in the thrall of the black box model of innovation. And this will be true *whatever* measure of innovation we choose to articulate and defend. This leads to my second suggestion.

2.) **From Public Goods to Public Choice**: We look at intellectual property through a lens constructed around the focal point of rational choice theory. We care deeply about the structure of incentives that it sets up, the possible free rider effects, moral hazards and so on. But we are curiously reticent in turning that lens back on the structure of incentives that intellectual property[30] policy sets up *in the policy-making process itself*. We should study the market for intellectual property *legislation* with the same care and rigor that we display in studying the market for intellectual property. To put it differently, one environmentalist critique of gene patents is that they create a powerful incentive for rights-holders to drive state scientific policy towards the utilization of the technologies in which the rights holders have exclusive rights, and do so quickly, before the patent term expires, regardless of whether the environmental consequences of those technologies are well understood. This kind of phenomenon ought to be well within the intellectual property scholar's field of inquiry.

---

[30]For a general background see MAXWELL L. STEARNS, PUBLIC CHOICE AND PUBLIC LAW; READINGS AND COMMENTARY (1997); Daniel A. Farber & Philip P. Frickey, The Jurisprudence of Public Choice, 65 TEX. L. REV. 873 (1987). For a similar suggestion about intellectual property policy see Arti Rai, *Addressing the Patent Gold Rush: the Role of Deference to PTO Patent Denials* 2 Wash. U. J.L. & Pol'y 199, 215, note 51 (2000).

More generally, we need models of the market for intellectual property legislation that can help us understand how to build structural protections of the public interest into the policy making process itself. Intellectual property is a legal device to avoid a "public goods" problem, meaning the problem of providing adequate incentives to produce goods which are non-rival and nonexcludable. But there are also a different variety of "public goods" in the market for intellectual property legislation. We know that legislation built around the self-interest of existing and aspirant monopolists will protect a variety of private goods, namely those of the firms and interests at the table. We know also that it will fail to protect certain kinds of interests—most notably those of large numbers of unorganized individuals with substantial collective, but low individual, stakes in the matters being discussed. It will fail to protect the interests of businesses that, as Jessica Litman puts it, are "outsiders who can't be at the negotiating table because their industries haven't been invented yet."[31] This will be even more true if those businesses are likely to challenge the business plans of the companies that *are* at the table. It will fail to protect the public domain, even as a source for future creativity, except in the limited cases where those around the table can clearly see that their *particular* form of production will work better with a rich commons-based system, rather than around a private property system tilted to give their particular form of production an advantage.

These are the dysfunctions that we know our market for intellectual property legislation will yield. Why then, do we not spend some of the time we currently invest trying to remedy dysfunctions in information and innovation markets, instead trying to think of remedies for the dysfunctions in our legislative process? How about an intellectual property ombudsman to represent the interests of the public and the public domain? Or a required "Information environment impact statement" by the GAO justifying each new grant of monopoly right in terms of its long term economic effects? Doubtless there are better ideas than these, but it is the focus on the policy making process that I wish to defend. Again, attention to the gene patenting controversy proves instructive. It may be that the complaints of the people on the bottom of the bus have more to teach those of us on the top than we would like to admit.

V Conclusion

I started by comparing the current progress of intellectual property expansion over the human genome to the English enclosure movement. As with the first enclosure movement, we are faced with a conflicting array of claims—claims about the essential common-ness of the property in question, or the moral and traditional reasons why it should not be shifted to individual ownership, claims about the effect of newfound property rights on environmental attitudes, claims about the limits of the marketplace, about economic effects, about encouragement of innovation and so on and so on. As with the first enclosure movement, some of these claims

---

[31] JESSICA LITMAN, DIGITAL COPYRIGHT (2001)

are essentialist or romantic or just analytically sloppy. As with the first enclosure movement, we have available to us a cooler language of economic incentives that promises a calmer and smoother discussion about the encouragement of progress. (Though the economic arguments for the second enclosure movement strike me as considerably weaker than those for the first.) And as with the first enclosure movement, to retreat from the full range of issues into those calmer, smoother waters would be a mistake, *even if we ended up supporting the furthest reaching extensions of property rights currently being proposed.*

In this chapter, I have tried to summarize—without endorsing—a broad range of criticisms of gene patenting. My point was *not* to say that this broad range of criticisms was correct; indeed I find many of them unpersuasive. Nor was it to reject the analytical structure of innovation economics that forms the backbone of intellectual property scholarship. Instead, I used the gene patenting debate as an example to make the larger point that intellectual property scholars are mistaken to write off most non-utilitarian criticisms as outside their purview, concentrating instead only on a positive exegesis of patent law and a restricted normative exegesis of which patent policy will best ensure continued research and development of commercially desirable products. Even if one is unconvinced by Rifkin or the environmentalists, in other words, when one focuses on the *pattern* of disciplinary exclusion, rather than the particular substantive arguments, it becomes apparent both how narrow and how contentious our disciplinary assumptions are. What's more, I argued, the major professional division among intellectual property scholars— that between maximalists and minimalists—has the strong, but probably unintended effect of helping to reinforce this narrowness over the methodological tenets of the discipline. The minimalists are those with the strongest incentive to challenge the expansion of intellectual property rights over the genome. Yet, because of the structure of the rhetoric in the discipline, they are also those most committed to the claim that only utilitarian arguments about the encouragement of future innovation are legitimate parts of the discourse. Minimalists are used to fighting off covert sweat of the brow claims, concealed appeals to natural right, and Hegelian notions of personality made manifest in expression—all deployed to argue that rightsholders should have their legally protected interests expanded yet again. Against these rhetorics, they insist on both constitutional and economic grounds that the reason to extend intellectual property rights can only be the promotion of innovation. When we turn to the question of whether there are non-economic reasons for *curtailing* the reach of patents over genetic material, their hard-wired reflex is to restrict their analysis to the same utilitarian domain. Beyond the debate between maximalists and minimalists, the gene patenting controversy throws other articles of faith and other shibboleths into doubt; from the various possible presumptions behind the idea of perpetual innovation to the assumed operation and dysfunction of the political process. Thus, I concluded, we have much to learn from the gene wars even though they may currently seem to yield more heat than light.

I think most intellectual property scholars will find this claim unconvincing. Told that they must supplement a discipline in which answers seem possible—even if disputed—with greater attention to debates in which they have much less familiarity and in which answers seem much more contested, they will understandably cavil. What's more, the intellectual biographies of many patent scholars mean that they find the intersection of a particular complex technology with a set of different innovation markets to be a much more personally hospitable terrain than the broader landscape I am suggesting.

I cannot prove that they are wrong, of course. For epistemological reasons, if nothing else, the proof of this pudding is in the eating. But I will offer two analogies to suggest that a broader debate about genetic patents within intellectual property scholarship would be of particular benefit.

First, consider the analogous effect of environmental thinking on both debates within, and the analytical structure of, tangible property scholarship.[32] The increase in the breadth of the analysis, the need for a multi-generational framework, the importance of externalities, the complexity and interrelationship of the natural systems within which schemes of public and private law land regulation intersect—all of these have been highlighted by the brush with environmental concerns. Disciplines do not develop in a vacuum; in this case the analytical toolkit itself has been changed by the encounter. Even the practitioner of a "just-the-facts-Ma'am" style of property scholarship, uncomfortable with the fuzziness and enthusiasm of all these tree-huggers, has seen his discipline improved in the process. Though consideration of environmental themes does not *logically require* us to be more self-conscious about our notions of "efficient" property policy, it has undoubtedly had that effect. There is, I would suggest, no reason to believe that the same thing would not be true of intellectual property scholarship as it confronts the very real questions posed by the consequences to environmental policy of patents on genetic material, including human genetic material. And this thought, in turn, suggests that in other areas the broader debate over genetic patents may have much to teach intellectual property scholars, in the same way that the debate over commodification changed property scholarship in the 80's and 90's. In particular, we intellectual property scholars need to grapple more thoroughly than we have with the limitations as well as the advantages of the utilitarian, progress-enhancing analytical framework we adopt. In other areas of scholarship, we are much more willing to point out the limits of utilitarian analysis, the weakness of a "willingness and ability to pay" model of social worth, the impact of wealth effects on the efficiency calculus, the intergenerational moral concerns hidden by the framework of growth economics, the problems with always assuming endogenous preferences, or exogenously determined rights. Just as we may have more to offer the debate over gene patents than a careful analysis of effects of exclusive rights at different points in the research stream, so too the broader debate over gene patents may have things to show us about the relative complacency of our methodological

---

[32] Carol Rose's work illustrates my point here particularly well.

assumptions.[33] Ironically, as I suggested earlier, this point may be particularly important, and particularly difficult, for the minimalists.

For a second analogy, consider the justice claims that have recently caused "access to essential medicines" to become a fundamental part of drug patent policy both domestically and internationally. Again, these are a set of issues that fit poorly within conventional intellectual property scholarship; but the arguments are not mere exhortations to take drugs away from companies and hand them over to the poor and the sick. The essential medicine questions are not simple, either economically or institutionally and—after some initial reticence—the academy now seems to be turning its eyes to the complicated points of treaty interpretation, regional institutional design, international price discrimination, and alternative patent regimes that this particular and real moment of human suffering forces us to think about. Can we really believe that our scholarly focus will be somehow *weaker* as a result of the forced encounter with claims of distributive justice and human rights? In fact, with any luck, the intensity of feeling about a particular controversy over AIDS drugs may actually force us to acknowledge the single greatest weakness behind a patent driven drug development policy; a patent driven system for drug development will, if working correctly, deliver drugs on which there is a high social valuation—measured in this case by ability and willingness to pay. To put it another way, to have a patent driven drug policy is to *choose* to deliver lots of drugs that deal with male-pattern baldness, but also with real and important diseases: rheumatoid arthritis, various cancers and heart disease. It is to choose *not to* have a system that delivers drugs for tropical diseases, or indeed for any disease which is suffered overwhelmingly by the national or global poor.

To say this is not to condemn drug patents; it is rather, to suggest precisely the two lines of inquiry I argued for in this article. First, if our goal is truly to help to eliminate human suffering, then we should spend more time thinking about alternative and supplementary ways of encouraging pharmaceutical innovation beyond the drug patent system. Second, when we talk about innovation and progress in the intellectual property system, we quickly and easily substitute some universal imagined ideal of Progress for the actual specific version of "progress" towards which our current distribution of entitlements and rights will push us. Many policies that might seem justified by the promotion of large "P" progress, might seem more questionable if they were instead pushing us towards the specific vision of progress held latently within the pattern of demand established by our current distribution of rights and wealth. To quote Amartya Sen, "there are plenty of Pareto optimal societies which would be perfectly horrible places to live."[34]

---

[33] I have struggled with this point particularly in my scholarship on the public domain. In that scholarship, I rely heavily on the basic tools of innovation economics. But I am always troubled when I read these lines from Aldo Leopold, from a chapter of Sand County Almanac entitled "Substitutes for a Land Ethic." "One basic weakness in a conservation system based wholly on economic motives is that most members of the land community have no economic value... When one of these non-economic categories is threatened, and if we happen to love it, we invent subterfuges to give it economic importance... It is painful to read those circumlocutions today." ALDO LEOPOLD, A SAND COUNTY ALMANAC 210-211 (1949).

[34] The source was an interview with an uncomprehending MSNBC interviewer at the time of Sen's receipt of the Nobel prize. The reader will have to trust the author's faulty memory for the precision of the quote.

Now if these lessons can be taught us in a concrete and unforgettable way by the debate over drug patents, is there any reason to believe that the larger debate over gene patents will offer us any less insight, or any less provocation? It *could* be, of course, that the end result would be exactly the same; perhaps all of us would find our conclusions unchanged, even if we were a little more critical about worshiping at the church of innovation, even if we clarified our definitions of that concept and of the notion of efficiency that underpins it, even if we broadened our scholarly focus to include the kind of institutional and environmentalist inquiries I suggest here, and made our discussion of commodification a little more similar to that which occurs in conventional property scholarship. Perhaps this change in methodology would leave our substantive positions unchanged, though I doubt it. Perhaps its effects would only be found in other areas, such as the essential medicines question, or the question of the goals of basic science policy, or the question of the redesign of the institutional framework through which intellectual property policy is made. But even if all that were true, the gene patenting debate could still teach intellectual property scholars a set of lessons we sorely need. At least, that is, if we have the courage to enter it.

V: CHALLENGES FOR THE FUTURE OF COMPUTATIONAL GENOMICS

The three chapters that comprise Section V describe some key challenges for computational genomics in the 21st century. In particular, these chapters focus on concerns related to three kinds of challenges: (1) the increased dependency on computational tools in genomic research and the impact that this will likely have for biology and for social issues affecting the ELSI Research Program;[1] (2) the need to establish clear ethical guidelines for health professionals in light of developments in bioinformatics; and (3) concerns about whether and how to proceed with nanotechnology research, which has significant implications for the future of genomics.

Challenges Involving Computational Tools and the ELSI Research Program

In Chapter 16, Francis Collins, Eric Green, Alan Guttmacher, and Mark Guyer outline a vision for the future of genomic research—in particular, a vision established by the National Human Genome Research Institute (NHGRI). This chapter originally appeared as an article in *Nature* in April 2003, commemorating two seminal events: the 50th anniversary of the discovery of the double helix by Watson and Crick, and the completion of the mapping of the human genome by an international group of scientists. Collins et al. begin with some remarks about the impact that future genomics research will likely have for biology. For example, the authors believe that as the structure of genomes continues to be elucidated, and as the "myriad encoded elements" are identified, further connections between genomics and biology will be made. These connections, in turn, will "accelerate the exploration of all realms of biological sciences." The authors also note, however, that very sophisticated computational tools and techniques will be needed to identify and analyze the functional genomic elements that will make this possible. These tools and techniques will consist of:

> ...increasingly powerful computational capabilities, including new approaches for tackling ever-growing and increasingly complex data sets and a suitably robust computational infrastructure for housing, accessing and analyzing those data sets.

---

[1] Recall our examination of the ELSI Research Program in Chapter 1.

Thus the future of genomics will become even more dependent on computers and the computational capabilities[2] made possible by them.[3]

Noting that many of the major benefits to be realized from genomics are in the area of health, Collins et al. point out that genomic research has also contributed to other aspects of society as well. For example, they note that genomics has created certain "opportunities" for research on social issues, helping us to understand more fully "how we define ourselves and each other." The authors also note that we can analyze the impact of genomics on concepts of "race, kinship, individual and group identity, health, disease and 'normality' for traits and behaviors." In particular, Collins et al. consider some future challenges pertaining to ethical, legal, and social implications (ELSI) research. They argue that the scope of issues currently considered in the ELSI framework will need to be expanded. They also believe that the ELSI research community will need to include individuals from minority groups and from other communities that may be disproportionately affected by the use or misuse of genetic information.

Challenges for Health Informatics Professionals

In Chapter 17, Kenneth Goodman examines some ethical concerns at the intersection of *health informatics* and genomics. He begins by noting that a plausible case can be made for the view that the two sciences that will have the greatest effect on 21st century health care are information technology and genetics. He also notes that while it is already recognized that informatics is significantly "reshaping the health professions," we are only beginning to appreciate the extraordinary risks and potential benefits of progress in the "human genome sciences."

Specifically, Goodman's chapter focuses on ethical and social issues that arise in the field of *clinical bioinformatics*. In his analysis of ethical aspects of this field, Goodman differentiates three categories of concern: accuracy and error, appropriate uses and users of digitized genetic information, and privacy and confidentiality.[4] Within the category of accuracy and error, Goodman examines four different kinds of issues, which he identifies as "risks to persons," "recanted linkage studies," "meta-analysis," and "decision support." In the category of appropriate uses and users, he considers some concerns relating to two important questions: (1) "Who should use clinical information"? (2) "In which contexts should that data be used"? Under the category of privacy and confidentiality, Goodman examines a tension between what he sees as two different, but mutually critical, kinds of needs regarding access to personal health information: (1) the need for "appropriate or author-

---

[2]See also Francis Collins' remarks on the critical nature of this dependency in an interview with Bradie Metheny in *Policy Roundup—State of Genomic Research,* available at www.laskerfoundation.org/homenews/genome/pround_gn.html.

[3]Recall our discussion in Chapter 1 of the dependency that has resulted because of the "computer-enhancing" roles that computing technologies have played in genomics.

[4]These categories are also used by the author in Goodman (1999).

ized access to personal information," and (2) the need to "prevent inappropriate or unauthorized access."

Goodman acknowledges that there are no clear standards to guide health professionals with respect to the diverse kinds of ethical questions that can arise in the field of clinical bioinformatics. Typically, health professionals are directed either to a set of principles commonly found in organizational guidelines (that might or might not exist) or to what he calls "best practice standards." But Goodman believes that these are inadequate for either one or both of the following reasons: (a) they tend to be so broad or overly simplistic that they cannot adequately guide behavior, or (b) they are so specific or detailed that they are too inflexible to be useful in diverse and unexpected cases. Goodman concludes his chapter by noting that clear and explicit guidelines are needed in order to strike an appropriate balance between these two extremes.

Nanotechnology and the Future of Computational Genomics Research

In Chapter 18, the closing chapter of Section V and of the book, John Weckert examines controversies surrounding research in nanotechnology, which has implications for the future of computational genomics research. *Nanotechnology*, a term coined by K. Eric Drexler in the 1980s, is a branch of engineering dedicated to the development of extremely small electronic circuits and mechanical devices built at the molecular level of matter. Current microelectricomechanical systems (or MEMS), tiny devices such as sensors embedded in conductor chips used in airbag systems to detect collisions, are one step away from the molecular machines envisioned in nanotechnology. Drexler (1991) believes that developments in this field will result in computers at the nano-scale, no bigger in size than bacteria, called *nanocomputers*. To appreciate the scale of future nanocomputers, we can imagine a mechanical or electronic device whose dimensions are measured in nanometers (billionths of a meter, or units of 10^{-9} meter). Ralph Merkle (1997) envisions nanocomputers having mass storage devices that can "store more than 100 billion bytes in a volume the size of a sugar cube." He also predicts that these nano-scale computers will be able to "deliver a billion instructions per second—a billion billion times more than today's desktop computers."

Which kinds of materials will be used to construct computing devices at the nano-level? Will they be biological? Nadrian Seeman (2004, p. 65) believes that DNA is a versatile component for making nanoscopic structures and devices, pointing out that materials could be constructed—i.e., "either made of DNA or made by it"—with structures that are precisely designed at the molecular level.[5] Seeman (p. 66) further argues that DNA is an "ideal molecule" for building nanometer-scale structures because strands of DNA can be:

---

[5]Seemen also points out the helix of DNA has a diameter of about two nanometers, and "it twists full circle every 3.5 nanometers or so, a distance of about 10 base pairs, which form the 'rungs' of DNA's ladder" (p. 65).

> ...programmed to self assemble into complex arrangements by producing the strands with the appropriate combinations of complementary bases, which preferentially bond together to form stretches of double helices... Nanometer-scale DNA machines can function by having parts of their structure change from one DNA conformation to another. These movements can be controlled by chemical means or by the use of special DNA strands.

So it would seem that the future of nanotechnology research and genomics research will intersect at some points and possibly converge at others. While there may be several opportunities at this juncture of research, we should also consider whether any serious challenges and possible risks may arise in pursuing research and development at the nano-level.

Nanotechnology's advocates are quick to point out many of the advantages that would likely result for the medical and health fields, as well as for the environment. For example, nano-particles inserted into bodies could diagnose diseases and directly treat diseased cells. Doctors could use nanomachines (or *nanites*) to make microscopic repairs on areas of the body that are difficult to operate on with conventional surgical tools. And with nanotechnology tools, the life signs of a patient could be better monitored. With respect to the environment, nanites could be used to clean up toxic spills, as well as to eliminate other kinds of environmental hazards. Nanites could also dismantle or "disassemble" garbage at the molecular level and recycle it again at the molecular level via "nanite assemblers."

Nanotechnology opponents, on the other hand, see things quite differently. Since all matter (inanimate objects and organisms) could theoretically be disassembled and reassembled by nanite assemblers and disassemblers, some worry about what would happen if strict "limiting mechanisms" were not built into those nanites. In other words, they worry that if nanomachines were created to be self-replicating and if there was a problem with their limiting mechanisms, they could multiply endlessly like viruses. Some also worry that nanite assemblers and disassemblers could be used to create weapons; and because guns, explosives, and electronic components of weapons could all be miniaturized, some critics (see, for example, Chen, 2002), worry that nanites themselves could be used as weapons.

Many of the claims and predictions regarding nanotechnology seem quite plausible; reputable scientists and researchers tend to agree with the projection that nanocomputers of some sort will be available sometime around the middle of the 21st century. Weckert suggests that because the evidence is so credible, it would be prudent for us to consider some of the ethical implications now, while there is still time to anticipate them. He also responds to some points raised in a controversial article by Bill Joy (2000), who has suggested that because developments in nanotechnology are threatening to make us an "endangered species," the only realistic alternative is to limit the development of that technology. Weckert believes that if Joy is correct about the dangers of nanotechnology, then we must seriously consider whether scientists should continue to engage in research in this area.

Furthermore, if nanotechnology research and development does continue, Weckert believes that we should determine *who* will be held morally responsible for outcomes that go awry.

Weckert does not examine any specific implications of nanotechnology research for genomics per se. However, he considers some overall benefits and harms that might develop from continued nanotechnology research in general, and he asks what our default presumption should be about this research continuing. That is, should there be what Weckert calls a "presumption in favour of freedom of research"? Or should the burden be on scientists to prove at the outset that research in a controversial area poses no threat?[6] Weckert suggests the following strategy for deliberating over this question:

> *If a* prima facie *case can be made that some research will likely cause harm . . . then the burden of proof should be on those who want the research carried out to show that it is safe.*

He goes on to say, however, that there should be:

> *. . . a presumption in favour of freedom until such time a* prima facie *case is made that the research is dangerous. The burden of proof then shifts from those opposing the research to those supporting it. At that stage the research should not begin or be continued until a good case can be made that it is safe.*

Has a *prima facie* or self evident case been made to show that nanotechnology poses a dangerous threat? If not, then in Weckert's scheme, there are no compelling grounds for halting nanotechnology research at this time. However, this decision could easily be reversed if future evidence were to strongly suggest that harm is more likely than not to result from that research. In that case, using Weckert's strategy, it would be permissible to restrict or even forbid nanotechnology research.

As research in nanotechnology proceeds, it would be prudent to establish an ethical framework to guide it. Some ethical guidelines for "molecular nanotechnology" have already been suggested by the Foresight Institute; and the U.S. government has created the National Nanotechnology Initiative (NNI) to monitor and guide research in this field. A fully developed and government-sponsored set of ethical standards for nanotechnology, similar to the ELSI standards that have guided the Human Genome Project, might provide an appropriate ethical standard for guiding nano-level research in general. In the case of genomics-specific research conducted at the nano-scale, some

---

[6] Arguments for how best to proceed in scientific research when there are concerns about harm to the public good are often examined in terms of what has become known as the "precautionary principle" in science. For a discussion of how this principle can be used to assess risks in research involving public health and the environment, see Foster (2002) and Wilson (2002).

special ELSI requirement could be articulated and then incorporated into an expanded set of guidelines for the ELSI Research Program.

Review Questions for Chapters in Section V
1. According to Collins, Green, Guttmacher, and Guyer, what implications will future developments in genomics likely have for the field of biology?
2. Why do Collins, Green, Guttmacher, and Guyer believe that greater "computational capabilities" will be needed to accomplish some of the goals and objectives identified in their vision of the future of genomics?
3. Identify some of the "opportunities" for research in the social sciences that Collins, Green, Guttmacher, and Guyer believe genomics has made possible.
4. Why do Collins, Green, Guttmacher, and Guyer believe that the scope of ELSI issues currently examined will need to be expanded in the future? Why do the authors also believe that more representatives from minority groups will need to be included in the expanded ELSI initiatives?
5. What does Kenneth Goodman mean by the expressions "health informatics" and "clinical bioinformatics"?
6. Which three categories of ethical concerns involving clinical bioinformatics does Goodman identify and examine?
7. Why does Goodman believe that "organizational guidelines" and "best practice standards" are not adequate as ethical guidelines for health informatics professionals?
8. What is nanotechnology? What are nanocomputers?
9. What kinds of challenges does research at the nano-level pose for computational genomics?
10. Describe the arguments John Weckert articulates for and against continued research and development in the field of nanotechnology.
11. Identify some of the ethical issues that could arise if nanotechnology research is allowed to continue.
12. What kinds of ethical guidelines for nanotechnology research, if any, should be established? How could the ethical guidelines established by the ELSI Research Program inform the framing of "nano-ethics" guidelines?

References
Chen, Andrew (2002). "The Ethics of Nanotechnology." Available at http://www.actionbioscience.org/newfrontiers/chen.html. Accessed 12/6/2004.
Collins, Francis S. (2004). Interview (with Bradie Metheny on "Key Facts About Genomic Research") in *Policy Roundup—State of Genomic Research.* Available at http://www.laskerfoundation.org/homenews/genome/pround_gn.html. Accessed 12/5/2004.
Drexler, K. Eric (1991). *Unbounding the Future.* New York: Quill.
Foster, Kenneth (2002). "The Precautionary Principle—Common Sense or Environmental Extremism," *IEEE Technology and Society*, Vol. 21, No. 4, pp. 8–13.
Goodman, Kenneth W. (1999). "Bioethics: Challenges Revisited," *MD Computing*, May/June, pp. 17–20.
Joy, Bill (2000). "Why the Future Doesn't Need Us," *Wired*, Vol. 8, No. 4.
Merkle, Ralph (1997). "It's a Small, Small, Small World," *Technology Review*, Vol. 25, February/March.

Seeman, Nadrian C. (2004). "Nanotechnology and the Double Helix," *Scientific American*, June, pp. 64–75.

Wilson, Richard (2002). "Is the Precautionary Principle a Principle?" *IEEE Technology and Society*, Vol. 21, No. 4, pp. 45–48.

Suggested Further Readings

Anderson, James G., and Kenneth W. Goodman, eds. (2002). *Ethics and Information Technology: A Case-Based Approach to a Health Care System in Transition*. New York: Springer.

Backus, Ann, Richard A. Spinello, and Herman T. Tavani, eds. (2004). *Genomics, Ethics, and ICT*. Special Issue of *Ethics and Information Technology* (Vol. 6, No. 2). Dordrecht, The Netherlands: Kluwer.

Christiani, David C., Richard R. Sharp, Gwen. W. Collman, and William A. Suk (2001). "Applying Genomic Technologies in Environmental Health Research: Challenges and Opportunities, *Journal of Occupational and Environmental Medicine*, Vol. 43, No. 6, pp. 526–533.

ELSI Research Program. National Human Genome Research Institute. Available at http://www.genome.gov/10001754. Accessed 1/27/2003.

Goodman, Kenneth W. (1996). "Ethics, Genomics, and Information Retrieval," *Computers in Biology and Medicine*, Vol. 26, No. 3, pp. 223–229.

Goodman, Kenneth W. (1998). "Bioethics and Health Informatics: An Introduction." In K. W. Goodman, ed. *Ethics, Computing, and Medicine: Informatics and the Transformation of Healthcare*. New York: Cambridge University Press, pp. 1–31.

Juengst, Eric (2004). "Face Facts: Why Human Genetics Will Always Provoke Bioethics," *Journal of Law, Medicine, and Ethics*, Vol. 32.

Marturano, Antonio (2002). "The Role of Metaethics and the Future of Computer Ethics," *Ethics and Information Technology*, Vol. 4, No. 1, pp. 71–78.

Moor, James H. (2001). "The Future of Computer Ethics: You Ain't Seen Nothin' Yet!" *Ethics and Information Technology*, Vol. 3, No. 2, pp. 89–91.

Sharp, Richard R., and J. Carl Barrett (2000). "The Environmental Genome Project: Ethical, Legal, and Social Implications," *Environmental Health Perspectives*, Vol. 108, No. 4.

Tavani, Herman T. (2004). *Ethics and Technology: Ethical Issues in an Age of Information and Communication Technology*. Hoboken, N. J.: John Wiley and Sons.

Taubes, Gary (2001). "Your Genetic Destiny for Sale," *Technology Review*, Vol. 104, No. 3, pp. 40–46.

Weckert, John (2001). "Computer Ethics: Future Directions," *Ethics and Information Technology*, Vol. 3, No. 2, pp. 93–96.

Weckert, John (2004). "Lilliputian Computer Ethics." In R. A. Spinello and H. T. Tavani, eds. *Readings in CyberEthics*. 2nd ed. Sudbury, MA: Jones and Bartlett, pp. 690–697.

Woodhouse, E. J. (2004). *Nanotechnology Controversies*. Special Issue of *IEEE Technology and Society*. (Vol. 23, No. 4). New York: IEEE.

16. A Vision for the Future of Genomics Research: A Blueprint for the Genomic Era

Francis S. Collins, Eric D. Green, Alan E. Guttmacher, and Mark S. Guyer[1]

The completion of a high-quality, comprehensive sequence of the human genome, in this fiftieth anniversary year of the discovery of the double-helical structure of DNA, is a landmark event. The genomic era is now a reality.

In contemplating a vision for the future of genomics research, it is appropriate to consider the remarkable path that has brought us here. Recognition of DNA as the hereditary material[2], determination of its structure[3], elucidation of the genetic code[4], development of recombinant DNA technologies[5,6], and establishment of increasingly automatable methods for DNA sequencing[7–10] set the stage for the Human Genome Project (HGP) to begin in 1990 (see also www.nature.com/nature/DNA50). Thanks to the vision of the original planners, and the creativity and determination of a legion of talented scientists who decided to make this project their overarching focus, all of the initial objectives of the HGP have now been achieved at least two years ahead of expectation, and a revolution in biological research has begun.

The project's new research strategies and experimental technologies have generated a steady stream of ever-larger and more complex genomic data sets that have poured into public databases and have transformed the study of virtually all life processes. The genomic approach of technology development and large-scale generation of community resource data sets has introduced an important new dimension into biological and biomedical research. Interwoven advances in genetics, comparative genomics, high-throughput biochemistry and bioinformatics are providing biologists with a markedly improved repertoire of research tools that will allow the functioning of organisms in health and disease to be analysed and comprehended at an unprecedented level of molecular detail. Genome sequences, the

---

This essay originally appeared in *Nature*, **422**: 1-13 (April). Copyright © 2003 by Nature Publishing Company. Reprinted by permission.

bounded sets of information that guide biological development and function, lie at the heart of this revolution. In short, genomics has become a central and cohesive discipline of biomedical research.

The practical consequences of the emergence of this new field are widely apparent. Identification of the genes responsible for human mendelian diseases, once a herculean task requiring large research teams, many years of hard work, and an uncertain outcome, can now be routinely accomplished in a few weeks by a single graduate student with access to DNA samples and associated phenotypes, an Internet connection to the public genome databases, a thermal cycler and a DNA-sequencing machine. With the recent publication of a draft sequence of the mouse genome[11], identification of the mutations underlying a vast number of interesting mouse phenotypes has similarly been greatly simplified. Comparison of the human and mouse sequences shows that the proportion of the mammalian genome under evolutionary selection is more than twice that previously assumed.

Our ability to explore genome function is increasing in specificity as each subsequent genome is sequenced. Microarray technologies have catapulted many laboratories from studying the expression of one or two genes in a month to studying the expression of tens of thousands of genes in a single afternoon[12]. Clinical opportunities for gene-based pre-symptomatic prediction of illness and adverse drug response are emerging at a rapid pace, and the therapeutic promise of genomics has ushered in an exciting phase of expansion and exploration in the commercial sector[13]. The investment of the HGP in studying the ethical, legal and social implications of these scientific advances has created a talented cohort of scholars in ethics, law, social science, clinical research, theology and public policy, and has already resulted in substantial increases in public awareness and the introduction of significant (but still incomplete) protections against misuses such as genetic discrimination (see www.genome.gov/PolicyEthics).

These accomplishments fulfil the expansive vision articulated in the 1988 report of the National Research Council, *Mapping and Sequencing the Human Genome*[14]. The successful completion of the HGP this year thus represents an opportunity to look forward and offer a blueprint for the future of genomics research over the next several years.

The vision presented here addresses a different world from that reflected in earlier plans published in 1990, 1993 and 1998 (refs 15–17). Those documents addressed the goals of the 1988 report, defining detailed paths towards the development of genome-analysis technologies, the physical and genetic mapping of genomes, and the sequencing of model organism genomes and, ultimately, the human genome. Now, with the effective completion of these goals, we offer a broader and still more ambitious vision, appropriate for the true dawning of the genomic era. The challenge is to capitalize on the immense potential of the HGP to improve human health and well-being.

The articulation of a new vision is an opportunity to explore transformative new approaches to achieve health benefits. Although genome-based analysis methods are rapidly permeating biomedical research, the challenge of establishing robust paths from genomic information to improved human health remains immense.

Current efforts to meet this challenge are largely organized around the study of specific diseases, as exemplified by the missions of the disease-oriented institutes at the US National Institutes of Health (NIH, www.nih.gov) and numerous national and international governmental and charitable organizations that support medical research. The National Human Genome Research Institute (NHGRI), in budget terms a rather small (less than 2%) component of the NIH, will work closely with all these organizations in exploring and supporting these biomedical research capabilities. In addition, we envision a more direct role for both the extramural and intramural programmes of the NHGRI in bringing a genomic approach to the translation of genomic sequence information into health benefits.

The NHGRI brings two unique assets to this challenge. First, it has close ties to a scientific community whose direct role over the past 13 years in bringing about the genomic revolution provides great familiarity with its potential to transform biomedical research. Second, the NHGRI's long-standing mission, to investigate the broadest possible implications of genomics, allows unique flexibility to explore the whole spectrum of human health and disease from the fresh perspective of genome science. By engaging the energetic and interdisciplinary genomics-research community more directly in health-related research and by exploiting the NHGRI's ability to pursue opportunities across all areas of human biology, the institute seeks to participate directly in translating the promises of the HGP into improved human health.

To fully achieve this goal, the NHGRI must also continue in its vigorous support of another of its vital missions—the coupling of its scientific research programme with research into the social consequences of increased availability of new genetic technologies and information. Translating the success of the HGP into medical advances intensifies the need for proactive efforts to ensure that benefits are maximized and harms minimized in the many dimensions of human experience.

A reader's guide

The vision for genomics research detailed here is the outcome of almost two years of intense discussions with hundreds of scientists and members of the public, in more than a dozen workshops and numerous individual consultations (see www.genome.gov/About/Planning). The vision is formulated into three major themes—genomics to biology, genomics to health, and genomics to society—and six crosscutting elements.

We envisage the themes as three floors of a building, firmly resting on the foundation of the HGP. For each theme, we present a series of grand challenges, in the spirit of the proposals put forward for mathematics by David Hilbert at the turn of the twentieth century[18]. These grand challenges are intended to be bold, ambitious research targets for the scientific community. Some can be planned on specific timescales, others are less amenable to that level of precision. We list the grand challenges in an order that makes logical sense, not representing priority. The challenges are broad in sweep, not parochial—some can be led by the NHGRI alone,

BOX 1 Resources

One of the key and distinctive objectives of the Human Genome Project (HGP) has been the generation of large, publicly available, comprehensive sets of reagents and data (scientific resources or 'infrastructure') that along with other new, powerful technologies, comprise a toolkit for genomics-based reserach. Genomic maps and sequences are the most obvious examples. Others include databases of sequence variation, clone libraries and collections of anonymous cell lines. The continued generation of such resources is critical, in particular:

- Genome sequences of key mammals, vertebrates, chordates, and invertebrates
- Comprehensive reference sets of coding sequences from key species in various formats, for example, full-length cDNA sequences and corresponding clones, oligonucleotide primers, and microarrays
- Comprehensive collections of knockouts and knock-downs of all genes in selected animals to accelerate the development of models of disease
- Comprehensive reference sets of proteins from key species in various formats, for example in expression vectors, with affinity tags and spotted onto protein chips
- Comprehensive sets of protein affinity reagents
- Databases that integrate sequences with curated information and other large data sets, as well as tools for effective mining of the data
- Cohort populations for studies designed to identify genetic contributors to health and to assess the effect of individual gene variants on disease risk, including a 'healthy' cohort
- Large libraries of small molecules, together with robotic methods to screen them and access to medicinal chemistry for follow-up, to provide investigators easy and affordable access to these tools.

whereas others will be best pursued in partnership with other organizations. Below, we clarify areas in which the NHGRI intends to play a leading role.

The six critically important crosscutting elements are relevant to all three thematic areas. They are: resources (Box 1); technology development (Box 2); computational biology (Box 3); training (Box 4); ethical, legal and social implications (ELSI, Box 5); and education (Box 6). We also stress the critical importance of early, unfettered access to genomic data in achieving maximum public benefit. Finally, we propose a series of 'quantum leaps', achievements that would lead to substantial advances in genomics research and its applications to medicine. Some of these may seem overly bold, but no laws of physics need to be violated to achieve them. Such leaps would have profound implications, just as the dreams of the mid-1980s about the complete sequence of the human genome have been realized in the accomplishments now being celebrated.

> **BOX 2 Technology development**
>
> The Human Genome Project was aided by several 'breakthrough' technological developments, including Sanger DNA sequencing and its automation, DNA-based genetic markers, large-insert cloning systems and the polymerase chain reaction. During the project, these methods were scaled up and made more efficient by 'evolutionary' advances, such as automation and miniaturization. New technologies, including capillary-based sequencing and methods for genotyping single-nucleotide polymorphisms, have recently been introduced, leading to further improvements in capacity for genomic analyses. Even newer approaches, such as nanotechnology and microfluidics, are being developed, and hold great promise, but further advances are still needed. Some examples are:
>
> - Sequencing and genotyping technologies to reduce costs further and increase access to a wider range of investigators
> - Identification and validation of functional elements that do not encode protein
> - *In vivo*, real-time monitoring of gene expression and the localization, specificity, modification and activity/kinetics of gene products in all relevant cell types
> - Modulation of expression of all gene products using, for example, large-scale mutagenesis, small-molecule inhibitors and knock-down approaches (such as RNA-mediated inhibition)
> - Monitoring of the absolute abundance of any protein (including membrane proteins, proteins at low abundance and all modified forms) in any cell
> - Improve imaging methods that allow non-invasive molecular phenotyping
> - Correlating genetic variation to human health and disease using haplotype information or comprehensive variation information
> - Laboratory-based phenotyping, including the use of protein affinity reagents, proteomic approaches and analysis of gene expression
> - Linking molecular profiles to biology, particularly pathway biology to disease.

I Genomics to biology

Elucidating the structure and function of genomes

The broadly available genome sequences of human and a select set of additional organisms represent foundational information for biology and biomedicine. Embedded within this as-yet poorly understood code are the genetic instructions for the entire repertoire of cellular components, knowledge of which is needed to unravel the complexities of biological systems. Elucidating the structure of genomes and identifying the function of the myriad encoded elements will allow connections to be made between genomics and biology and will, in turn, accelerate the exploration of all realms of the biological sciences.

For this, new conceptual and technological approaches will be needed to:

- Develop a comprehensive and comprehensible catalogue of all of the components encoded in the human genome.
- Determine how the genome-encoded components function in an integrated manner to perform cellular and organismal functions.
- Understand how genomes change and take on new functional roles.

BOX 3 Computational biology

Computational methods have become intrinsic to modern biological research, and their importance can only increase as large-scale methods for data generation become more prominent, as the amount and complexity of the data increase, and as the questions being addressed become more sophisticated. All future biomedical research will integrate computational and experimental components. New computational capabilities will enable the generation of hypotheses and stimulate the development of experimental approaches to test them. The resulting experimental data will, in turn, be used to generate more refined models that will improve overall understanding and increase opportunities for application to disease. The areas of computational biology critical to the future of genomics research include:

- New approaches to solving problems, such as the identification of different features in a DNA sequence, the analysis of gene expression and regulation, the elucidation of protein structure and protein-protein interactions, the determination of the relationship between genotype and phenotype, and the identification of the patterns of genetic variation in populations and the processes that produced those patterns
- Reusable software modules to facilitate interoperability
- Methods to elucidate the effects of environmental (non-genetic) factors and of gene-environment interactions on health and disease
- New ontologies to describe different data types
- Improved database technologies to facilitate the integration and visualization of different data types, for example, information about pathways, protein structure, gene variation, chemical inhibition and clinical information/phenotypes
- Improved knowledge management systems and the standardization of data sets to allow the coalescence of knowledge across disciplines.

Grand Challenge I-1: Comprehensively identify the structural and functional components encoded in the human genome

Although DNA is relatively simple and well understood chemically, the human genome's structure is extraordinarily complex and its function is poorly understood. Only 1–2% of its bases encode proteins[7], and the full complement of protein-coding sequences still remains to be established. A roughly equivalent amount of the non-coding portion of the genome is under active selection[11], suggesting that it is also

> **BOX 4 Training**
>
> Meeting the scientific, medical and social/ethical challenges now facing genomics will require scientists, clinicians and scholars with the skills to understand biological systems and to use that information effectively for the benefit of humankind. Adequate training capacity will be required to address the following needs:
>
> - **Computational skills** As biomedical research is becoming increasingly data intensive, computational capability is increasingly becoming a critical skill.
> - **Interdisciplinary skills** Although a good start has been made, expanded interactions will be required between the sciences (biology, computer science, physics, mathematics, statistics, chemistry and engineering), between the basic and the clinical sciences, and between the life sciences, the social sciences and the humanities. Such interactions will be needed at the individual level (scientists, clinicians and scholars will need to be able to bring relevant issues, concerns and capabilities from different disciplines to bear on their specific efforts), at a collaborative level (reserachers will need to be able to participate effectively in interdisciplinary research collaborations that bring biology together with many other disciplines) and at the disciplinary level (new disciplines will need to emerge at the interfaces between the traditional disciplines).
> - **Different perspectives** Individuals from minority or disadvantaged populations are significantly under-represented as both researchers and participants in genomics research. This regrettable circumstance deprives the field of the best and brightest from all backgrounds, narrows the field of questions asked, can lessen sensitivity to cultural concerns in implementing research protocols, and compromises the overall effectiveness of the research. Genomics can learn from successful efforts in training individuals from under-represented populations in other areas of science and health (see, for example, www.genome.gov/Pages/Grants/Policies/ActionPlanGuide).

functionally important, yet vanishingly little is known about it. It probably contains the bulk of the regulatory information controlling the expression of the approximately 30,000 protein-coding genes, and myriad other functional elements, such as non-protein-coding genes and the sequence determinants of chromosome dynamics. Even less is known about the function of the roughly half of the genome that consists of highly repetitive sequences or of the remaining non-coding, non-repetitive DNA.

The next phase of genomics is to catalogue, characterize and comprehend the entire set of functional elements encoded in the human and other genomes. Compiling this genome 'parts list' will be an immense challenge. Well-known classes of functional elements, such as protein-coding sequences, still cannot be accurately predicted from

> **BOX 5   Ethical, legal and social implications (ELSI)**
>
> Today's genomics research and applications rest on more than a decade of valuable investigation into their ethical, legal and social implications. As the application of genomics to health increases along with its social impact, it becomes ever more important to expand on this work. There is an increasing need for focused ELSI research that directly informs policies and practices. One can envisage a flowering of 'translational ELSI research' that builds on the knowledge gained from prior and forthcoming 'basic ELSI research', which would provide knowledge for direct use by researchers, clinicians, policy-makers and the public. Examples include:
>
> - The development of models of genomics research that use attention to these ELSI issues for enhancing the research, rather than viewing such issues as impediments
> - The continued development of appropriate and effective genomics research methods and policies that promote the highest levels of science and of protecting human subjects
> - The establishment of crosscutting tools, analogous to the publicly accessible genomic maps and sequence databases that have accelerated other genomics research (examples of such tools might include searchable databases of genomic legislation and policies from around the world, or studies of ELSI aspects of introducing clinical genetic tests)
> - The evaluation of new genetic and genomic tests and technologies, and effective oversight of their implementation, to ensure that only those with confirmed clinical validity are used for patient care.

sequence information alone. Other types of known functional sequences, such as genetic regulatory elements, are even less well understood; undoubtedly new types remain to be defined, so we must be ready to investigate novel, perhaps unexpected, ways in which DNA sequence can confer function. Similarly, a better understanding of epigenetic changes (for example, methylation and chromatin remodelling) is needed to comprehend the full repertoire of ways in which DNA can encode information.

Comparison of genome sequences from evolutionarily diverse species has emerged as a powerful tool for identifying functionally important genomic elements. Initial analyses of available vertebrate genome sequences[7,11,19] have revealed many previously undiscovered protein-coding sequences. Mammal-to-mammal sequence comparisons have revealed large numbers of homologies in non-coding regions[11], few of which can be defined in functional terms. Further comparisons of sequences derived from multiple species, especially those occupying distinct evolutionary positions, will lead to significant refinements in our understanding of the functional importance of conserved sequences[20]. Thus, the generation of additional genome sequences from several well-chosen species is crucial to the functional characterization of the human genome (Box 1). The generation of such large sequence data

> **BOX 6 Education**
>
> Marked health improvements from integrating genomics into individual and public health care depend on the effective education of health professionals and the public about the interplay of genetic and environmental factors in health and disease. Health professionals must be knowledgeable about genomics to use the outcomes of genomics research effectively. The public must be knowledgeable to make informed decisions about participation in genomics research and to incorporate the findings of such research into their own health care. Both groups must be knowledgeable to engage profitably in discussion and decision-making about the societal implications of genomics.
>
> Promising models for genomics and genetics education exist (see, for example, www.nchpeg.org), but they must be expanded and new models developed. We have entered a unique 'educable era' regarding genomics; health professionals and the public are increasingly interested in learning about genomics, but its widespread application to health is still several years away. For genomics-based health care to be maximally effective once it is widely feasible, and for members of society to make the best decisions about the uses of genomics, we must take advantage now of this unique opportunity to increase understanding. Some examples are:
>
> - Health professionals vary, both individually and by discipline, in the amount and type of genomics education that they require. So multiple models of effective genomics-related education are needed.
> - Print, web and video educational products that the public can consume when actively seeking genomic information should be created and made easily available.
> - The media are crucial sources of information about genomics and its societal implications. Initiatives to provide the media with greater understanding of genomics are needed.
> - High-school students will be both the users of genomic information and the genomics researchers of the future. Especially as they educate all sectors of society, high-school educators need information and materials about genomics and its implications for society, to use in their classrooms.

sets will benefit from further advances in sequencing technology that yield significant cost reductions (Box 2). The study of sequence variation within species will also be important in defining the functional nature of some sequences (see Grand Challenge I-3).

Effective identification and analysis of functional genomic elements will require increasingly powerful computational capabilities, including new approaches for tackling ever-growing and increasingly complex data sets and a suitably robust computational infrastructure for housing, accessing and analysing those data sets (Box 3). In parallel, investigators will need to become increasingly adept in dealing with this treasure trove of new information (Box 4). As a better understanding of genome function is gained, refined computational tools for *de novo* prediction of the identity and behaviour of functional elements should emerge[21].

Complementing the computational detection of functional elements will be the generation of additional experimental data by high-throughput methodologies. One example is the production of full-length complementary DNA (cDNA) sequences (see, for example, www.mgc.nci.nih.gov and www.fruitfly.org/EST/full.shtml). Major challenges inherent in programmes to discover genes are the experimental identification and validation of alternate splice forms and messenger RNAs expressed in a highly restricted fashion. Even more challenging is the experimental validation of functional elements that do not encode protein (for example, regulatory regions and non-coding RNA sequences). High-throughput approaches to identify them (Box 2) will be needed to generate the experimental data that will be necessary to develop, confirm and enhance computational methods for detecting functional elements in genomes.

Because current technologies cannot yet identify all functional elements, there is a need for a phased approach in which new methodologies are developed, tested on a pilot scale and finally applied to the entire human genome. Along these lines, the NHGRI recently launched the Encyclopedia of DNA Elements (ENCODE) Project (www.genome.gov/Pages/Research/ENCODE) to identify all the functional elements in the human genome. In a pilot project, systematic strategies for identifying all functionally important genomic elements will be developed and tested using a selected 1% of the human genome. Parallel projects involving well-studied model organisms, for example, yeast, nematode and fruitfly, are ongoing. The lessons learned will serve as the basis for implementing a broader programme for the entire human genome.

Grand Challenge I-2: Elucidate the organization of genetic networks and protein pathways and establish how they contribute to cellular and organismal phenotypes

Genes and gene products do not function independently, but participate in complex, interconnected pathways, networks and molecular systems that, taken together, give rise to the workings of cells, tissues, organs and organisms. Defining these systems and determining their properties and interactions is crucial to understanding how biological systems function. Yet these systems are far more complex than any problem that molecular biology, genetics or genomics has yet approached. On the basis of previous experience, one effective path will begin with the study of relatively simple model organisms[22], such as bacteria and yeast, and then extend the early findings to more complex organisms, such as mouse and human. Alternatively, focusing on a few well-characterized systems in mammals will be a useful test of the approach (see, for example, www.signaling-gateway.org).

Understanding biological pathways, networks and molecular systems will require information from several levels. At the genetic level, the architecture of regulatory interactions will need to be identified in different cell types, requiring, among other things, methods for simultaneously monitoring the expression of all genes in a cell[12]. At the gene-product level, similar techniques that allow *in vivo*, real-time measurement of protein expression, localization, modification and activity/kinetics will be needed (Box 2). It will be important to develop, refine and scale up techniques that modulate gene expression, such as conventional gene-knockout methods[23], newer knock-down approaches[24] and small-molecule inhibitors[25] to establish the temporal and cellular expression pattern of individual proteins and to determine the functions of those proteins. This is a key first step towards assigning all genes and their products to functional pathways.

The ability to monitor all proteins in a cell simultaneously would profoundly improve our ability to understand protein pathways and systems biology. A critical step towards gaining a complete understanding of systems biology will be to take an accurate census of the proteins present in particular cell types under different physiological conditions. This is becoming possible in some model systems, such as microorganisms[26]. It will be a major challenge to catalogue proteins present in low abundance or in membranes. Determining the absolute abundance of each protein, including all modified forms, will be an important next step. A complete interaction map of the proteins in a cell, and their cellular locations, will serve as an atlas for the biological and medical explorations of cellular metabolism[27] (see www.nrcam.uchc.edu, for example). These and other related areas constitute the developing field of proteomics.

Establishing a true understanding of how organized molecular pathways and networks give rise to normal and pathological cellular and organismal phenotypes will require more than large, experimentally derived data sets. Once again, computational investigation will be essential (Box 3), and there will be a greatly increased need for the collection, storage and display of the data in robust databases. By modelling specific pathways and networks, predicting how they affect phenotype, testing hypotheses derived from these models and refining the models based on new experimental data, it should be possible to understand more completely the difference between a 'bag of molecules' and a functioning biological system.

Grand Challenge I-3: Develop a detailed understanding of the heritable variation in the human genome

Genetics seeks to correlate variation in DNA sequence with phenotypic differences (traits). The greatest advances in human genetics have been made for traits associated with variation in a single gene. But most phenotypes, including common diseases and variable responses to pharmacological agents, have a more complex origin, involving the interplay between multiple genetic factors (genes and their products) and nongenetic factors (environmental influences). Unravelling such complexity will require both a complete description of the genetic variation in the human genome and the development of analytical tools for using that information to understand the genetic basis of disease.

Establishing a catalogue of all common variants in the human population, including single-nucleotide polymorphisms (SNPs), small deletions and insertions, and other structural differences, began in earnest several years ago. Many SNPs have been identified[28], and most are publicly available (www.ncbi.nlm.nih.gov/SNP). A public collaboration, the International HapMap Project (www.genome.gov/Pages/Research/HapMap), was formed in 2002 to characterize the patterns of linkage disequilibrium and haplotypes across the human genome and to identify subsets of

SNPs that capture most of the information about these patterns of genetic variation to enable large-scale genetic association studies. To reach fruition, such studies need more robust experimental (Box 2) and computational (Box 3) methods that use this new knowledge of human haplotype structure[29].

A comprehensive understanding of genetic variation, both in humans and in model organisms, would facilitate studies to establish relationships between genotype and biological function. The study of particular variants and how they affect the functioning of specific proteins and protein pathways will yield important new insights about physiological processes in normal and disease states. An enhanced ability to incorporate information about genetic variation into human genetic studies would usher in a new era for investigating the genetic bases of human disease and drug response (see Grand Challenge II-1).

Grand Challenge I-4: Understand evolutionary variation across species and the mechanisms underlying it

The genome is a dynamic structure, continually subjected to modification by the forces of evolution. The genomic variation seen in humans represents only a small glimpse through the larger window of evolution, where hundreds of millions of years of trial-and-error efforts have created today's biosphere of animal, plant and microbial species. A complete elucidation of genome function requires a parallel understanding of the sequence differences across species and the fundamental processes that have sculpted their genomes into the modern-day forms.

The study of inter-species sequence comparisons is important for identifying functional elements in the genome (see Grand Challenge I-1). Beyond this illuminating role, determining the sequence differences between species will provide insight into the distinct anatomical, physiological and developmental features of different organisms, will help to define the genetic basis for speciation and will facilitate the characterization of mutational processes. This last point deserves particular attention, because mutation both drives long-term evolutionary change and is the underlying cause of inherited disease. The recent finding that mutation rates vary widely across the mammalian genome[11] raises numerous questions about the molecular basis for these evolutionary changes. At present, our understanding of DNA mutation and repair, including the important role of environmental factors, is limited.

Genomics will provide the ability to substantively advance insight into evolutionary variation, which will, in turn, yield new insights into the dynamic nature of genomes in a broader evolutionary framework.

Grand Challenge I-5: Develop policy options that facilitate the widespread use of genome information in both research and clinical settings

Realization of the opportunities provided by genomics depends on effective access to the information (such as data about genes, gene variants, haplotypes, protein structures, small molecules and computational models) by a wide range of potential users, including researchers, commercial enterprises, healthcare providers, patients and the public. Researchers themselves need maximum access to the data as soon as possible (see 'Data release', below). Use of the information for the development of therapeutic and other products necessarily entails consideration of the complex issues of intellectual property (for example, patenting and licensing) and commercialization. The intellectual property practices, laws and regulations that affect genomics must adhere to the principle of maximizing public benefit, but must also be consistent with more general and longer-established intellectual property principles. Further, because genome research is global, international treaties, laws, regulations, practices, belief systems and cultures also come into play.

Without commercialization, most diagnostic and therapeutic advances will not reach the clinical setting, where they can benefit patients. Thus, we need to develop policy options for data access and for patenting, licensing and other intellectual property issues to facilitate the dissemination of genomics data.

II Genomics to health

Translating genome-based knowledge into health benefits

The sequencing of the human genome, along with other recent and expected achievements in genomics, provides an unparalleled opportunity to advance our understanding of the role of genetic factors in human health and disease, to allow more precise definition of the non-genetic factors involved, and to apply this insight rapidly to the prevention, diagnosis and treatment of disease. The report by the US National Research Council that originally envisioned the HGP was explicit in its expectation that the human genome sequence would lead to improvements in human health, and subsequent five-year plans reaffirmed this view[15-17]. But how this will happen has been less clearly articulated. With the completion of the original goals of the HGP, the time is right to develop and apply large-scale genomic strategies to empower improvements in human health, while anticipating and avoiding potential harm.

Such strategies should enable the research community to achieve the following:

- Identify genes and pathways with a role in health and disease, and determine how they interact with environmental factors.
- Develop, evaluate and apply genome-based diagnostic methods for the prediction of susceptibility to disease, the prediction of drug response, the early detection of illness and the accurate molecular classification of disease.
- Develop and deploy methods that catalyse the translation of genomic information into therapeutic advances.

Grand Challenge II-1: Develop robust strategies for identifying the genetic contributions to disease and drug response

For common diseases, the interplay of multiple genes and multiple non-genetic factors, not a single allele, usually dictates disease susceptibility and response to treatments. Deciphering the role of genes in human health and disease is a formidable problem for many reasons, including impediments to defining biologically valid phenotypes, challenges in identifying and quantifying environmental exposures, technological obstacles to generating sufficient and useful genotypic information, and the difficulties of studying humans. Yet this problem can be solved. Vigorous development of crosscutting genomic tools to catalyse advances in understanding the genetics of common disease and in pharmacogenomics is needed. Prominent among these will be a detailed haplotype map of the human genome (see Grand Challenge I-3) that can be used for whole-genome association studies of all diseases in all populations, as well as further advances in sequencing and genotyping technology to make such studies feasible (see 'Quantum leaps', below).

More efficient strategies for detecting rare alleles involved in common disease are also needed, as the hypothesis that alleles that increase risk for common diseases are themselves common[30] will probably not be universally true. Computational and experimental methods to detect gene–gene and gene–environment interactions, as well as methods allowing interfacing of a variety of relevant databases, are also required (Box 3). By obtaining unbiased assessments of the relative disease risk that particular gene variants contribute, a large longitudinal population-based cohort study, with collection of extensive clinical information and ongoing follow-up, would be profoundly valuable to the study of all common diseases (Box 1). Already, such projects as the UK Biobank (www.ukbiobank.ac.uk), the Marshfield Clinic's Personalized Medicine Research Project (www.mfldclin.edu/pmrp) and the Estonian Genome Project (www.geenivaramu.ee) seek to provide such resources. But if the multiple population groups in the United States and elsewhere in the world are to benefit fully and fairly from such research (see Grand Challenge II-6), a large population-based cohort study that includes full representation of minority populations is also needed.

Grand Challenge II-2: Develop strategies to identify gene variants that contribute to good health and resistance to disease

Most human genetic research has traditionally focused on identifying genes that predispose to illness. A relatively unexplored, but important, area of research focuses on the role of genetic factors in maintaining good health. Genomics will facilitate further understanding of this aspect of human biology and allow the identification of gene variants that are important for the maintenance of health, particularly in the presence of known environmental risk factors. One useful research resource would be a 'healthy cohort', a large epidemiologically robust group of individuals (Box 1) with unusually good health, who could be compared with cohorts of individuals with diseases and who could also be intensively studied to reveal alleles protective for conditions such as diabetes, cancer, heart disease and Alzheimer's disease. Another promising approach would be rigorous examination of genetic variants in individuals at high risk for specific diseases who do not develop them, such as sedentary, obese smokers without heart disease or individuals with HNPCC mutations who do not develop colon cancer.

Grand Challenge II-3: Develop genome-based approaches to prediction of disease susceptibility and drug response, early detection of illness, and molecular taxonomy of disease states

The discovery of variants that affect risk for disease could potentially be used in individualized preventive medicine—including diet, exercise, lifestyle and pharmaceutical intervention—to maximize the likelihood of staying well. For example, the discovery of variants that correlate with successful outcomes of drug therapy, or with unfortunate side effects, could potentially be rapidly translated into clinical practice. Turning this vision into reality will require the following: (1) unbiased determination of the risk associated with a particular gene variant, often overestimated in initial studies[31]; (2) technological advances to reduce the cost of genotyping (Box 2; see 'Quantum leaps', below); (3) research on whether this kind of personalized genomic information will actually alter health behaviours (see Grand Challenge II-5); (4) oversight of the implementation of genetic tests to ensure that only those with demonstrated clinical validity are applied outside of the research setting (Box 5); and (5) education of health care professionals and the public to be well-informed participants in this new form of preventive medicine (Box 6).

The time is right for a focused effort to understand, and potentially to reclassify, all human illnesses on the basis of detailed molecular characterization. Systematic analyses of somatic mutations, epigenetic modifications, gene expression, protein expression and protein modification should allow the definition of a new molecular taxonomy of illness, which would replace our present, largely empirical, classification schemes and advance both disease prevention and treatment. The reclassification of neuromuscular diseases[32] and certain types of cancer[33] provides striking initial examples, but many more such applications are possible.

Such a molecular taxonomy would be the basis for the development of better methods for the early detection of disease, which often allows more effective and less costly treatments. Genomics and other large-scale approaches to biology offer the potential for developing new tools to detect many diseases earlier than is currently feasible. Such 'sentinel' methods might include analysis of gene expression in circulating leukocytes, proteomic analysis of body fluids, and advanced molecular analysis of tissue biopsies. An example would be the analysis of gene expression in peripheral blood leukocytes to predict drug response. A focused effort to use a genomic approach to characterize serum proteins exhaustively in health and disease might also be highly rewarding.

Grand Challenge II-4: Use new understanding of genes and pathways to develop powerful new therapeutic approaches to disease

Pharmaceuticals on the market target fewer than 500 human gene products[34]. Even though not all of the 30,000 or so human protein-coding genes' will have products targetable for drug development, this suggests that there is an enormous untapped pool of human gene-based targets for therapeutic intervention. In addition, the new understanding of biological pathways provided by genomics (see Grand Challenge I-2) should contribute even more fundamentally to therapeutic design.

The information needed to determine the therapeutic potential of a gene generally overlaps heavily with the information that reveals its function. The success of imatinib mesylate (Gleevec), an inhibitor of the BCR-ABL tyrosine kinase, in treating chronic myelogenous leukaemia relied on a detailed molecular understanding of the disease's genetic cause[35]. This example offers promise that therapies based on genomic information will be particularly effective. Grand Challenge I-1 describes the 'functionation' of the genome, which will increasingly be the critical first step in the development of new therapeutics. But stimulating basic scientists to approach biomedical problems with a genomic attitude is not enough. A therapeutic mindset, lacking in much of academic biomedical research and training, must be explicitly encouraged, and tools developed and provided for its implementation.

A particularly promising example of the gene-based approach to therapeutics is the application of 'chemical genomics'[25]. This strategy uses libraries of small molecules (natural compounds, aptamers or the products of combinatorial chemistry) and high-throughput screening to advance understanding of biological pathways and to identify compounds that act as positive or negative regulators of individual gene products, pathways or cellular phenotypes. Although the pharmaceutical industry applies this approach widely as the first step in drug development, few academic investigators have access to this methodology or are familiar with its use.

Providing such access more broadly, through one or more centralized facilities, could lead to the discovery of a host of useful probes for biological pathways that would serve as new reagents for basic research and/or starting points for the development of new therapeutic agents (the 'hits' from such library screens will generally require medicinal chemistry modifications to yield therapeutically usable compounds).

Also needed are new, more powerful technologies for generating deep molecular libraries, especially ones tagged to allow the ready determination of precise molecular targets. A centralized database of screening results should lead to further important biological insights. Generating molecular probes for exploring the basic biology of health and disease in academic laboratories would not supplant the major role of biopharmaceutical companies in drug development, but could contribute to the start of the drug development pipeline. The private sector would doubtless find many of these molecular probes of interest for further exploration through optimization by medicinal chemistry, target validation, lead compound identification, toxicological studies and, ultimately, clinical trials.

Academic pursuit of this first step in drug development could be particularly valuable for the many rare mendelian diseases, in which often the gene defect is known but the small market size limits the private sector's motivation to shoulder the expense of effective pharmaceutical development. Such translational research in academic laboratories, combined with incentives such as the US Orphan Drug Act, could profoundly increase the availability of effective treatments for rare genetic diseases in the next decade. Further, the development of therapeutic approaches to single-gene disorders might provide valuable insights into applying genomics to reveal the biology of more common disorders and developing more effective treatments for them (in the way that, for example, the search for compounds that target the presenilins has led to general therapeutic strategies for late-onset Alzheimer's disease[36]).

Grand Challenge II-5: Investigate how genetic risk information is conveyed in clinical settings, how that information influences health strategies and behaviours, and how these affect health outcomes and costs

Understanding how genetic factors affect health is often cited as a major goal of genomics, on the assumption that applying such understanding in the clinical setting will improve health. But this assumption actually rests on relatively few examples and data, and more research is needed to provide sufficient guidance about how to use genomic information optimally for improving individual or public health.

Theoretically, the steps by which genetic risk information would lead to improved health are: (1) an individual obtains genome-based information about his/her own health risks; (2) the individual uses this information to develop an individualized prevention or treatment plan; (3) the individual implements that plan; (4) this leads to improved health; and (5) healthcare costs are reduced. Scrutiny of these assumptions is needed, both to test them and to determine how each step could best be accomplished in different clinical settings.

Research is also required that critically evaluates new genetic tests and interventions in terms of parameters such as benefits, access and cost. Such research should be interdisciplinary and use the tools and expertise of many fields, including genomics, health education, health behaviour research, health outcomes research, healthcare delivery analysis, and healthcare economics. Some of these fields have historically paid little attention to genomics, but high-quality research of this sort could provide important guidance in clinical decision-making—as the work of several disciplines has already been helpful in caring for people with an increased risk of colon cancer as a result of mutations in PAP or HNPCC[37].

Grand Challenge II-6: Develop genome-based tools that improve the health of all

Disparities in health status constitute a significant global issue, but can genome-based approaches to health and disease help to reduce this problem? Social and other environmental factors are major contributors to health disparities; indeed, some would question whether heritable factors have any significant role. But population differences in allele frequencies for some disease-associated variants could be a contributing factor to certain disparities in health status, so incorporating this information into preventive and/or public-health strategies would be beneficial. Research is needed to understand the relationship between genomics and health disparities by rigorously evaluating the diverse contributions of socioeconomic status, culture, discrimination, health behaviours, diet, environmental exposures and genetics.

It is also important to explore applications of genomics in the improvement of health in the developing world (www3.who.int/whosis/genomics/genomics_report.cfm), where both human and non-human genomics will play significant roles. If we take malaria as an example, a better understanding of human genetic factors that influence susceptibility and response to the disease, and to the drugs used to treat it, could have a significant global impact. So too could a better understanding of the malarial parasite itself and of its mosquito vector, which the recently reported genome sequences[38,39] should provide. It will be necessary to determine the appropriate roles of governmental and non-governmental organizations, academic institutions, industry and individuals to ensure that genomics produces clinical benefits for resource-poor nations, and is used to produce robust local research expertise.

To ensure that genomics research benefits all, it will be critical to examine how genomics-based health care is accessed and used. What are the barriers to equitable access, and how can they be removed? This is relevant not only in resource-poor nations, but also in wealthier countries where segments of society, such as indigenous populations, the uninsured, or rural and inner city communities, have traditionally not received adequate health care.

III Genomics to Society

Promoting the use of genomics to maximize benefits and minimize harms

Genomics has been at the forefront of giving serious attention, through scholarly research and policy discussions, to the impact of science and technology on society. Although the major benefits to be realized from genomics are in the area of health, as described above, genomics can also contribute to other aspects of society. Just as the HGP and related developments have spawned new areas of research in basic biology and in health, they have also created opportunities for research on social issues, even to the extent of understanding more fully how we define ourselves and each other.

In the next few years, society must not only continue to grapple with numerous questions raised by genomics, but must also formulate and implement policies to address many of them. Unless research provides reliable data and rigorous approaches on which to base such decisions, those policies will be ill-informed and could potentially compromise us all. To be successful, this research must encompass both 'basic' investigations that develop conceptual tools and shared vocabularies, and more 'applied', 'translational' projects that use these tools to explore and define appropriate public-policy options that incorporate diverse points of view.

As it has in the past, such research will continue to have important ramifications for all three major themes of the vision presented here. We now address research that focuses on society itself, more than on biology or health. Such efforts should enable the research community to:

- Analyse the impact of genomics on concepts of race, ethnicity, kinship, individual and group identity, health, disease and 'normality' for traits and behaviours.
- Define policy options, and their potential consequences, for the use of genomic information and for the ethical boundaries around genomics research.

Grand Challenge III-1: Develop policy options for the uses of genomics in medical and non-medical settings

Surveys have repeatedly shown that the public is highly interested in the concept that personal genetic information might guide them to better health, but is deeply concerned about potential misuses of that information (see www.publicagenda.org/issues/pcc_detail.cfm?issue_type=medical_research&list=7). Topping the list of concerns is the potential for discrimination in health insurance and employment. A significant amount of research on this issue has been done[40], policy options have been published[41-43], and many US states have now passed anti-discrimination legislation (see www.genome.gov/Pages/PolicyEthics/Leg/StateIns

and www.genome.gov/Pages/PolicyEthics/Leg/StateEmploy). The US Equal Employment Opportunity Commission has ruled that the Americans with Disabilities Act should apply to discrimination based on predictive genetic information[44], but the legal status of that construct remains in some doubt. Although an executive order protects US government employees against genetic discrimination, this does not apply to other workers. Thus, many observers have concluded that effective federal legislation is needed, and the US Congress is currently considering such a law.

Making certain that genetic tests offered to the public have established clinical validity and usefulness must be a priority for future research and policy making. In the United States, the Secretary's Advisory Committee on Genetic Testing extensively reviewed this area and concluded that further oversight is needed, asking the Food and Drug Administration to review new predictive genetic tests prior to marketing (www4.od.nih.gov/oba/sacgtl reports/oversight_report.pdf). That recommendation has not yet been acted on; meanwhile, numerous websites offering unvalidated genetic tests directly to the public, often combined with the sale of 'nutraceuticals' and other products of highly questionable value, are proliferating.

Many issues currently swirl around the proper conduct of genetic research involving human subjects, and further work is needed to achieve a satisfactory balance between the protection of research participants from harm and the ability to conduct clinical research that benefits society as a whole. Much effort has gone into developing appropriate guidelines for the use of stored tissue specimens (www.georgetown.edu/research/nrcbl/nbac/hbm.pdf), for community consultation when conducting genetic research with identifiable populations (www.nigms.nih.gov/news/reports/community_consultation.html #exec), and for the consent of non-examined family members when conducting pedigree research (www.nih.gov/sigs/bioethics/nih_third_party_rec.html), but confusion still remains for many investigators and institutional review boards.

The use of genomic information is not limited to the arenas of biology and of health, and further research and development of policy options is also needed for the many other applications of such information. The array of additional users is likely to include the life, disability and long-term care insurance industries, the legal system, the military, educational institutions and adoption agencies. Although some of the research informing the medical uses of genomics will be useful in broader settings, dedicated research outside the health care sphere is needed to explore the public values that apply to uses of genomics other than for health care and their relationship to specific contextual applications. For example, should genetic information on predisposition to hyperactivity be available in the future to school officials? Or should genetic information about behavioural traits be admissible in criminal or civil proceedings? Genomics also provides greater opportunity to understand ancestral origins of populations and individuals, which raises issues such as whether genetic information should be used for defining membership in a minority group.

Because uses of genomics outside the healthcare setting will involve a significantly broader community of stakeholders, both research and policy development in this area must involve individuals and organizations besides those involved in the medical applications of genomics. But many of the same perspectives essential to research and policy development for the medical uses of genomics are also essential. Both the potential users of non-medical applications of genomics and the public need education to understand better the nature and limits of genomic information (Box 6) and to grasp the ethical, legal and social implications of its uses outside health care (Box 5).

Grand Challenge III-2: Understand the relationships between genomics, race and ethnicity, and the consequences of uncovering these relationships

Race is a largely non-biological concept confounded by misunderstanding and a long history of prejudice. The relationship of genomics to the concepts of race and ethnicity has to be considered within complex historical and social contexts.

Most variation in the genome is shared between all populations, but certain alleles are more frequent in some populations than in others, largely as a result of history and geography. Use of genetic data to define racial groups, or of racial categories to classify biological traits, is prone to misinterpretation. To minimize such misinterpretation, the biological and sociocultural factors that interrelate genetics with constructs of race and ethnicity need to be better understood and communicated within the next few years.

This will require research on how different individuals and cultures conceive of race, ethnicity, group identity and self-identity, and what role they believe genes or other biological factors have. It will also require a critical examination of how the scientific community understands and uses these concepts in designing research and presenting findings, and of how the media report these. Also necessary is widespread education about the biological meaning and limitations of research findings in this area (Box 6), and the formulation and adoption of public-policy options that protect against genomics-based discrimination or maltreatment (see Grand Challenge III-1).

Grand Challenge III-3: Understand the consequences of uncovering the genomic contributions to human traits and behaviours

Genes influence not only health and disease, but also human traits and behaviours. Science is only beginning to unravel the complicated pathways that underlie such attributes as handedness, cognition, diurnal rhythms and various behavioural characteristics. Too often, research in behavioural genetics, such as that regarding sexual orientation or intelligence, has been poorly designed and its findings have been communicated in a way that oversimplifies and overstates the role of genetic factors. This has caused serious problems for those who have been stigmatized by the suggestion that alleles associated with what some people perceive as 'negative' physiological or behavioural traits are more frequent in certain populations. Given this history and the real potential for recurrence, it is particularly important to gather sufficient scientifically valid information about genetic and environmental factors to provide a sound understanding of the contributions and interactions between genes and environment in these complex phenotypes.

It is also important that there be robust research to investigate the implications, for both individuals and society, of uncovering any genomic contributions that there may be to traits and behaviours. The field of genomics has a responsibility to consider the social implications of research into the genetic contributions to traits and behaviours, perhaps an even greater responsibility than in other areas where there is less of a history of misunderstanding and stigmatization. Decisions about research in this area are often best made with input from a diverse group of individuals and organizations.

Grand Challenge III-4: Assess how to define the ethical boundaries for uses of genomics

Genetics and genomics can contribute understanding to many areas of biology, health and life. Some of these human applications are controversial, with some members of the public questioning the propriety of their scientific exploration. Although freedom of scientific inquiry has been a cardinal feature of human progress, it is not unbounded. It is important for society to define the appropriate and inappropriate uses of genomics. Conversations between diverse parties based on an accurate and detailed understanding of the relevant science and ethical, legal and social factors will promote the formulation and implementation of effective policies. For instance, in reproductive genetic testing, it is crucial to include perspectives from the disability community. Research should explore how different individuals, cultures and religious traditions view the ethical boundaries for the uses of genomics—for instance, which sets of values determine attitudes towards the appropriateness of applying genomics to such areas as reproductive genetic testing, 'genetic enhancement' and germline gene transfer.

Implementation: the NHGRI's role

The vision for the future of genomics presented here is broad and deep, and its realization will require the efforts of many. Continuation of the extensive collaboration between scientists and between funding sources that characterized the HGP will be essential. Although the NHGRI intends to participate in all the research areas discussed here, it will need to focus its efforts to use its finite resources as effectively as possible. Thus, it will take a major role in some areas, actively collaborate in others, and have only a supporting role in yet others. The NHGRI's priorities and areas of emphasis will also evolve as milestones are met and new opportunities arise.

The approach that has characterized genomics and led to the success of the HGP—an initial focus on technology development and feasibility studies, followed by pilot efforts to learn how to apply new strategies and technologies efficiently on a larger scale, and then implementation of full-scale production efforts—will continue to be at the heart of the NHGRI's priority-setting process. The following are areas of high interest, not listed in priority order.

Large-scale production of genomic data sets: The NHGRI will continue to support genomic sequencing, focusing on the genomes of mammals, vertebrates, chordates and invertebrates; other funders will support the determination of additional genome sequences from microbes and plants. With current technology, the NHGRI could support the determination of as much as 45–60 gigabases of genomic DNA sequence, or the equivalent of 15–20 human genomes, over the next five years. But as the cost of sequencing continues to decrease, the cost/benefit ratio of sequence generation will improve, so that the actual amount of sequencing done will be greatly affected by the development of improved sequencing technology.

The decisions about which genomes to sequence next will be based on the results of comparative analyses that reveal the ability of genomic sequences from unexplored phylogenetic positions to inform the interpretation of the human sequence and to provide other insights. Finally, the degree to which any new genomic sequence is completed—finished, taken to an advanced draft stage or lightly sampled—will be determined by the use for which the sequence is generated. And, of course, the NHGRI's sequencing programme will maintain close contact with, and take account of the plans and output of, other sequencing programmes, as has happened throughout the HGP.

A second data set ready for production-level effort is the human haplotype map (HapMap). This project, a collaboration between the NHGRI, many other NIH institutes, and four international partners, is scheduled for completion within three years. The outcome of the International HapMap Project will significantly shape the future direction of the NHGRI's research efforts in the area of genetic variation.

Pilot-scale efforts The NHGRI has initiated the ENCODE Project to begin the development of the human genome 'parts list'. The first phase will address the application and improvement of existing technologies for the large-scale identification of coding sequences, transcription units and other functional elements for which technology is currently available. When the results of the ENCODE Project show evidence of efficacy and affordability at the pilot scale, consideration will be given to implementing the appropriate technologies across the entire human genome.

Technology development Many areas of critical importance to the realization of the genomics-based vision for biomedical research require new technological and methodological developments before pilots and then large-scale approaches can be attempted. Recognizing that technology development is an expensive and high-risk undertaking, the NHGRI is nevertheless committed to supporting and fostering technology development in many of these crucial areas, including the following.

DNA sequencing. There is still great opportunity to reduce the cost and increase the throughput of DNA sequencing, and to make rapid, cheap sequencing available more broadly. Radical reduction of sequencing costs would lead to very different approaches to biomedical research.

Genetic variation. Improved genotyping methods and better mathematical methods are necessary to make effective use of information about the structure of variation in the human genome for identifying the genetic contributions to human diseases and other complex traits.

The genome 'parts list'. Beyond coding sequences and transcriptional units, new computational and experimental approaches are needed to allow the comprehensive determination of all sequence-encoded functional elements in genomes.

Proteomics. In the short term, the NHGRI expects to focus on the development of appropriate, scalable technologies for the comprehensive analysis of proteins and protein machines in human health and in both rare and complex diseases.

Pathways and networks. As a complement to the development of the genome 'parts list' and increasingly effective approaches to proteome analysis, the NHGRI will encourage the development of new technologies that generate a synthetic view of genetic regulatory networks and interacting protein pathways.

Genetic contributions to health, disease and drug response. The NHGRI will place a high priority on creating and applying new crosscutting genomics tools, technologies and strategies needed to identify the genetic bases of medically relevant phenotypes. Research on the genetic contributions to rare and common diseases, and to drug response, will typically involve biological systems and diseases of primary interest to other NIH institutes and other funding organizations. Accordingly, the NHGRI expects that its involvement in this area of research will often be implemented through partnerships and collaborations. The NHGRI is particularly interested in stimulating research approaches to the identification of gene variants that confer disease resistance and other manifestations of 'good health'.

Molecular probes, including small molecules and RNA-mediated interference, for exploring basic biology and disease. Exploration of the feasibility of expanding chemical genomics in the academic and public sectors, particularly with regard to the establishment of one or more centralized facilities, will be pursued by the NHGRI in partnership with others.

Databases Another type of community resource for the biological and biomedical research communities is represented by databases (Box 3). But their support represents a potentially significant problem. Funding agencies, reflecting the interest of the research community, tend to prefer to use their research funds to support the generation of new data, and the ongoing need for continued and increasing support for the data archives and robust access to them is often given less attention. Both the scientific community and the funding agencies must recognize that investment in the creation and maintenance of effective databases is as important a component of research funding as data generation. The NHGRI has been a major source of support for several major genetics/genomics-oriented databases, including the Mouse Genome Database (www.informatics.jax.org/mgihome/MGD/aboutMGD.shtml), the *Saccharomyces* Genome Database (genome-www.stanford.edu/Saccharomyces), FlyBase (flybase.bio.indiana.edu), WormBase (www.wormbase.org) and Online Mendelian Inheritance in Man (www.ncbi.nlm.nih.gov/omim). The NHGRI will continue to be a leader in exploring effective solutions to the issues of integrating, displaying and providing access to genomic information.

Ethical, legal and social research The NHGRI's ELSI research activities will increasingly focus on fundamental, widely relevant, societal issues. The community of scholars and researchers working in these social fields, as well as the scope of issues being explored, need to be expanded. The ELSI research community must include individuals from minority and other communities that may be disproportionately affected by the use or misuse of genetic information. New mechanisms for promoting dialogue and collaboration between the ELSI researchers and genomic and clinical researchers need to be developed; such examples might include structural rewards for interdisciplinary research, intensive summer courses or mini-fellowships for cross-training, and the creation of centres of excellence in ELSI studies to allow sustained interdisciplinary collaboration.

Longitudinal population cohort(s) This promising research resource will be so broadly applicable, and will require such extensive funding that, although the NHGRI might have a supporting role in design and oversight, success will demand the involvement and support of many other funding sources.

Non-genetic factors in health and disease A consequence of an improved definition of the genetic factors underlying human health and disease will be an improvement in the recognition and definition of the environmental and other non-genetic contributions to those traits. This is another area in which the NHGRI will be involved through the development of new strategies and by forming partnerships.

Use of genomic information to improve health care The NHGRI will catalyse collaboration between the diverse scholarly disciplines whose joint efforts will be necessary for research on the best ways for patients and health care providers to make effective use of personalized genetic information in the improvement of health. The NHGRI will also strive to ensure that research in this area is informed by, and extends knowledge of, the societal implications of genomics.

Improving the health of all people It will be important for the NHGRI to support research that explores how to ensure that genomic information is used, to the extent that such information is relevant, to reduce global health disparities. That will include a vigorous effort to increase the representation of minorities in the ranks of genomics researchers. But the full solution of the health disparities problem can only come about through a committed and sustained effort by governments, medical systems and society.

Policy development The NHGRI will continue to help facilitate public-policy development in the area of genetic/genomic science. Effective policy development will require attention to those issues for which it could have the greatest impact on the policy agenda and could help to facilitate genomic science. The NHGRI will also focus on issues that would assist the public in benefiting from genomics, such as privacy of genetic information, access to genetics services, direct-to-consumer/providers marketing, patenting and licensing of genetic information, appropriate treatment of human participants in research, and standards, usefulness and quality in genetic testing.

Data release

An important lesson of the HGP has been the benefit of immediately releasing data from large-scale sequencing projects, as embodied in the Bermuda principles (www.gene.ucl.ac.uk/hugolbermuda.htm). Some other large-scale data production projects have followed suit (such as those for full-length cDNAs and single-nucleotide polymorphisms), to the benefit of the scientific community. Scientific progress and public benefit will be maximized by early, open and continuing access to large data sets and by ensuring that excellent scientists are attracted to the task of producing more resources of this sort. For this system to continue to work, the producers of community-resource data sets have an obligation to make the results of their efforts rapidly available for free and unrestricted use by the scientific community, and resource users have an obligation to recognize and respect the important contribution made by the scientists who contribute their time and efforts to resource production.

Although these principles have been generally realized in the case of genomic DNA sequencing, they have not been for many other types of community-resource projects (structural biology coordinates or gene expression data, for example). The development of effective systems for achieving the rapid release of data without restrictions and for providing continued widespread access to materials and research tools should be an integral component of the planning and development of new community resources. The scientific community should also develop incentives to support the voluntary release of such data before publication by individual investigators, by appropriately rewarding and protecting the interests of scientists who wish to share their data with the community in such a generous manner.

Quantum leaps

It is interesting to speculate about potential revolutionary technical developments that might enhance research and clinical applications in a fashion that would rewrite entire approaches to biomedicine. The advent of the polymerase chain reaction, large-insert cloning systems and methods for low-cost, high-throughput DNA sequencing are examples of such advances that have already occurred.

During the course of the NHGRI's planning discussions, other ideas were raised about analogous 'technological leaps' that seem so far off as to be almost fictional but which, if they could be achieved, would revolutionize biomedical research and clinical practice.

The following is not intended to be an exhaustive list, but to provoke creative dreaming:

- the ability to determine a genotype at very low cost, allowing an association study in which 2,000 individuals could be screened with about 400,000 genetic markers for $10,000 or less;
- the ability to sequence DNA at costs that are lower by four to five orders of magnitude than the current cost, allowing a human genome to be sequenced for $1,000 or less;
- the ability to synthesize long DNA molecules at high accuracy for $0.01 per base, allowing the synthesis of gene-sized pieces of DNA of any sequence for between $10 and $10,000;
- the ability to determine the methylation status of all the DNA in a single cell; and the ability to monitor the state of all proteins in a single cell in a single experiment.

Conclusions

Preparing a vision for the future of genomics research has been both daunting and exhilarating. The willingness of hundreds of experts to volunteer their boldest and best ideas, to step outside their areas of self-interest and to engage in intense debates about opportunities and priorities, has added a richness and audacity to the outcome that was not fully anticipated when the planning process began.

To the extent that this article captures the sense of excitement of the new discipline of genomics, it is to their credit. A complete list of the participants in this planning process can be found at www.genome.gov/About/Vision/Acknowledgements.

A final word is appropriate about the breadth of the vision articulated here. A choice had to be made between portraying a broad view of the future of genomics research and focusing more narrowly on the specific role of the NHGRI. Recognizing that researchers and the public are more interested in the promise of the field than about the funding source responsible, we have focused here on the broad landscape of scientific opportunity. We have, however, identified the areas that are particularly appropriate for leadership by the NHGRI throughout this article. These are generally research areas that are not specific to a particular disease or organ system, but have broader biomedical and/or social implications. Yet even in those instances, the word 'partnership' appears numerous times intentionally. We expect to have partnerships not only with other public funding sources, such as the other 26 NIH institutes and centres, but also with many other governmental agencies, private foundations and private-sector organizations. Indeed, public-private partnerships, such as the SNP Consortium, the Mouse Sequencing Consortium and the International HapMap Project, provide powerful new models for the generation of public data sets with immediate and far-reaching value. Thus, many of the most exciting opportunities in genomics research cross traditional boundaries of specific disease definitions, classically defined scientific disciplines, funding sources and public versus private enterprise. The new era will flourish best in an environment where such traditional boundaries become ever more porous.

Although the opportunities described here are thought to be highly achievable, the formal initiation of specific programmes will require more detailed analysis. The relative priorities of each component must be addressed in the light of limited resources to support research. The NHGRI plans to release a revised programme announcement and other grant solicitations later this year, providing more specific guidance to extramural researchers about plans for the implementation of this vision. Furthermore, in genomics research, we have learned to expect the unexpected. From past experience, it would be surprising (and rather disappointing) if biological, medical and social contexts did not change in unpredictable ways. That reality requires that this vision be revisited on a regular basis.

In conclusion, the successful completion this month of all of the original goals of the HGP emboldens the launch of a new phase for genomics research, to explore the remarkable landscape of opportunity that now opens up before us. Like Shakespeare, we are inclined to say, "what's past is prologue" (The Tempest, Act II, Scene 1). If we, like bold architects, can design and build this unprecedented and noble structure, resting on the firm bedrock foundation of the HGP, then the true promise of genomics research for benefiting humankind can be realized.

"Make no little plans; they have no magic to stir men's blood and probably will themselves not be realized. Make big plans; aim high in hope and work, remembering that a noble, logical diagram once recorded will not die, but long after we are gone will be a living thing, asserting itself with evergrowing insistency" (attributed to Daniel Burnham, architect).

Acknowledgements

The formulation of this vision could not have happened without the thoughtful and dedicated contributions of a large number of people. The authors were greatly assisted by Kathy Hudson, Elke Jordan, Susan Vasquez, Kris Wetterstrand, Darryl Leja and Robert Nussbaum. A subcommittee of the National Advisory Council for Human Genome Research, including Wylie Burke, William Gelbart, Eric Juengst, Maynard Olson, Robert Tepper and David Valle, provided a critical sounding board for draft versions of this document. We also thank Aravinda Chakravarti, Ellen Wright Clayton, Raynard Kington, Eric Lander, Richard Lifton and Sharon Terry for serving as working-group chairs at the meeting in November 2002 that refined this document. Finally, we thank the hundreds of individuals who participated as workshop planners and/or participants during this 18-month process.

References

1. On behalf of the US National Human Genome Research Institute. Endorsed by the National Advisory Council for Human Genome Research, whose members are Vickie Yates Brown, David R. Burgess, Wylie Burke, Ronald W. Davis, William M. Golbart, Eric T. Juengst, Bronya J. Keats, Raju Kucherlapati, Richard P. Lifton, Kim J. Nickerson, Maynard V. Olson, Janet D. Rowley, Robert Tepper, Rohert H. Waterston and Tadataka Yamada.
2. Avery, O. T., MacLeod, C. M. & McCarty, M. Studies of the chemical nature of the substance inducing transformation of pneumococcal types. Induction of transformation by a desoxyribonudeic acid fraction isolated from *Pneumococcus* Type III. *J. Exp. Med.* 79, 137–158 (1944).
3. Watson, J. D. & Crick, F. H. C. Molecular structure of nucleic acids: A structure for deoxyribose nucleic acid. *Nature* 171, 737 (1953).
4. Nirenberg, M. W. The genetic code: II. *Sci. Am.* 208, 80–94 (1963).
5. Jackson, D. A., Symons, R. H. & Berg, P. Biochemical method for inserting new genetic information into DNA of Simian Virus 40: circular SV40 DNA molecules containing lambda phage genes and the galactose operon of *Escherichia coli. Prof. Natl. Acad. Sci. USA* 69, 2904–2909 (1972).
6. Cohen, S. N.,Chang, A. C., Boyer, H. W. & Helling, R. B. Construction of biologically functional bacterial plasmids *in vitro. Proc. Natl. Acad. Sci. USA* 70, 3240–3244 (1973).
7. The International Human Genome Sequencing Consortium. Initial sequencing and analysis of the human genome. *Nature* 409, 860–921 (2001).
8. Sanger, F. & Coulson, A. R. A rapid method for determining sequences in DNA by primed synthesis with DNA polymerase. *J. Mol. Biol.* 94, 441–448 (1975).
9. Maxam, A. M. & Gilbert, W. A new method for sequencing DNA. *Proc. Natl. Acad. Sr. USA* 74, 560–564 (1977).
10. Smith, L. M. et al. Fluorescence detection in automated DNA sequence analysis. *Nature* 321, 674–679 (1986).
11. The Mouse Genome Sequencing Consortium. Initial sequencing and comparative analysis of the mouse genome. *Nature* 420, 520–562 (2002).
12. The Chipping Forecast II. *Nature Genet.* 32, 461–552 (2002).

13. Guttmacher, A. E. & Collins, F. S. Genomic medicine—A primer. *N. Engl. J. Med.* 347, 1512–1520 (2002).
14. National Research Council. *Mapping and Sequencing the Human Genome* (National Academy Press, Washington DC, 1988).
15. US Department of Health and Human Services, US DOE. *Understanding Our Generic Inheritance. The US Human Genome Project The First Five Years.* NIH Publication No. 90-1590 (National Institutes of Health, Bethesda, MD, 1990).
16. Collins, F. & Galas, D. A new five-year plan for the US Human Genome Project. *Science* 262, 43–46 (1993).
17. Collins, F. S. *et al.* New goals for the US Human Genome Project: 1998–2003. *Science* 282, 682–689 (1998).
18. Hilbert, D. Mathematical problems. *Bull. Am. Math. Soc.* 8, 437–479 (1902).
19. Aparicio, S. *et al.* Whole-genome shotgun assembly and analysis of the genome of *Fugu rubripes. Science* 297, 1301–1310 (2002).
20. Sidow, A. Sequence first. Ask questions later. *Cell* 111, 13–16 (2002).
21. Zhang, M. Q. Computational prediction of enkaryotic protein coding genes. *Nature Rev. Genet.* 3, 698–709 (2002).
22. Banerjee, N. & Zhang, M. X. Functional genomics as applied to mapping transcription regulatory networks. *Curr. Opin. Microbiol.* 5, 313–317 (2002).
23. Van der Werden, L., Adams, D. I. & Bradley, A. Tools for targeted manipulation of the mouse genome. *Physiol. Genomics* II, 133–164 (2002).
24. Hannon, G. I. RNA interference. *Nature* 418, 244–251 (2002).
25. Stockwell, B. R. Chemical genetics: Ligand-based discovery of gene function. *Nature Rev. Genet.* 1, 116–125 (2000).
26. Gavin, A. C. *et al.* Functional organization of the yeast proteome by systematic analysis of protein complexes. *Nature* 415, 141–147 (2002).
27. Tyson, I. I., Chen, K. & Novak, B. Network dynamics and cell physiology. *Nature Rev. Mol. Cell Biol.* 2, 908–916 (2001).
28. Sachidanandam, R. et al. A map of human genome sequence variation containing 1.42 million single nucleotide polymorphisms. *Nature* 409, 928–933 (2001).
29. Gabriel, S. B. *et al.* The structure of haplotype blocks in the human genome. *Science* 296, 2225–2229 (2002).
30. Reich, D. E. & Lander, E. S. On the allelic spectrum of human disease. *Trends Genet.* 17, 502–510 (2001).
31. Hirschhorn, I. N., Lohmueller, K., Byrne, E. & Hirschhorn, K. A comprehensive review of genetic association studies. *Genet. Med.* 4, 45–61 (2002).
32. Wagner, K. R. Genetic diseases of muscle. *Neurol. Clin.* 20, 645–678 (2002).
33. Golub, T. R. Genomic approaches to the pathogenesis of hematologic malignancy. *Curr. Opin. Hematol.* 8, 252–261 (2001).

34. Drews, I. & Ryser, S. The role of innovation in drug development. *Nature Biotechnol.* 15, 1318–1319 (1997).
35. Druker, B. J. Imatinib alone and in combination for chronic myeloid leukemia. *Semin. Hematol.* 40, 50–8 (2003).
36. Selkoe, D. I. Alzheimer's disease: genes, proteins, and therapy. *Physiol. Rev.* 81, 741–66 (2001).
37. Lynch, H. T. & de la Chapelle, A. Genomic medicine: hereditary colorectal cancer. *N. Engl. J. Med.* 348, 919–932 (2003).
38. Gardoer, M. J. *et al.* Genome sequence of the human malaria parasite *Plasmodium falciparum. Nature* 419, 498–511 (2002).
39. Holt, R. A. *et al.* The genome sequence of the malaria mosquito *Anopheles gambiae. Science* 298, 129–149 (2002).
40. Anderlik, M. R. & Rothstein, M. A. Privacy and confidentiality of genetic information: What rules for the new science? *Annu. Rev. Genom. Hum. Genet.* 2, 401–433 (2001).
41. Hudson, K. L., Rothenberg, K. H., Andrews, L. B., Kahn, M. I. E. & Collins, F. S. Genetic discrimination and health-insurance—An urgent need for reform. *Science* 270, 391–393 (1995).
42. Rothenberg, K. *et al.* Genetic information and the workplace: Legislative approaches and policy challenges. *Science* 275, 1755–1757 (1997).
43. Fuller, B. P. *et al.* Policy forum: Ethics–privacy in genetics research. *Science* 285, 1359–1361 (1999).
44. Miller, P. S. Is there a pink slip in my genes? *J. Health Care Law Policy* 3, 225–265 (2000).

17. Bioinformatics: Challenges at the Frontier

Kenneth W. Goodman

One could make a very good argument for the case that information technology and genetics are the sciences that will have the greatest effect on 21st century health care. It is already clear that informatics is reshaping the health professions; and we are just now coming to terms with the extraordinary risks and potential benefits of progress in the human genome sciences.

Now all we need to do is figure out what to make of their intersection, that is, the intersection of health informatics and genomics. Add in ethical and social issues, and we create one of the greatest intellectual and practical challenges in the history of science.

Bioinformatics, or the use of information technology to acquire, store, manage, share, analyze, represent and transmit genetic data, has blossomed in the past decade. The term is most often used by scientists who are using computers to sequence and otherwise analyze the genomes of humans and other species. But if we like, we can stipulate that bioinformatics includes applications as pedestrian as the use of a personal computer to store the results of genetic tests ("Patient X has the BRCA1 gene"), as well as the use of intelligent machines to link physiological traits with a database to diagnose genetic maladies, to predict clinical correlations, to conduct research, and so forth.

At any rate, it is clear that the future of genomics will be built upon vast amounts of computer-based information. From data acquisition via microarrays or "gene chips" to diagnosis using genetic decision support systems to research based on data mining of very large databases and warehouses, social and ethical issues raised by the genomic sciences will stimulate clinicians and scientists at least as much as those raised by clinical informatics.

Moreover, the broad and rapid growth of bioinformatics presents exciting opportunities and challenges not only for clinicians and scientists, but also for society, individuals, policy-making bodies and governments (Goodman 1996).

While bioinformatics raises many issues for human subjects research, we can confine ourselves here to more clinical concerns, and group them under the label

---

This essay, which draws from material in *MD Computing* (1999; 16, 3:17–20), appears for the first time in *Ethics, Computing, and Genomics*. Copyright © 2005 by Kenneth W. Goodman. Printed by permission.

"clinical bioinformatics." (It is well to note, though, that once human genetic information is stored on a computer, it becomes much easier to study it—and in some cases the distinction between clinical and research issues will become quite narrow.) Let us then organize the ethical and social issues raised by clinical bioinformatics into the following categories: (1) accuracy and error (2) appropriate uses and users of digitized genetic information, and (3) privacy and confidentiality.

Accuracy and Error

We have already learned from clinical health informatics that accuracy and error avoidance raise ethical issues. These are often related to evolving standards of care. If there are emerging or established standards for database management, for instance, then a system that relies on a database will be more or less useful, reliable and safe, according as the database is appropriately maintained, tested, augmented, and so on. The reason to link error and ethics is that errors, however unintentional, can lead to harms. Whether a harm constitutes a *wrong* is one of the main challenges of morality.

The challenge here is how simultaneously to nurture the growth of an exciting new science while ensuring that patients and their families are not harmed or wronged.

Several current and future issues related to the accuracy of bioinformatics systems can be identified:

Risks to persons—To the extent that we can expect more and more frequent computer-aided discoveries of the genetic loci of human diseases, then errors can pose or increase risks to public health and even the well-being of individual humans. Patients may also be at risk when computers are used to predict the expression of future genetic maladies. The risks may be psychological and will likely vary depending on whether there is a treatment or cure for a given malady. The role of genetic counselors will loom large here.

Recanted linkage studies—Preliminary or unreplicated linkage studies are sometimes recanted or re-evaluated. Erroneous linkage analyses can throw colleagues off the track and, perhaps more importantly, cause unnecessary psychological trauma for individuals who fear they may be affected. In the case of purported linkages that correlate with race or ethnicity, there is the added risk of producing social stigma. This is perhaps especially the case in neurogenetics and psychiatric genetics.

Meta-analysis—It is exciting to observe the emergence of meta-analysis in genomics. In this research technique, the results of previous studies are aggregated and reanalyzed by statistical software in hopes of achieving statistical significance or adequate sample sizes. This raises ethical issues by virtue, in part, of doubts about the quality of included data and the validity of inferences based on diversity of data. These doubts are important when meta-analytic results are applied to patient care (Goodman 1998a).

Decision support—Diagnostic and decision support systems are well known to raise ethical issues in clinical medicine. But there is yet no critical analysis of decision support for genetic diagnoses, in which, for example, clinical information, photographic material, pedigree, and gene localization data are analyzed by computers. The growth of genomic data bases and the increasing availability of genetic information at the clinical level suggest that decision support systems are a ripe source for ethical and social inquiry (Miller and Goodman 1998; Goodman 1998b).

Appropriate Uses and Users

Questions concerning who should use clinical information systems and in what contexts have been shown to raise interesting and important ethical issues and, indeed, many of these have been recognized for decades (Miller, Schaffner and Meisel 1985); we should expect that genetic data processing will elicit related concerns and pose new problems.

For instance, suppose a physician or nurse begins including genetic data in patient charts, uses those data to predict the likelihood of clinical manifestations and correlations, and then employs those analyses to refer patients to genetic counselors. The first question to be answered is, Was this novel use undertaken with the patients' consent? Because genetic information can frighten or alarm patients in ways that other health and medical data do not, we need to ask if the patients knew genetic data was being gathered and stored for clinical purposes. In the absence of a treatment or cure for a particular genetic malady, it is not unreasonable for any individual patient to prefer *not* to know a genetic diagnosis or prognosis. So, consent seems to be a crucial gate through which the physician or nurse must pass before using these data "for the patient's sake."

The weight of valid or informed consent seems to many to be greater here—that is, in the land of genetics—than for more familiar kinds of clinical decision support.

Or suppose that individuals' genetic data were being collected by governments, managed care organizations or other third-party payers with the goal of shaping or adjusting risk pools or coverage eligibility. The difference between evidence-based actuarial calculations and discrimination can be very thin, indeed. To the extent that computers are used for these tasks—and how else could such tasks be undertaken?—it will be essential for individuals, institutions and society to agree on ethically optimized strategies for clinical bioinformatics applications.

Now we must ask *who* should use a genetic diagnostic or prognostic system. It was just implied that there might a problem with a particular use—to determine health benefits, say. Does this further suggest that bioinformatics tools should not be used by certain entities?

Consider that individual physicians, nurses, genetic counselors or psychologists might use computer systems to improve patient care, as well as for less worthy purposes. Does it follow that certain users—in addition to uses—might be problematic?

One way to think about this is by asking if the user is employing a computer in a task not normally within his or her competence. The idea is that if you are unable or untrained to perform certain tasks without a computer, then it is inappropriate to suppose the computer can somehow imbue you with the skills necessary to the job. For instance if a physician or nurse does not normally render genetic diagnoses, it is unwise to suppose that she or he acquires competence via the machine.

In fact, it is more than unwise—it is a patent mistake. Computers can improve our skills at many tasks but rarely, if ever, give us new professional skills or abilities. Therefore, an appropriate use of a genetic decision support system, for instance, will be to assist adequately trained professionals, not to replace them or to bring them "up to speed" in domains in which they lack basic skills (cf. Miller 1990).

This point must be clearly understood: Computers can be outstanding educational tools in bioinformatics as elsewhere, but there is a difference between acquiring a skill and presuming its existence. We have learned from "ordinary" clinical computing that humans practice medicine and nursing but computers do not. This is a lesson well worth applying to bioinformatics.

Privacy and Confidentiality

The electronic storage of genetic information replicates a tension already familiar in health informatics: the tension between (1) the need for appropriate or authorized access to personal information, and (2) the need to prevent inappropriate or unauthorized access. Striking a balance between these two imperatives is an exciting but sometimes vexing challenge.

Privacy and confidentiality are potentially threatened when individual genetic data are maintained or transmitted using computers. The threats include bias and discrimination, personal stigma (as opposed to population or subgroup stigma), psychological stress, and tensions within families, among other risks. The difficulties posed by expectations of privacy and confidentiality are well explored in regard to the electronic patient record (see Alpert 1998, for a review), but we do not yet know whether the inclusion of genetic data adds to or alters those difficulties.

Specifically, our objective is to determine if and in what way bioinformatics raises ethical issues that are distinct from ethics and genetics and, depending on what is found, either to adapt existing conceptual and pedagogic tools or provide new ones. The key means by which we plan to meet these objectives are the successful development of ethically optimized guidelines (for organizations that maintain data bases, for IRBs, etc.) and model curricula in ethics and bioinformatics (for students and professionals).

Striking a Balance

The need for organizational policies, best-practice standards, and/or guidelines is widespread in the human sciences. Because the thrust of the proposed research is at the intersection of three vast areas of inquiry and practice—genetics, computing, and ethics—the challenge we face is extraordinary: Guidelines and standards often fail because they are either so broad or simplistic that they cannot adequately guide behavior, or are so specific or detailed that they are too inflexible to be useful in diverse and unexpected cases.

There is therefore a need to strike a balance between these two shortcomings. In striking that balance, the successful completion of this project would provide a very useful tool for organizations.

As to educational materials, it is worth observing that research ethics curricula usually overlook issues in bioinformatics. If we are correct in anticipating that the future of genetic research will be inextricably linked to information-processing technologies, then this oversight is, or will be, quite serious. Indeed, we may well conclude from our inquiry that there are larger nets to cast and that we should extend our emphasis on genetics to include all biology and medicine and the changes mediated by information technology.

Acknowledgments

An earlier version of this chapter appears in *MD Computing* 1999; 16 (3): 17-20. I am grateful to Springer Verlag (*www.springeronline.com*) for permission to reuse material from that article.

References

Alpert SA. (1998) Health care information: Access, confidentiality, and good practice. In K.W. Goodman, ed., *Ethics, Computing and Medicine: Informatics and the Transformation of Health Care*, Cambridge and New York: Cambridge University Press, 75–101.

Goodman KW. (1996) Ethics, Genomics, and Information Retrieval. *Computers in Biology and Medicine* 26:223–229.

Goodman KW. (1998a) Meta-analysis: Conceptual, ethical and policy issues. In K.W. Goodman, ed., *Ethics, Computing and Medicine: Informatics and the Transformation of Health Care*, Cambridge and New York: Cambridge University Press, 139–167.

Goodman KW. (1998b) Ethical and Legal Issues in Use of Decision Support Systems. In E. Berner, ed., *Decision Support Systems.* New York: Springer Verlag: 217–233.

Miller R., Goodman KW. (1998) Ethical challenges in the use of decision-support software in clinical practice. In K.W. Goodman, ed., *Ethics, Computing and Medicine: Informatics and the Transformation of Health Care*, Cambridge and New York: Cambridge University Press, 1998, 102–115.

Miller RA. (1990) Why the standard view is standard: People, not machines, understand patients' problems. *Journal of Medicine and Philosophy* 15: 581–91.

Miller RA, Schaffner KF, and Meisel A. (1985) Ethical and legal issues related to the use of computer programs in clinical medicine. *Annals of Internal Medicine* 102: 529–36.

18. The Control of Scientific Research: The Case of Nanotechnology[2]

John Weckert

1. Introduction

Over the past hundred years or so, developments in science and technology have radically altered our world. Some of these developments have been unambiguous boons—millions of people are alive today as a result of breakthroughs in understanding and treating infectious diseases. Other developments, however, have been less obviously beneficial—it is far from clear that the world is a safer or better place as a result of the advances in nuclear technology, for example, whether for peaceful or military purposes. In any case, attitudes to science and technology have also altered in this period. These attitudes were once highly optimistic. Now there is a good deal of scepticism about the alleged benefits of much scientific research or even fear of its potentially damaging effects. With these changes of attitude have come calls for greater oversight and control over scientific research, even, in some cases, for a halt to research. Over twenty-five years ago, for instance, Joseph Weizenbaum claimed that there are some kinds of research in which computer scientists should not engage, saying that:

> There are ... two kinds of computer applications that either ought not be undertaken at all, or, if they are contemplated, should be approached with utmost caution. The first kind I would call simply obscene. These are ones whose very contemplation ought to give rise to feelings of disgust in every civilized person. The proposal I have mentioned, that an animal's visual system and brain could be coupled to computers,

---

This essay originally appeared in the *Australian Journal of Professional and Applied Ethics*, **3**: 29-44. Copyright © 2001 by the Centre for Applied Philosophy and Public Ethics, Charles Sturt University, Australia. Reprinted by permission.

> is an example. ... I would put all projects that propose to substitute a computer system for a human function that involves interpersonal respect, understanding, and love in the same category. ... The second kind ... is that which can easily be seen to have irreversible and not entirely foreseeable side effects. (Weizenbaum, 1984, pp. 268–270)

Within the area of information technology, recent developments in nanotechnology and quantum computing have reignited the issue of constraining or even halting research. These technologies have the potential to radically change information technology. If they are successful, and there are signs that they will be, computers will become very, very small, very, very fast, and have an enormous amount of memory relative to computers of today. This is creating excitement in some quarters, but anxiety in others. Speaking of nanotechnology, in a recent and much publicised article, Bill Joy wrote that:

> It is most of all the power of destructive self-replication in genetics, nanotechnology, and robotics (GNR) that should give us pause. ... In truth we have had in hand for years clear warnings of the dangers inherent in widespread knowledge of GNR technologies—of the possibility of knowledge alone enabling mass destruction. ... The only realistic alternative I see is relinquishment: to limit development of the technologies that are too dangerous, by limiting our pursuit of certain kinds of knowledge. (Joy, 2000, pp. 248, 254)

Given the variety of benefits promised by nanotechnology in, for example, medicine, the environment, and information technology (see next section), Joy's claim seems a little strong. This paper will discuss a few concerns, to see what the appropriate reaction to this technology is. Are the worries enough to give his call for a halt to research any plausibility?[3]

Before proceeding to examine these claims that some research in computing should not take place, we need to look at nanotechnology and quantum computing, to see what dangers there might be. It must be noted too, that others have recognised potential dangers, and a set of guidelines has already been proposed (Foresight, 2000).

2. Nanotechnology and Quantum Computing

Nanotechnology is relatively new, although it was first mooted in 1959 by Richard Feynman in a lecture entitled "There's plenty of room at the bottom" (Feynman, 1959). Nanotechnology involves the manipulation of individual atoms and molecules in order to build structures. One nanometre is one billionth of a metre (three to five atoms across), and "nanoscience and nanotechnology generally refer to the world as it works on the nanometer scale, say, from one nanometer to several hundred nanometers" (NSTC, 1999, p.1). This technology is important because it enables very small things to be built, and gives great control over the properties of materials constructed "from the bottom up." This technology is claimed to have enormous benefits in a variety of fields. Nanoparticles inserted into the body could diagnose and cure diseases, building materials could adapt to the weather conditions, cheap and clean energy could be produced, sensoring devices could become much more sensitive, and computers much faster and have more memory. (For discussions in layman's terms, see the 2001 articles by Wilson, Luntz, Martin, Wallace and Harvey.) While these benefits have not yet accrued, they are not mere speculation. The theories underlying the proposed applications are, it is argued, scientifically based. The technology for manipulating atoms individually is available, and as long ago as 1989 IBM physicists produced the IBM logo by manipulating atoms.

The main interest in this paper is the relationship of nanotechnology with computing. This technology will enable the development of computers on a nano scale, and quantum computers. These computers will be much more powerful than current ones. According to one researcher, it is expected that a quantum computer would be able to perform some computations that could not be performed by all the current computers on the planet linked together, before the end of the universe (Simmons, 2000). Computers will also become very small. For example, at one end of the spectrum there would be devices that incorporate "nanoscale computers and several binding sites that are shaped to fit specific molecules, [that] would circulate freely throughout the body ..."(Merkle, 1997). At the other end "we should be able to build mass storage devices that can store more than 100 billion billion bytes in a volume the size of a sugar cube, and massively parallel computers of the same size that can deliver a billion billion instructions per second—a billion times more than today's desktop computers" (Merkle, 1997). Smalley is a little more cautious, but still estimates that the efficiency of computers will be increased by a factor of millions (Smalley, 1999). More cautious still are Steane and Reiffel, who write "Depending on the success of nanotechnology, miniaturization will increase computing power by somewhere between 1,000 and 100,000 times the current speed—and that is all" (Steane and Reiffel, 2001, p. 43). They admit that quantum computers will be faster, but "In many situations ... because of the way they must be built, will perform much slower than classical computers" (p. 44). In addition to this increase in power, there could be "detecting devices so sensitive that they could pick up the equivalent of the drop of a pin on the other side of the world" (Davies, 1996 p. vii).

This technology, if it develops in the predicted matter, will facilitate some interesting developments. Monitoring and surveillance will become very easy, particularly when the new computer and communication technologies are linked. People with microscopic implants will be able to be tracked using Geographic Positioning Systems (GPSs) just as cars can be now, only more efficiently. One need never be lost again! Other implants could increase our memory, reasoning ability, sight, hearing, and so on. Some argue that the distinction between humans and machines may no longer be useful (Kurzweil, 1999). And virtual reality will be indistinguishable from reality itself.

Before considering some of the ethical issues raised by these possible developments, it is worth asking just how credible are the claims. Many of the claims are quite credible, it seems. Many are being made by researchers with impressive records in computer science, and research is being supported by reputable universities. More importantly, some progress has been made. It is already possible to manipulate individual atoms, as stated earlier, nanotubes have been developed, and some progress has been made on building simple quantum computers. Theoretically nanotechnology can work, and progress has been made on building some applications. There is enough evidence to suggest that nanotechnology and quantum computing are possible, probably sometime around the middle of this century, and so it is worth looking at any ethical questions that they might raise. (For discussions of current and recent research see articles in *Scientific American*, September, 2001; *Science*, 290, 2000; *IoN*, 2001; NSTC, 1999; NAI, 2001. See also Benenson, et al, 2001.)

3. Ethical questions

While nanotechnology has potential benefits and dangers in a wide variety of areas, for example in health and in the environment as previously mentioned, we will consider here just some potential dangers in the computing area. That there are also benefits in this area is not being questioned.

There are at least two sets of issues. One set concerns existing problems which will be exacerbated by the miniaturisation of computers. This miniaturisation will be accompanied by an increase in power (computers that are much faster and with much more memory), and by much more sensitive input devices. The second set concerns potentially new problems, problems which as yet have not arisen, at least not in any significant way.

3.1. Exacerbated problems

It is likely that most existing ethical problems arising from the use of computers will be exacerbated. Easier and faster copying onto ever smaller devices will make protecting intellectual property more difficult. There will be more worries about Internet content as that content becomes more realistic and more difficult to control. It is likely however that one of the greatest impacts will be on privacy and data protection, and the potential for control.

3.1.1. Privacy

Privacy and data protection problems will be enormously increased. Vast databases that can be accessed at very high speeds will enable governments, businesses and so on to collect, store and access much more information about individuals than is possible today. In addition, the capacity for data mining, the exploration and analysis of very large amounts of data for the purpose of discovering meaningful and useful rules and patterns, will increase dramatically. And perhaps most importantly, the monitoring and surveillance of workers, prisoners, and in fact, the population in general, will be greatly enhanced with the use of small, powerful computers and new sensoring devices for input. GPSs will be able to specify the location of individuals, cameras with neural nets (or other learning technologies), will be able to pick out unusual behaviour in crowds, or just on the streets, and notify authorities. (There is a continuum of course between problems arising from current technology and those to be a result to future nanotechnology.) According to a US National Defense Research Institute report:

> *Various threats to individual privacy include pervasive sensors, DNA "fingerprinting" genetic profiles that indicate disease predispositions, Internet-accessible databases of personal information, and other information technology threats. (RAND, 2000, p.39)*

While all of these possibilities have benefits, for example for safety and efficiency, the possibility is also opened for large scale control of individuals, either by governments, employers or others with authority or power. There will be a need for the reassessment of privacy legislation, the use of personal information by governments and corporations, and guidelines and legislation for the use of monitoring devices.

3.2. New problems
3.2.1. Artificial intelligence

If machines are developed that behave in much the same way as humans do, in a wide variety of contexts, the issue will arise of whether or not they are things with moral rights and responsibilities. Consideration would need to be given as how they should be treated. If they behaved like us, would we be justified in treating them differently? With nanotechnology facilitating much faster processing and much more sensitive input devices, systems that behave like humans will become more feasible. If they behave like we do in a variety of ways, it would not matter if they were not *really* intelligent or *really* had minds, whatever that might mean. We give animals the benefit of the doubt with respect to the experience of pain (and other humans, in fact), so there may be no good reasons for not likewise treating machines.

3.2.2. Bionic humans

Chip implants in humans that enhance various of the senses, memory and perhaps even other capacities such as reasoning ability and creativity, may blur the distinction between human and machine. We already have spectacles, hearing aids, cochlear implants, hair implants, skin grafts, tooth implants, pace makers, transplants, and so on, so why should these more advanced implants matter? Perhaps there is a difference between helping people to be "normal", that is, correcting a deficiency, and making a normal person a "super human". But it would need to be spelt out just why this is the case.

3.2.3. Virtual reality

Virtual reality systems will improve to the point where it may become difficult to tell the difference between "real" and "virtual" reality. There may be no apparent difference between really hang gliding and doing so virtually (except that crashing may be less painful!). Suppose that we could have virtual sex with anyone of our choosing, without his or her permission, or knowledge. Would this constitute rape, or be immoral for some other reason? And what about virtual murder or other crimes?

4. The control of research

If Weizenbaum and Joy are right then there is perhaps some research in nanotechnology and quantum computing that computer scientists ought not do, and if they do, they can be held morally responsible for the consequences of that research. Or so it would seem. There are, however, a number of issues which need to be sorted out before we can be confident in affirming this. One concerns the differences between pure research, technological development, and the use to which that development is put. The first question here is whether there is any pure research that should not be undertaken. This issue is often discussed under the heading of "forbidden knowledge"; is there any knowledge that we should not attempt to discover? A related question is whether there is any technology that should not be developed, and the further one of limits to the uses of that technology. However, first it will be useful to consider a few instances of scientific research being "controlled".

4.1. Some cases

The control of scientific research is fraught with danger, and therefore should not be contemplated lightly. There are of course a variety of ways in which control can be exerted, some of which seem more objectionable than others. Research controlled by State ideology seems particularly dangerous, illustrated well by the Lysenko affair in the USSR. Geneticists who opposed Lysenko's Lamarkian views on heredity were actively persecuted, with unfortunate consequences both for them and for Soviet agriculture. Examples such as this, and the role of science, especially eugenics, in Nazi Germany, certainly highlight dangers with State control of, or just interference in, scientific research.

The situation, however, is not a simple one in which the correct way is just a matter of complete freedom for scientists, whatever that means. At one extreme there is the situation in which the State sacks, imprisons or worse, dissident scientists who will not "toe the party line", as happened in the Lysenko case. But influence is not usually so blatant. Even in the Lysenko affair there were factors which are not entirely foreign to us, as can be seen in this quotation:

> In 1931 and 1932, [a number of scientists]... lost their positions at the Communist Academy. There was increasing pressure to abandon basic research that was unlikely to lead to immediate practical measures that would advance Soviet agriculture and there were strong implications that research in "pure science" was tantamount to sabotage. (Sheehan, 1993)

Pressure to undertake certain kinds of research, for example applied research, can be exerted by the allocation of funding, and by the attitudes of "big names" in the field. If funding is only or mainly available for a certain kind of research, that area will obviously grow at the expense of others. The Nazis employed this means to encourage research into eugenics (Kevles and Hood, 1992, p. 8). And if "big names" attack an area or approach, it discourages activity in that area or approach.

While this sort of control is undoubtedly less painful and more acceptable to individual scientists, it is not clear that it is much better for the scientific enterprise as a whole. In both cases research is channelled in a direction that suits the aims of the authorities. This is illustrated by two cases in the history of research in Artificial Intelligence (AI).

In 1973 Sir James Lighthill, in the Lighthill Report, reported to the Science Research Council of the United Kingdom that the field of AI was making little progress and that it should no longer be funded (Crevier, 1993). As a result, funding was stopped and research groups disbanded. In this case a government commissioned report that was acted upon stifled research in a field which has proven to be useful, for quite a number of years. The second case is different, because it was not government driven, but the effect was similar. Artificial neural network research began in the late 1950's and appeared to have some potential. However, in 1969 Marvin Minsky and Seymour Papert published a book that pointed out a basic limitation of the technology to that time, essentially a neural network of just two layers. This book had the effect of almost completely halting research on artificial neural networks until the early 1980's (Crevier, 1993, pp. 106-107; Minsky and Papert, 1969).

Three models for the control of scientific research are apparent in the cases above. In the first, control by government is direct and explicit. Just get rid of scientists engaging in the wrong kind of research. In the second, control is exercised more subtly through the distribution of research funds. This approach allows research to proceed if alternative funding is available, or if it is of such a nature that not much funding is required. However, in practice it does have the effect of stifling certain kinds of research, as the Lighthill case illustrates. The third model is different, because the driving force was not the government, but two leading figures in the AI field. They simply "got it wrong". Artificial neural network research was not leading to a dead end as they believed, and this has been amply demonstrated by developments since that research resumed.

This discussion perhaps just states the obvious, namely, that control can be exerted on scientific research in various ways. But it also highlights the fact that there is not a simple dichotomy between State control on the one hand, and scientific freedom on the other. In all of the cases mentioned, potentially fruitful research was stifled, and not always by State intervention. This of course does not show that intervention by the State is good, or that it is easily justified. What it does indicate is that scientists themselves are not always good judges of potentially fruitful research programmes. Perhaps at times they are too closely involved to be able to "see the wood for the trees". Additionally, they may have strong vested interests in the collapse of one type of research (I am not suggesting that Minsky and Papert were in this category). It is not obvious that State intervention is always worse than leaving it up to the scientists, or that it is always more harmful to the scientific enterprise.

4.2. Forbidden knowledge

Earlier a distinction was made between knowledge, the technology developed from that knowledge, and the uses to which the knowledge is put. Certain uses of knowledge (or the technology based on the knowledge) ought to be avoided if those uses cause harm. The emphasis for the moment is on the knowledge itself. The knowledge must also be distinguished from the method of gaining that knowledge. Clearly, certain methods for gaining knowledge are wrong, for example those which cause harm. (In particular situations it might be that there is some greater good that makes some degree of harm, both in the gaining of knowledge and in its use, permissible.) The question of whether the knowledge itself ought to be forbidden is, or seems to be, quite a different matter. The knowledge is neither morally good nor bad in the way that the methods or uses might be.

It is interesting to note in this context the view of Antoine Arnauld in the *Port Royal Logic*, written in the seventeenth century:

> ... to err in science is not a grave matter, for science has but little bearing on the conduct of life. It would doubtless be beneficial now to consider what leads men to make false judgments in general and particularly in morals and in important and frequently discussed civil matters. (Arnauld, 1964, p. 265)

On one reading this is dramatically at odds with current thinking. Few would now say that "science has little bearing on life". On another reading however, it is not so different. Science itself, it might be argued, does have little bearing. What does have a bearing is the use to which it is put, and it is here that the moral and civil matters come into play. If science does affect life little, it would seem that there is no justification, indeed no point, in attempting to control it. Whether or not this is true, is the question of this section.

It is difficult to make sense of the claim that there is some knowledge that is morally bad in itself regardless of any consequences. It is certainly difficult to find any examples. It is easy to find examples of knowledge that have harmful consequences. Joy himself seems to acknowledge this when he says that knowledge of nanotechnology should be limited in order to limit the technologies that would be developed from that knowledge. If knowledge has harmful consequences should it be forbidden? Not always, because many types of knowledge can be put to both beneficial and harmful uses, and we do not want to automatically rule out the beneficial.

At this point it is worth looking briefly at Somerville's argument in *The Ethical Canary*, because, perhaps, the discussion should not be couched just in consequentialist terms. Her two basic principles are these:

> ... we have a profound respect for life, in particular human life ... and we must act to protect the human spirit—the intangible, invisible, immeasurable reality that we need to find meaning in life and to make life worth living—that deeply intuitive sense of relatedness or con-

nectedness to the world and the universe in which we live (Somerville, 2000, xi–xii).

She suggests that if scientific research violates either one of these principles, then it ought to be avoided even if it has some beneficial consequences. Her second principle has particular relevance for nanotechnology and quantum computing, although she does not discuss these fields. An argument, similar in some ways, is presented by Lenman. Talking about having some of our needs met by computers rather than being personally involved in the meeting of those needs, he writes:

> Human lives ... are structured around the project of meeting human needs and our sense of the value in our lives is informed by the quality of our participation in this project. When we find ourselves disengaged from this project our sense of this value is accordingly undermined. ... The project of meeting human needs is not the same as and cannot without loss be replaced by the project of simply having our needs met. (Lenman 2001, p.10)

We will return to these comments in the conclusion, and now consider some more consequentialist arguments.

To help to clarify matters, the consequences of knowledge can be divided into two groups, physical and mental. Physical consequences are uses to which the knowledge is put. The knowledge that is of most concern is that which will almost certainly be used in harmful ways. It is hard to imagine, for example, that the knowledge, both theoretical and technical, necessary for the development of weapons will not be used to harm people. It is similarly difficult to believe that the knowledge required for the development of sophisticated surveillance devices will not be used for illegitimate control. There may be no way to prevent the harm without preventing the acquisition of the knowledge in the first place. A case could be made that that knowledge is not the fit subject of research and ought to be forbidden. Mental consequences do not necessarily involve uses. Some knowledge is such that simply knowing it has negative consequences on life. Nicholas Rescher puts it this way:

> Some information is simply not safe for us to have—not because there is something wrong with its possession in the abstract, but because it is the sort of thing we humans are not well suited to cope with. There are various things we simply ought not to know. If we did not have to live our lives amidst a fog of uncertainty about a whole range of matters that are actually of fundamental interest and importance to us, it would no longer be a human mode of existence that we would live. Instead, we would become a being of another sort, perhaps angelic, perhaps machine-like, but certainly not human. (Resher, 1978, p. 9)

Suppose that as a result of research in IT it became known how to build machines which in behaviour were indistinguishable from humans, and moreover, that it was obvious that these machines were purely deterministic and without freewill. If we knew this, we would obviously have to see ourselves in a new light. Would we, in our present stage of evolution, be able to cope? If the GNR technology discussed by Joy develops in the manner that he fears, would we be able to continue to live happy and satisfying lives? If some knowledge has profound effects on the way we see ourselves, should it be forbidden? It seems that here, just as in the case of knowledge that almost inevitably leads to harm, a plausible argument can be made for forbidding it.

Volkman (2001) argues that there is a kind of technological determinism. While it is not logically impossible to stop technological development, "it is part of our nature as humans that we will not give up on knowledge and technology." If this is true then of course there is not much point in saying that we ought to do so. Many things seem to be part of human nature in the sense that humans have engaged in these activities over the ages and across cultures, but we do not necessarily see them as acceptable for that reason alone. Rape and murder have been unfortunately common, but are not for that reason thought to be acceptable in civilised society. At least parts of our "nature" can be overcome, and some ought to be. Perhaps the quest of technological development at all costs is in this category. But there is more to Volkman's technological determinism than this. It has an ethical component: it is "ethically impossible to recognize a means to improvement and not pursue it." This seems to mean that it is always unethical not to pursue a development that will improve the human lot in some way, despite any risks. The reason for this is that it is hubris to suppose that we can tell whether or not the benefits of any particular technological development are worth the risks. Therefore, according to Volkman "it is ultimately wrong to pursue the moratorium on technology advocated by Bill Joy" (Volkman, 2001, p. 361). It is however, not clear why this conclusion follows. If it is hubris to suppose that we can tell if some technology will have more benefits or harms for humans, it would seem no more reasonable or ethical to pursue the benefits than to avoid the harms. The reasonable and ethical strategy would surely be to assess the potential harms and benefits on the basis of what we know, and act on that. It is not obvious yet that Joy is wrong.

It has been argued that it can be justifiable to restrict or prohibit research, but who should do this restricting or prohibiting? According to Johnson (1996, pp. 197-217), while it is justifiable to prohibit knowledge (or certain kinds of research), there is nobody to do the forbidding in the way that God was able to forbid Adam and Eve from eating the fruit in the Garden of Eden. However, we must seek the potential forbidder!

Four possibilities are considered by Marcello Pera: the State, committees of experts, the scientists themselves, and the people. He rejects state intervention for a number of reasons, the most plausible being that it is a threat to individual freedom. The committee of experts is rejected because there are no experts in ethics and because it would lower the sense of responsibility of scientists, and he also rejects the scientists themselves, because they would serve their own interests. He concludes that "the supervision of science should be granted to everyone" (Pera, 1989, pp. 58–72).

Pera's conclusion is hardly satisfactory given that we are not told how "everyone" would make the decisions. The only possibilities seem to be numerous referenda, which would be impracticable, or through the State, which he rejects. The most plausible suggestions are the State or the scientists themselves. There are problems with State intervention, but with appropriate legislation these can be minimised. There are also problems with self-regulation by the scientists, but again despite Pera, this could work. With respect to the State, Singer writes:

> In general, the social consequences of attempting to put some areas of knowledge beyond reach will be worse than the consequences of gaining the knowledge, and doing our best to see that it is used responsibly. (Singer, 1996, p. 229)

Agazzi also has worries about state intervention in science:

> If we were to accept social control of science, meaning total planning of scientific research to make it completely goal-oriented, we would also have to accept two highly undesirable consequences. First, we would have to limit the freedom of science ... Second, ... we could not avoid the question: "Who will determine the goals?" (Agazzi, 1989, pp. 215–216)

A strong statement for the freedom of science from political control is supplied by David Baltimore:

> First, the criteria determining what areas to restrain inevitably express certain sociopolitical attitudes that reflect a dominant ideology. Such criteria cannot be allowed to guide scientific choices. Second, attempts to restrain directions of scientific inquiry are more likely to be generally disruptive of science than to provide the desired specific restraints.(Baltimore, 1979, p. 41)

A number of arguments are offered to support these claims. First is what Baltimore calls the "Error of Futurism", that is the supposition that we can predict the consequences of any research accurately enough to make any sensible decisions. The second argument is a version of one of John Stuart Mill's. Freedom of speech and expression allows the development of new ideas, increases the choices in life and generally renews and vitalises life and makes it richer. A third argument is that repression in scientific research is likely to lead to repression in other areas and so will increase fear rather than strength in society. A fourth argument is based on the unpredictability of science. Even if some research is not allowed, the knowledge to which it may have led might emerge from other research quite unexpectedly.

This argument is aimed at pure or basic research, and Baltimore admits that the further one moves toward applications of research, the weaker these arguments become. If these arguments hold for pure research, then all of the responsibility for undertaking worthwhile research rests on the scientists themselves, which is where, according to Baltimore and others, it ought to rest. It must be noted here that it does not necessarily follow that the scientists' responsibilities extend to the *uses* to which the knowledge is put. They create or generate the knowledge from their research but others decide how it is to be used. Whether this separation of responsibilities is ultimately sustainable is another matter, but for the moment we will accept it.

We now return to Baltimore's arguments that research should not be externally controlled. His first argument, the "Error of Futurism," is that prediction is too unreliable to provide the basis for any restrictions on research. Consider for example Weizenbaum's prediction that research into speech recognition could have no useful consequences (Weizenbaum, 1984, p. 271). It now appears to be an important tool in Human Computer Interface design for users with certain disabilities. Prediction is certainly fraught with danger, however we often must base our actions on predictions, on what we believe may happen, and it is not clear why the case of research should be any different (but see Brown's argument in the following section).

The second argument is that freedom of speech and expression allows the development of new ideas, thereby increasing life's choices and generally making it richer. This is true, and this form of freedom is undoubtedly an important good, but it is not the only one, and can be in conflict with others. In general we are restricted in the performance of actions which will, or are likely to, harm others. Again it is unclear why research should be treated differently. The third argument is that repression in scientific research is likely to lead to repression in other areas and so will increase fear rather than strength in society. While we can agree that repression is not good, many things are restricted, or repressed in a civilised society, for example, certain research using human subjects and driving under the influence of alcohol, but these restrictions would surely reduce rather than increase fear. If probable harm is as closely associated with knowledge as suggested earlier, there is no reason why pure research should be treated differently from other aspects of life. The final argument was that, because of the unpredictability of science, the undesired knowledge might emerge unexpectedly from research other than that which was disallowed. This is true but not to the point. While it may not be possible to ensure that some undesirable knowledge will be discovered, it is almost certainly possible to reduce that probability.

It is then permissible or even obligatory on occasions to restrict or forbid research on the ground that mental or physical harm is likely to result from it. However, this should not be done lightly, because freedom in this area is important, not only for the good of science but also for the good of society. There should be a presumption in favour of freedom of research. If research is to be restricted the burden of proof should be on those who want to restrict it. However, there seems to be a conflicting intuition here. If a prima facie case can be made that some particular research will most likely cause harm, either mentally or physically, then the burden of proof should be on those who want the research carried out to demonstrate that it is safe. The situation then appears to be this. There is a presumption in favour of freedom until such time as a prima facie case is made that the research is dangerous. The burden of proof then shifts from those opposing the research to those supporting it. At that stage the research should not begin or be continued until a good case has been made that it is safe.

The conclusion to this point then is that the case against the State having a role in the control of scientific research has not been made, but that such control has dangers and it should not be embraced lightly. The argument so far has focussed primarily on pure research, because that is where it is most difficult to make a case for control. However, it is not obviously much different in the case of technological development, one of the fruits of pure research. Just as a scientist can say that he or she is just adding to knowledge and therefore has no responsibility for the use to which that knowledge is put, so technologists can say that they are just developing tools, and it is not up to them how those tools are used.

5. Nanotechnology and quantum computing research

It has been argued in this paper that nanotechnology and quantum computing do raise some worrying ethical questions. While there seems to be potential for great benefit, it is not benefit unalloyed, and some of the potential problems were outlined earlier. It has also been argued that there are cases in which it could be justified to halt certain types of research. The question here is whether research into nanotechnology and quantum computing is in this category.

One scientist who thinks that it is not is Ralph Merkle, who says of Joy's view that "it seems to embrace one of the worst strategies we could adopt" Merkle, 2001). He supports his belief with a number of arguments: (1) there cannot be informed public discussion without active research; (2) research can only be stopped if we know about it; (3) others will do it even if we do not; and (4) we would be turning our backs on possibly saving lives and raising living standards (Merkle, 2001).

While it is true that public discussion could be more informed if there were an active research programme, this is not a very strong argument if that research is dangerous, or likely to be. Nor is it true that without that research there can be no informed discussion. Enough scientists are working in related areas, in physics and in the biological sciences, to have a fairly good understanding of what the potential is. The second argument is true in the sense that something cannot be stopped unless it has first of all been started, but that is not so interesting. Presumably what is meant is that unless we have been working on something we will not know enough about to have good reasons for not doing it. But this is not quite right, for the same reason that (1) was not. We can know enough about an activity to know that it is potentially dangerous and ought to be avoided, without actually doing it, particularly when much is known about related activities. The third reason, that others will do it even if we do not, is a little more worrying. While in many cases this type of argument is good, in this case it does seem to have some validity. On this point Merkle writes: "The only practical result [of halting research] would be to give less scrupulous nations the economic and military advantages of developing this technology before we did". While this is not a noble reason, it does carry some weight.

If it is known that nanotechnology can be used to develop more sophisticated weapons, and if it is known that others are developing this technology, there does seem to be some argument for doing the same, if only to develop defences against those possible weapons. Exploring this vexed issue would take us too far afield. It is enough to point out here that certain kinds of research, for example into biological weapons, is not pursued (at least overtly), even though "less scrupulous nations" might. Rather, efforts are made to stop every one doing so. Perhaps nanotechnology should be in this category. He also has a variant of this argument:

> This [Joy's] approach suffers from major problems: telling researchers not to research nanotechnology and companies not to build it when there are vast fortunes to be made, glory to be won, and national strategic interests at stake either won't work, or will push research underground where it can't be regulated while depriving anyone who actually obeys the ban of the many benefits nanotechnology offers. (Merkle 2001)

Brown gives an argument with a twist, against Joy. He objects that people such as Joy predict the future of technology without taking into account the human and social factors, and calls it "tunnel vision". "It excludes all the other factors that come into play as technologies develop. In particular, it excludes the social factors that always shape and redirect technology", and this, he says, makes prediction much more difficult (Brown, 2001, p. 30). This seems true, but the twist comes when it is realised that it is partly because of critics like Joy that the development is tempered by human concerns. Brown himself recognises this, and talks of self-*unfulfilling* prophecies (p.30). Therefore, while people like Joy might be wrong in their predictions, they are wrong *because* they made them.

Finally there is the argument that much good could flow from the research and therefore it ought to be undertaken. This too has some force, but the potential good needs to be weighed against the potential harm, and it is not obvious that the former would outweigh the latter, although the converse is not obvious either.

6. Conclusion

Given the quite fundamental changes that these technologies could facilitate, it is not enough to merely consider the potential benefits and harms as they might apply to life as we know it now. The issue is more one of the kind of life that we want. Can we, and do we want to, live with artificial *intelligences*? We can happily live with fish that swim better than we do, with dogs that hear better, hawks that see and fly better, and so on, but things that can reason better seem to be in a different and altogether more worrying category. Do we want to be "super human" relative to our current abilities, with implants that enhance our senses, our memory and our reasoning ability? What would such implants do to our view of what it is to be human? Does it matter if our experiences are "real" or not, that is, if they are had in a virtual world or in the real one? Would there be any sense in that distinction? These are all big questions that cannot be answered here, but the suggestion is that they are the important ones when considering the future of research into nanotechnology and quantum computing. These questions seem related to Somerville's second principle, that of protecting the human spirit, and also to Lenman's argument regarding the value given to human life by participating in the satisfaction of certain human needs. Perhaps research into nanotechnology and quantum computing does not harm the human spirit or reduce the value in human life, but much more examination of the issues is warranted.

Notes

1. I wish to thank the editor of the *Australian Journal of Professional and Applied Ethics* for his helpful comments.
2. An earlier abbreviated version of this chapter was published in James H. Moor and Terrell Ward Bynum, editors, *CyberPhilosophy: The Intersection of Philosophy and Computing*, Blackwell, Malden, Mass. 2002.
3. One worry that will not be considered here is that posed by self-replication, and "grey goo". This would certainly be a major problem if it were technically feasible, however there is some dispute about the possibility (see Ashley, 2001, Smalley, 2001, Whitesides, 2001).

References

Agazzi, E. (1989), Responsibility: the genuine ground for the regulation of a free science, in Shea, W.R. and Sitter, B. eds. *Scientists and Their Responsibilities*, Watson Publishing International, Canton, MA. 203–219.

Arnauld, A. (1964), *The Art of Thinking: Port-Royal Logic*, Bobbs-Merrill, Indianapolis.

Ashley, S. (2001), Nanobot construction crews, *Scientific American*, September, 76–77.

Baltimore, D. (1979), Limiting science: a biologist's perspective, in G. Holton and R.S. Morrison (eds) *Limits of Scientific Inquiry*, W.W. Norton, New York, 37–45.

Benenson, Y. Paz-Elizur, T., Adar, R., Keinan, E., Livneh, Z., and Shapiro, E., (2001), Programmable and autonomous computing machine made of biomolecules, *Nature*, 414, 22 November, 430–434.

Brown, J.S. (2001), Don't count society out: a response to Bill Joy, National Science Foundation, *Societal Implications of Nanoscience and Nanotechnology*, 30–36.

Crevier, D. (1993), *AI: The Tumultuous History of the Search for Artificial Intelligence*, Basic Books, New York.

Davies, P (1996), Forward in Milburn, Gerard, *Quantum Technology*, Allen and Unwin, St. Leonards, NSW.

Feynman, R. (1959), There's Plenty of Room at the Bottom, An Invitation to Enter a New Field of Physics, talk given on December 29th at the annual meeting of the American Physical Society the at California Institute of Technology Online at *http://www.zyvex.com/nanotech/feynman.html*.

Foresight, (2000), Foresight Guidelines on Molecular Nanotechnology *http://www.foresight.org/guidelines/pr001.html*

Harvey, E. (2001), Can nanotechnology deliver our dreams?, *Australasian Science*, June, 26–29.

IoN, (2001), The Institute of Nanotechnology *http://www.nano.org.uk/*

Joy, B. (2000), Why the future doesn't need us, *Wired* magazine, 8.04.

Kelves, D.J. and Hood, L. (1992), eds. *The Code of Codes: Scientific and Social Issues in the Human Genome Project*, Harvard University Press, Cambridge, Mass.

Kevles, D.J. (1993), Out of eugenics, in Kevles, D.J. and Hood, L. (1993), eds. *The Code of Codes: Scientific and Social Issues in the Human Genome Project,* Harvard University Press, Cambridge, Mass. 3-36.

Kurzweil, R. (1999), *The Age of Spiritual Machines: When Computers Exceed Human Intelligence,* Allen and Unwin, St. Leonards, NSW.

Lenman, J. (2001), On becoming redundant or what computers shouldn't do, *Journal of Applied Philosophy*, 18, 1-11.

Luntz, S. (2001), Photosynthetic factories and nanosponges, *Australasian Science*, June, 30-32.

Martin, D. (2001), Our body's nanosensors, *Australasian Science*, June, 33-35.

Merkle, R. (1997), It's a small, small, small world, MIT Technology Review, Feb/Mar. 25. Online at *http://www.techreview.com/articles/fm97/merkle.html.*

Merkle, R. (2001), Nanotechnology: what will it mean? From the January 2001 issue of *IEEE Spectrum* http://www.zyvex.com/Publications/articles/Spectrum.html.

Minsky, M. and Papert, S. (1969), *Perceptrons—An Introduction to Computational Geometry*, MIT Press.

NAI (2001), Proceedings and Outcomes Report of a 30 March Workshop on Nanotechnology in Australian Industry http://www.isr.gov.au/industry/advancedmanufacturing/NanoWorkshop Report.pdf

NSTC, (1999),Nanotechnology: Shaping the World Atom by Atom, report of National Science and Technology Council (NSTC), Committee on Technology, The Interagency Working Group on Nanoscience, Engineering and Technology, online at *http://itri.loyola.edu/nano/IWGN.Public. Brochure/.*

Pera, M. (1989), Should science be supervised, and if so, by whom? In Shea, W.R. and Sitter, B. eds. *Scientists and Their Responsibilities*, Watson Publishing International, Canton, MA. 58-72.

Rescher, N. (1987), *Forbidden Knowledge and Other Essays on the Philosophy of Cognition*, D. Reidel.

Science, 290, (2000), special issue on nanotechnology.

Sheehan, H. (1993), *Marxism and the Philosophy of Science: A Critical History* Humanities Press International, New Jersey. Extracts available at *http://www.dcu.ie/~comms/hsheehan/ lysenko.htm*

Simmons, M. (2000), Faster, smaller, smarter, talk at Small Things, Big Science: Nanotechnology, *Horizons of Science forum*, University of Technology, 23 November.

Singer, P. (1996), Ethics and the limits of scientific freedom, *Monist*, 79, 218-229.

Smalley, R.E. (1999), Nanotechnology, Prepared Written Statement and Supplemental Material of R. E. Smalley, Rice University, June 22, online at *http://www.house.gov/science/smalley_062299.htm.*

Smalley, R.E. (2001), Of chemistry, love and nanobots, *Scientific American*, September, 68-69.

Somerville, M. (2000), *The Ethical Canary: Science, Society and the Human Spirit*, Viking, Ringwood, Vic.

Steane, A.M., and Rieffel, E.G. (2001), Beyond bits: the future of quantum information processing, *Computer,* January, 38–45.

Volkman, R. (2001), Playing God: technical hubris in literature and philosophy, *Proceedings of the Fifth International Conference on The Social and Ethical Impacts of Information and Communication Technologies, Ethicomp20001,* Gdansk, June 18-20, Volume 1, 350–361.

Wallace, G. (2001), Artificial muscles stronger than steel, *Australasian Science*, June, 36–37.

Weizenbaum, J. (1984), *Computer Power and Human Reason: From Judgement to Calculation*, Penguin Books, Harmondsworth.

Whitesides, G.M. (2001), The once and future nanomachine, *Scientific American*, September, 70–75.

Wilson, M. (2001), Nanomachines: the new industrial revolution, *Australasian Science*, June, 26–29.

Index

A
Abortion, 40, 44
Access, to genetic information, 77–78
  adverse effects from, 229
  controls, 164
  creating problems of, 78–79
  healthcare workers' views of, 146–148
  incentives for, 228–230
  information inequity and, 78
  legitimate need for, 115–116
  medical records, 122–124, 146–148
  restrictions on, 117–118
  for scientists, 228–230
  solving problems of, 78–79
Accountability principle, 156
Accuracy, of clinical bioinformatics, 318–319
Acquisition, of intangible property, 201–203
Actants, 253
ADA (Americans with Disabilities Act), 138
Adenine, 110
Administrative proceedings, HHS Rule for disclosures in, 144
Advanced capitalism, 241
American Medical Association (AMA), genetic patent policy, 226
Americans with Disabilities Act (ADA), 138
Anglo-American systems, of intellectual property, 199–200
Animal rights, 40, 44
Anti-genetic exceptionalism, 20–23
Anti-germination technology, 56–58
Aristotelian middle ground, 36
Artificial intelligence, 327
Artificial neural networks, 329
Associative privacy, 101
Authors, ownership rights of, 204
Automobile seat-belt interlocks, 61–66
Autonomy
  contractual, disclosure and, 70–71
  control of genetic material and, 220–221
  genetic property rights and, 201, 227
  informed consent and, 11, 69–71
  law and, 67–69
  private bargaining and, 67

B
BAC to BAC sequencing, 79, 237
Banking, disclosures for, 145
BERAC (Budget and Environmental Research Advisory Council), 13
Berne Convention Standards of 1989, 199

Bioinformatics
   clinical. *See* Clinical bioinformatics
   content/methodology of, 9, 10t
   definition of, 8, 317
Biological controls, programmable, 70
   vs. technical controls, 66–67
   vs. technological controls, 63–64
Biological information, science of, 241
Biomedical research, genetic data resources, ownership of, 222–223
Bionic humans, 327
Bio-prospecting, 220–221, 230
Biotechnology industry
   property rights for genetic materials and, 222–226
   technological costs, 240–241
Blackboxing, 253
BRCA 1, 112
BRCA 2, 112
Budget and Environmental Research Advisory Council (BERAC), 13
Bush administration, 132–133

C

Categorical Imperative, Kantian, 47, 48, 51
Categorial privacy, 181–183
Causal relations, inferring, 159
Celera, 250, 252
Chemical genomics, 302
Chicago economics model, 269–270
"Clickwrap" agreements, 59
Clinical bioinformatics, 280
   accuracy of, 318–319
   appropriate use/users, 319–320
   definition of, 317–318
   guidelines/standards for, 320–321
   privacy/confidentiality, 320
Codes of Conduct, 127
Codons, 110
Coercion, of research subjects, protection against, 176
"Cohen Theorem," 63, 66
Collection limitation principle, 156

Commercialization, deCODE Genetics case and, 17–18
Common heritage of mankind, gene patenting and, 259–260
Common morality
   applications of, 50–51
   avoidance of harm and, 43
   definition of, 28
   function of, 33
   general agreement on, 34
   goal of, 43
   impartiality and, 40–41, 53
   moral judgments and, 42–43
   moral theories and, 34–36
   philosophical accounts of, 36
   public nature of, 53
   as public system, 41–42
   rationality and, 39
   scope of, 43–44
   vs. Kantian system, 52–53
   vs. other moral systems, 51–53
Communitarism, 242
Community consent, 177
Compensation, for genetic material donors, 229–230
Computational biology
   content/methodology of, 9, 10
   definition of, 8
   genomics research and, 292, 295–296
   techniques, 23–24
Computational genomics
   computational aspects of, 9
   content/methodology of, 9, 10
   definition of, 8–9
   ELSI framework applications, 14–18
   ethical aspects of, 11–18
   ethical issues, special considerations for, 18–24
   genomic aspects of, 9–10
   research, nanotechnology and, 281–283
   vs. bioinformatics, 7–8
   vs. genome informatics, 9–11
Computational tools, ELSI Research Program and, 279–280

Computer-assisting role, of computer technology, 6, 23
Computer databases. *See* Databases
Computer-enhancing role, of computer technology, 6–7, 23
Computer networks, 6
Computers, electronic medical records and, 116
Computer science, theoretical connections with genetics/genomics, 5–7
Computer technology, privacy rights and, 115
Conceptual muddles, 23–24
Confidentiality
  clinical bioinformatics and, 320
  computational genomics and, 11–12
  deCODE Genetics case and, 17
  of genetic information, 214
  group, 181–183
  of research subjects, 177–178
Connecticut, Griswold v., 101, 114
Consent issues. *See also* Informed consent
  deCODE Genetics case and, 17
  medical consent, 68–69
  policies, 12
Consequences, of knowledge, 330–335
Consequentialism. *See* Utilitarianism
Constitution, technological controls and, 63–64
Consumer preference, 269–270
Content management technology, 58–60
Contracts, 251
Contractual theory, 67–69
Control, of scientific research, nanotechnology and, 328–335
Control/restricted access theory, 102, 115–116
Control theory, 101–102
Convergence of technologies, 75
Copyleft, 248
Copyright Act, 62
Copyrights
  in digital environments, 200

laws, 58–59
prevention of patent control, 252
"sweat and brow" claims, 260–261
Coroner disclosure, HHS Rule for, 144
Corporate ownership, of genetic information in computer databases, 188–189
Counseling, 130
Crackers, 239
Cryptography, 163–164
Cultural beliefs, 38
Cystic fibrosis, 111
Cytosine, 110

D
Data
  genomics. *See* Genomics, data
  personal. *See* Personal data
  production, large-scale, 310–311
  protection, 100
  sharing, 12
Databases
  access, deCODE Genetics case and, 16–17
  accuracy of, 12
  knowledge discovery in, 78–79, 170
  NHGRI, 309
  ownership of genetic information in, 188–189
  protecting personal information in, 178–179
  searchable, 77
Data mining
  algorithms for, 103
  applications of, 175
  computerized techniques, 171
  decreasing risks of, 163–164
  for group profile creation, 160–161, 171–172
  Human Genome Project and, 171
  informed consent and, 181–183
  mathematical algorithms for, 161
  for profile creation, 78–79, 157–158
  risks of, 163–164
  shotgun method, 171
  technology/tools for, 103–104, 167

vs. knowledge discovery in databases, 170
Data quality principle, 156
Deception, 44, 47, 176
Decision support, clinical bioinformatics and, 318–319
deCODE Genetics, Inc.
  cross-referencing of information, 1, 167
  database ownership and, 188–189, 230, 231
  ESLI concerns, 16–18, 19
  exclusive access and, 229
  historical aspects, 15–16
  Icelandic healthcare database and, 215–218
  informed consent and, 178
Democratic government, morally acceptable alternatives and, 43
Deontological ethical theories, 27–28, 52–53
Deoxyribose nucleic acid. *See* DNA materials
Department of Energy (DOE), 13
Department of Health and Human Services, privacy standards, 141–146
Depression, from genetic disclosure, 38–39
Detection of disease, 301
Determinants, of disease, 153–154
Digital divide, 29
Digital media, content licensing of, 58–59
Digital Millennium Copyright Act (DMCA), 60, 256
Dignity, 11
Disclosure
  to government health data systems/health directories, 144
  for health oversight purposes, 144
  to law enforcement, 144
  mandated, 70–71
  minimum necessary, 143

Discrimination
  based on genetic information, 109, 117–118, 304–305
  from group profiles, 159–160
  pathologicization of difference, 92–93
  vulnerability to, 170
Disease. *See also specific diseases*
  determinants of, 153–154
  genetic components of, 299–300
  genetic contributions to, 309
  non-genetic factors in, 310
  resistance, genetic components of, 300
  risk factors, 301
  susceptibility, 300–301
Disease-causing genes, 21, 91, 215
Disinterest, 242
Distributional group profiles, 173–175
DMCA (Digital Millennium Copyright Act), 60, 256
DNA materials
  dispersal, informed consent for, 227
  nanotechnology and, 281–282
  observation methods for, 236
  as patentable subject matter, 223
  samples, 213
  structure of, 110
DNA sequencing, technology development, NHGRI and, 308
DOE (Department of Energy), 13
Doe v. SEPTA, 138–139
Dosage, internal and biologically effective, 88–89
Drug response, genetic contributions to, 299–300, 309
Dynamic negotiation system, 130–133

E
Early biological effects, detection of, 90
Economic factors
  in genetic materials donations, 221–226
  of intellectual property rights, 263–265
Education, genomics research and, 295

Efficiency, contractual theory and, 67
EGP. *See* Environmental Genome Project
ELSI Research Program
   application to computational genomic research, 14–18
   computational tools and, 279–280
   deCODE case and, 16–18, 19
   historical aspects of, 12–14
   NHGRI activities, 309–310
   privacy and, 99, 183, 184
   program areas, 13
   recommendations from, 94–95
ENCODE Project, 308
Environmental criticisms, of gene patenting, 259
Environmental exposures, responses, genetic variability of, 85
Environmental Genome Project (EGP), 168–170
   ELSI challenges and, 14, 30
   initiatives of, 87–88, 168
   recommendations for, 93–94
Environmental genomics
   definition of, 168
   research subjects, risks of, 168–169
Environmental health research
   recommendations for, 93–94
   relevance of Human Genome Project, 86–88
Environmental policy, property law and, 267
Epidemiology
   definition of, 153
   in information age, 154–155
"Error of Futurism," 333–334
ESTs (expressed sequence tags), patents on, 190, 223, 252, 263
Ethical, Legal and Social Implications Research Program. *See* ELSI Research Program
Ethical issues
   in computational genomics, 11–18
      special considerations for, 18–24
   definition of, 11
   genomic uses, 307
   in nanotechnology, 326–328
Ethnicity
   disease-related polymorphisms and, 92
   genomics and, 306
Eugenics, 61
European Biology Directive, 190
European Union Directive on Privacy, 100, 124–125, 126, 156, 228
Experimental methods, 153
Experimentation, 171
Expressed sequence tags (ESTs), patents on, 190, 223, 252, 263
External perspective, 174

F

Factor analysis, 148
Fair information principles, 140–141, 148, 156
Fairness, 11, 13, 16
Faith-based objections, to gene patenting, 258, 265, 266
Federal agencies, medical records protections, 138
Federal Trade Commission (FTC), 140
Firm, concept of, 250–251
Fragile X syndrome, 111
Freedom of contract, 68
Freedom of speech/expression, 334
Free software movement. *See* Open source movement
Frequency data, 146, 147, 151–152
FTC (Federal Trade Commission), 140
Funding of research, allocation of, 328–329

G

"Gay gene," 117–118
Gene-environment interactions, 29–30
   lung cancer and, 88–91
   regulatory implications, 93–94
Gene machines, 110–112
Genes
   causative, identifying for specific diseases, 14–15
   definition of, 10

highly penetrant, 21
Gene splicing techniques, 56
Gene therapy
  development of, 301–302
  future of, 58
  techniques, government regulation of, 207–208
Genetic code, regulation by, 62
Genetic control elements, 57–58
Genetic counselors, 45, 47, 50
Genetic discrimination, 92
Genetic exceptionalism
  alternative analysis of, 21–23
  argument against, 20–21
  definition of, 1–2, 19, 213–214
  thesis, evaluation of, 20
Genetic information
  access to. *See* Access, to genetic information
  acquiring, 79–81
  appropriate use/users, 319–320
  confidentiality of, 214
  cross-referenced, 177–178
  discovery of, 205–206
  discrimination and. *See* Discrimination
  free flow of, 248
  freely shared, 205
  from Human Genome Project, 79–81
  increased, researchers and, 154–155
  from indigenous people, 220–221, 228
  lack of forgetfulness on, 158
  "loosely," 20
  loss of control over, 227
  meaning of, 80
  misappropriation/misuse of, 227–228
  misconceptions, 21
  ownership of. *See* Ownership, of genetic information
  privacy, need for strong protection and, 116–118
  privacy and, 91–92
  public ownership of, 204
  secondary use of, 179–181
  significance of, 81–82
  trivial linked to sensitive information, 158
  usage policies, 298–299
  uses for, 22–23
Genetic Information Nondiscrimination Act, 100
Genetic material, patentability of, 223–224
Genetic mutations, 90
Genetic networks, 296–297
Genetic Privacy Act (GPA), 218–219
Genetic programming, autonomous consent and, 69
Genetic risk information
  clinical effects of, 302–303
  knowledge of, 112
Genetics
  medical benefits of, 112–113
  theoretical connections with computer science, 5–7
Genetic susceptibility, to lung cancer, 90–91
Genetic testing
  deCODE Genetics case and, 17
  genetic exceptionalism and, 19–20
  for PKU, 112
Genetic variability
  in response to environmental exposures, 85
  technology development, NHGRI and, 308
Genome. *See also* Human Genome Project
  component identification, NHGRI and, 308
  definition of, 10
  human, 236
    heritable variation in, 297–298
    identifying structural/functional components of, 292–296
    meaning of, 236–237
    structure/function, elucidation of, 291–299
Genome informatics, 1
  content/methodology of, 9, 10t

definition of, 9
vs. computational genomics,
  9–11, 10t
Genome sequences, 287–288
  identifying, 6
  inter-species comparisons,
    294–295, 298
Genomics
  data
    access to, 77–78
    for health care improvements,
      310
    large-scale production of,
      307–308
  definition of, 10
  as public good, 76
  race/ethnicity and, 306
  theoretical connections with computer science, 5–7
Genomics research
  computational biology and, 292
  education and, 295
  future of, 287–312
  health benefits of, 299–304
  informed consent in, 176–181
  NHGRI mission and, 289
  resources and, 290
  technology development and, 291
  training and, 293
GeoCities case, 140
Germ-line alteration of humans, 58
Gleevec (imatinib mesylate), 301–302
Global relevance, 29
GNU Public License, 248, 251–252
Golden Rule, 47, 48
Government. *See also* State government
  control of research and, 329, 333
  democratic, morally acceptable alternatives and, 43
  genetic information usage, 319
  health data systems, disclosures to, 144
  paternalism of, 68–69
GPA (Genetic Privacy Act), 218–219
Griswold v. Connecticut, 101, 114

Group characteristics, inferring on individual group members, 159
Group profiles, 155–156
  data mining for, 157–158, 171–172
  definition of, 172
  distributional, 161–163, 173–175
  effects of using, 158–160
  nondistributional, 161–163, 173–175
  observational methods for, 157
  from personal data, 172–173
  risks of, 159–160
  use, reliability of, 104, 160–163
Groups
  confidentiality of, 181–183
  socially identifiable, 170
  stigma of, 12
GSTM1 polymorphism, 90
Guanine, 110
Guaymi cell line, patenting of, 220

H
Hacker ethic, 248–249
Hackers of code, 191, 239
Haemophilus influenzae, 111
Harm
  avoiding, rationality as, 36–39
  causing, moral rules and, 45
  irrationality and, 37–38
  morally relevant features and, 48–50
  rankings of, 37–38
  suffered by moral agents, 43–44
HDOs (health database organizations), 123
Health
  genetic contributions to, 309
  genomic information and, 310
  non-genetic factors in, 310
Health care improvements, genomic information for, 310
Health database organizations (HDOs), 123
Health directories, disclosures to, 144
Health improvement, genomic tools for, 303–304
Health informatics, 280

Health informatics professionals,
    challenges for, 280–283
Health Information Privacy Model
    Act, 139
Health Insurance Portability and
    Accountability Act of 1996
    (HIPAA), 124–127
  enactment of, 99, 137
  genetic test results and, 218
  HHS reguations and, 142–143
  purpose of, 103
HGP. *See* Human Genome Project
HHS Rule, 137, 141–142
  disclosures
    to government health data systems/health directories, 144
    to law enforcement, 144
  goals of, 142
  limitations/protections in, 142–143
  minimum disclosure required, 145–146
  permissible uses/disclosures, 143–145
  statutory background, 142
HIPPA. *See* Health Insurance
    Portability and
    Accountability Act of 1996
Hippocrates, 153–154
HUGO (Human Genome
    Organization), 76, 77, 79, 190
Human artifacts, 55
Human Genome Organization
    (HUGO), 76, 77, 79, 190
Human Genome Project (HGP)
  as bioinformatics project, 237–238
  as blackbox, 253
  data mining and, 171
  genetic information from, 79–81
  goal of, 111, 168
  meaning/function of genes and, 236
  morality and, 33
  origin of, 287
  relevance to environmental health
    research, 86–88
Human research, 305
Humans, germ-line alteration of, 58

Human subject research, bioinformatics issues. *See* Clinical
    bioinformatics
Human traits/behaviors, genomic
    contributions to, 306–307
Huntington's disease
  gene carriers, 91
  gene locus, 15, 111
  genetic exceptionalism and,
    19–20, 21
  genetic knowledge, reactions to,
    38–39
  predictive testing, 117
  withholding information on, 50
Hypotheses, 154–155

I
Iceland
  deCODE Genetics, Inc. and, 15–16
  as homogeneous population, 15–16
Icelandic healthcare database,
    deCODE Genetics, Inc. and,
    215–218, 228
Imatinib mesylate (Gleevec), 301–302
Impartiality, 39–41
  common morality and, 40–41, 53
  definition of, 35–36, 39–40
  moral rules and, 40–41
  moral rules *vs.* ideals and, 46
  toward minimal group, 40–41
Incentives, for access to genetic
    information, 228–230
Indigenous people, genetic material
    from, 220–221, 228
Individual participation principle, 156
Individual profiles, 104, 155–156
Individuals, ownership of genetic
    information in computer
    databases, 188–189
Inference controls, 164
Inference errors, 159
Informatics
  definition of, 8
  pace of change in, 247
Information. *See* Genetic information
Informational privacy, 101

Informational technology development, "free market" approach, 65–66
Information flow controls, 164
Information technology, 78–79
Information withholding, moral acceptability of, 43
Informed consent
  autonomy and, 11, 69–71
  cognitive attitudes and, 179–180
  data mining and, 181–183
  to disclosure of information, 130
  ethical issues, 176
  financial value of genetic material and, 230
  in genomics research context, 176–181
  medical, 70
  opacity of, 180–181
  principle of, 179
  privacy and, 227–228
  protecting personal information in computer databases, 178–179
  standards for, 169
  valid, 17, 176, 179
  vs. presumed consent, 177
Innovation, perpetual, 269–272
Innovation policy, genetic patents and, 261–262
Institutional Review Boards (IRBs), 30, 169
Insurance companies, data-mining applications, 175
Intellectual property (intangible property)
  features of, 197–199
  legislation, 272
  Lockean model of, 200–208
  policy
    limits of, 262–268
    minimalists vs. maximalists, 263–265, 274
    for research and clinical settings, 298–299
    as public good, 272–273
  rights, 187
  absence of, 263
  for developmental coordination, 252–253
  inequality argument against, 206–208
  intellectual debates over, 265–268
  Lockean model of, 200–208
  open source philosophy and, 240
  prospect theory of, 251
  protection of, 257
  scholarship, 268–277
  social nature of, 204–206
  utilitarian incentives-based argument, 199–200
Interference in personal affairs, protection against, 101
Internal perspective, 174
International HapMap Project, 297–298, 308
International law, 228
Inter-species sequence comparisons, 294–295, 298
Intrusion
  non-governmental into personal lives, 114
  protection against, 101
Inventors, ownership rights of, 204
Involuntary commitment, moral disagreements on, 37
IRBs (Institutional Review Boards), 30, 169
Irrationality, 35–38

J
Jehovah's Witnesses, blood transfusions and, 38
Judicial proceedings, HHS Rule for disclosures in, 144
Justifications
  for intangible property rights, 200–208
  for violating moral rules, 47–48

K

Kant, Immanuel
   Categorical Imperative, 47, 48, 51
   deontological ethical theories, 27–28, 52–53
KDD (knowledge discovery in databases), 78–79, 170
Kennedy-Kassebaum health insurance reform bill, "Administrative Simplification" clause, 135
Killing, as violation of moral rule, 43, 44
Knowledge
   consequences of, 330–335
   genome-based, translating to health benefits, 299–304
   from societal interaction, 205–206
Knowledge discovery in databases (KDD), 78–79, 170

L

Labor theory of acquisition, 200–201
Law, autonomy and, 67–69
Law enforcement, disclosures, 144
Legal perspective
   on genetic engineering advances, 28–29
   on privacy-enhancing proprietary rights, 218–220
   on protection of medical records privacy, 140
"Lex genetica," 62–64, 71
Licensing scheme, compulsory, 229
Limitation principle, 156
Linkage studies, recanted, 318
Linux software, 252
Locke, John, 187, 200–201
Lockean model of property, 187
Lock-out technology, 61–62
Low-penetrance genes, 91
Lung cancer, gene-environment interaction and, 88–91
Lying, 47, 52

M

Mapping, of human genome, 111
Market failure, 67–68
Market interaction, benefits from, 204
Market value, 205–206
Medical consent, 68–69, 70
Medical data, personally identifiable, 123
Medical examiners, disclosure by, HHS Rule for, 144
Medical information
   centralized data system for, 126
   gathering, 110
   guidelines for, 124–127
Medical records
   abuses of, 138
   access to, 122–124, 147
   accuracy of, 146–147
   authorized users of, 124
   disclosures, ADA, 138–139
   electronic, 110, 116
   genetic information privacy, need for strong protection and, 116–118
   legal protections for privacy, 140
   owner of, 123
   privacy, impact of technology on, 139–140
   secondary use, unauthorized, 147
Medical research, disclosures for, 145
Medicine, genetic revolution in, 88
Mendelian diseases, 224
Messenger RNA, 110
Meta-analysis, 318
Microarray technologies, 288
Mill, John Stuart, 51. See also Utilitarianism
Mitosis, 110
M13 libraries, 79, 237
Modeling, computational, 6–7
Modesty, methodological, 267
Molecular biology, future of, 82
Monogenic susceptibility, 90
Monopolies, intellectual property rights and, 263–264

Moore v. Regents of California,
    222–223, 231
Moral agents, harm suffered by, 43–44
Moral agreement, areas of, 34
Moral disagreements
    basis for, 37, 43–44
    on involuntary commitment, 37
    limits on, 43
    settling, 40
Moral ideals, 45–46
Moral judgments
    common morality and, 36, 42–43
    disagreement in, 33
Morally acceptable alternatives,
    selection of, 43
Morally relevant features, 19, 48–50, 52
Moral rules
    "do not kill," 43, 44
    impartiality and, 40–41
    listing of, 45
    violations of, 43, 46–47
        consequences of, 51
        justification for, 44, 47–48
        morally relevant features and,
            48–50
        public allowance for, 49–50
    vs. moral ideals, 46
Moral systems
    gene patenting and, 258–259
    justified, 45–51
    morally relevant features and, 48–50
    vs. common morality, 51–53
Moral theory, 34–36

N
NACHGR (National Advisory Council
    for Human Genome
    Research), 13
Nanocomputers, 281
Nanotechnology
    control of research and, 328–335
    convergence of technology and, 75
    definition of, 324
    ethical issues, 326–328
    future of computational genomics
        research and, 281–283
    quantum computing and, 325–326
    research, 335–336
National Advisory Council for
    Human Genome Research
    (NACHGR), 13
National Committee on Health
    Information and Privacy,
    126–127
National Human Genome Research
    Institute (NHGRI), 13, 279
    ELSI research activities, 309–310
    goals of, 289
    role of, 307–310
National Institute of Environmental
    Health Sciences (NIEHS). *See*
    Environmental Genome
    Project
National Institutes of Health (NIH),
    13, 252
National Research Council, *Mapping
    and Sequencing the Human
    Genome,* 288–289
Natural rights theory, 187, 208
Networks, genetic regulatory, 309
New Genetics, 241
NHGRI (National Human Genome
    Research Institute), 289
NIEHS (National Institute of
    Environmental Health
    Sciences). *See* Environmental
    Genome Project
NIH (National Institutes of Health),
    13, 252
Nondistributional group profiles,
    173–175
Nontransparent correlations, 175
Normative protections, 19, 23–24
Nucleotides (bases), 110

O
Observation methods
    for creating group profiles, 157
    data mining and, 171

for DNA materials, 236
in epidemiology, 153
OECD (Organization for Economic Cooperation and Development), 156
Office of Health and Human Services (HHS), privacy standards. *See* HHS Rule
Office of Technology Assessment (OTA), *Protecting Privacy in Computerized Medical Information,* 176
Opacity, of consent, 180–181
Openness principle, 156
Open source movement
  adoption in genetics field, 239–240
  description of, 238–239
  for genetics and software research, 77, 241–242
  tenets of, 247–248
Oregon, Genetic Privacy Act of 1999, 219, 231
Organization for Economic Cooperation and Development (OECD), 156
Organized scepticism, 242
Orphan Drug Act, 302
OTA (Office of Technology Assessment), *Protecting Privacy in Computerized Medical Information,* 176
Outcomes, unanticipated, 23–24
Oversight activities, disclosure for, 144
Ownership, of genetic information
  arguments on, 258–262
  in computer databases, 187, 188–189
  privacy and, 214–215, 217

P

PAH-DNA adducts, 88–89, 90
PAHs (polycyclic aromatic hydrocarbons), 88–89
Pain, 44, 48
Pareto-based principle, Lockean, 201–203, 207
Patentable subject matter, nature of, 260–261
Patent Act of 1790, 199
Patent and Trade Organization (PTO), 190
Patenting-and-publish strategy, 242
Patent protections, validity/scope on genetic material, 224
Patents
  AMA policy on, 226
  arguments against, 258–262, 265–266
  copyrights and, 252
  criticisms of, 274
  curtailing, reasons for, 274
  effect on research efforts, 222
  on expressed sequence tags, 190, 223, 252, 263
  on gene sequences, 240
  on genetic material, 223–224, 231
  for Guaymi cell line, 220–221
  ill-effects on research, 225
  industry from, 268
  justification on normative basis, 225
  on non-personal genetic data, 187, 190–192
  public domain and, 256–257
  on research tools, 200
  utilitarian perspective on, 225–226
Patent specialists, 262–263
Paternalism
  autonomy and, 67–68
  stigma of, 69
Patient information, gathering, 146
Payment, disclosures for, 145
Personal data
  group profiles from, 172–173
  ownership of, 217
  in personal profiles, 156–157
  property rights, 219–220
  protecting in computer databases, 178–179
  secondary uses of, 103–104
  sensitivity of, 109
Personal reasons/beliefs, 37–38
Pharmacogenetics, 87

Phenylketonuria (PKU), 112
PHRAP, 79
Physicians
  genetic information usage, 319
  medical consent and, 68–69
  withholding of information from patient, 43
PKU (phenylketonuria), 112
Plant Variety Protection Act, 56–57, 62
Policy vacuums, 23
Polycyclic aromatic hydrocarbons (PAHs), 88–89
Polygenic disorders, 224
Polygenic susceptibility, 90
Polymorphisms
  genetic, 87
  risk-conferring, 92–93
Popper-Hempel model, 236
Population cohorts, large, 310
Population genomics
  basic assumption of, 14
  definition of, 14
  strategy for, 14
Postgenome era, 81–82
p53 tumor-suppressor gene, 90
Pre-competitive information, 190
Presumed consent, 177
Prima facie claims to control, 201, 203
Privacy
  clinical bioinformatics and, 320
  computer-related concerns, 101
  concept of, 11–12, 100–103
  corporate self-regulation, 127–128
  deCODE case and, 16
  as ESLI Research Program areas, 13
  evolution of concept, 100–101
  genetic information and, 91–92, 116–118
  informed consent and, 227–228
  loss of, 121–122
  of medical records, 138–141
  nanotechnology and, 326–327
  nature of, 113–116
  need for strong protection and, 116–118
  ownership of genetic information and, 214–215
  protection, 129
    categorial, 181–183
    collective, 181–183
    for genetic information, 117–118
    hybrid approach, 128–133
    individual, 181–183
    legal, 117–118, 156–157
    state level, 139
  rights, 113–116, 216, 227
  right to, 113–116
  value of, 121–122
Privacy Act of 1974, 101, 138
Private bargaining, autonomy and, 67
Private groups, medical records protections, 138
Private lawmaking, legal/regulatory interventions, 64–66
Probes, molecular, 309
Profiles, 155–156, 173–175
"Programmable biological artifacts," 28–29
Promises, breaking, 44, 52
Property rights
  biotechnology industry and, 222–226
  copyleft uses, 248
  economic factors in genetic materials donations, 221–226
  for healthcare databases, 215–218
  to personal data, 217–220
  privacy-enhancing, legal support for, 218–220
  self-determination and, 220–221
  in source materials, 230
  vested in subject, 219
Prospect theory of intellectual property rights, 251
Protein pathways, 296–297
Proteomics, NHGRI and, 308
Psychological privacy, 101
Psychosocial risks, of environmental genomics research subjects, 168

PTO (Patent and Trade Organization), 190
Public allowance, for moral rules violations, 49–50
Public domain capture problem, 249
Public goods, genomic data as, 76
Public health activities, HHS Rule and, 143
Public health regulatory policies, implications of gene-environment interactions, 93–94
Public knowledge, group profiles and, 160
Public policy development, 29, 63, 310
Public system
 common morality as, 35, 41–42
 definition of, 41
Purpose specification principle, 156

Q
Quantum computing
 nanotechnology and, 325–326
 research in, 335–336

R
Race
 disease-related polymorphisms and, 92–93
 genomics and, 306
Rationality
 accounts of, 38–39
 as avoiding harms, 36–39
 common morality and, 39
 definition of, 35, 38–39
 justification for violating moral rules, 48
 self-interest and, 39
Red Hat, 250, 252
Regents of California, Moore v., 222–223, 231
Reliability, of group profile usage, 160–163
Religious beliefs, decision-making and, 38
Religious faith, objections to gene patenting and, 258, 265, 266

Research
 biomedical, ownership of genetic data resources, 222–223
 control, nanotechnology and, 328–335
 ELSI. *See* ELSI Research Program
 environmental health, 86–88, 93–94
 funding, allocation of, 328–329
 genomics. *See* Genomics research
 medical, disclosures for, 145
 in nanotechnology, 335–336
 property rights for genetic materials and, 222–226
 in quantum computing, 335–336
 restrictions/prohibitions, 332–335
 source material, privatization of, 222–223
Researchers, increased information and, 154–155
Resources, for genomics research, 290
Respect, 11
Restricted access theory, 101, 102
Ribonucleic acid (RNA), 110
Rights of sources, 213, 230, 260
Risks
 accuracy of bioinformatic systems and, 318
 to persons, 280
RNA (ribonucleic acid), 110
Roe v. Wade, 101, 114
Rule sets, technologically embedded, 61–63

S
Safeguards principle, 156
"Salient differences," 14
Scholarship, intellectual property, 268–277
Scientific practice, norms of, 241–242
Scientific research perspective, 29–30
Scientists, ownership of genetic information in computer databases, 188–189
Scripts, technological, 64–71
Seat-belt interlocks, automobile, 61–62, 64–65, 66

Secondary use
  of genetic information, 179–181
  of personal information, 103–104
  unauthorized, of medical records, 147
Security, 129
Seed licensing, 56–57
Self-determination, proprietary rights and, 220–221
Self-interest, rationality and, 39
Self-ownership, 198
SEPTA, Doe v., 138–139
Sequencing, of human genome, 111–112
Shotgun sequencing method, 6, 79, 171, 237–238
Shrinkwrapped licenses, 59, 70
Simulations, computational, 6–7
Single-nucleotide polymorphisms (SNPs), 297
Socially identifiable groups, 170
Social values, 207
Society, intangible property rights and, 204–206
Software "lock-out" systems, 59
Sovereignty, 201
Spirituality, gene patenting and, 258
Standardization process, 250
State government
  genetic privacy legislation, 219
  intervention, autonomy and, 67–68
  privacy protection, 139
Stigmatism
  of genomic research, 11
  vulnerability to, 170
Suffering, 45
Susceptibility
  to carcinogens, assessing variation in, 90–91
  to lung cancer, 90–91
Susceptibility genes, 21, 87, 91

T
Tangible property, vs. intellectual property, 197–199
Tay-Sachs disease, 111

Technical controls, vs. biological controls, 66–67
Technological controls, 61–63, 66–67
Technological scripts, 64–71
Technology
  development
    genomics research and, 291
    knowledge, usage of, 330–335
    "tunnel vision" and, 336
  development, NHGRI and, 308–309
  impact on medical records privacy, 139–140
  quantum leaps in, 311
  unintended consequences of, 148
Third parties, medical data and, 130
Thymine, 110
Trade secrecy, 251
"Tragedy of the commons," 222, 255
Training, genomics research and, 293
Transparent correlations, 175, 176
Triplet coding, 110

U
UHICA (Uniform Health Care Information Act), 139
Uniform Computer Information Transaction Act (UCITA), 59, 60
Uniform Health Care Information Act (UHICA), 139
"Unique health identifier," 125, 128
United States Department of Energy (DOE), 13
UPOV, 56
Utilitarian incentives-based argument, of intellectual property, 187, 199–200
Utilitarianism (consequentialism)
  criticisms of, 27
  definition of, 27
  gene patenting and, 266–267
  Greatest Happiness Principle, 51
  just, 226–228
  morally relevant features and, 52
  vs. Kantian system, 52–53

V

Values
- embedded technological, 70
- intangible property rights and, 201–203, 204
- technologically embedded, 29

Violations of moral rules, 46–47

Virtual reality systems, 327–328

Voluntariness, of informed consent, 176

W

Withholding information, justification for, 47

X

X-rays diffraction technique, 236